Some other books by the authors:

Cosmos (C.S.)

Contact (C.S.)

Murmurs of Earth: The Voyager Interstellar Record (C.S., A.D. *et al.*)

A Famous Broken Heart (A.D.)

Broca's Brain (C.S.)

The Dragons of Eden (C.S.)

The Cold and the Dark: The World After Nuclear War (C.S. *et al.*)

COMET

Painting by Kim Poor.

COMET

CARL SAGAN · ANN DRUYAN

RANDOM HOUSE
NEW YORK

Library of Congress Cataloging in Publication Data
Sagan, Carl, 1934–
    Comet.

    Bibliography: p.
    Includes index.
    1. Comets.   2. Halley's comet.   I. Druyan, Ann,
1949–   .   II. Title.
QB721.S34   1985   523.6   85-8308
ISBN 0-394-54908-2

Manufactured in the United States of America

Design: Robert Aulicino
Composed by: Typographic Images

DEDICATION

To Shirley Arden,
for more than a decade of friendship
and hard work well done.
With our love and admiration.

Contents

INTRODUCTION

BEFORE THE EARTH WAS FORMED, THERE WERE comets here. Afterwards, and for all subsequent eons, comets have graced our skies. But until very recently, the comets performed without an audience; there was, as yet, no consciousness to wonder at their beauty. This all changed a few million years ago, but it was not until the last ten millennia or so that we began to make permanent records of our thoughts and feelings. Ever since, comets have left a good deal more than dust and gas in their wakes; they have trailed images, poetry, questions, and insights. In this book, we have sought to rediscover those trails, to explore our present understanding of the comets, and to speculate on what else may be possible.

We were inspired by the 1985/86 return of one of the Earth's most brilliant (and punctual) visitors, Halley's Comet. In the upper right-hand corner of most pages of this book is a single frame of a movie. It illustrates the motion of Halley's Comet in its actual orbit around the Sun during this particular passage. You hold the pages between the thumb and fingers of your right hand and flip from front to back. The moving dot is the comet, the curve is its elliptical orbit, and the dates when Halley's Comet appears in those positions is indicated just below. Notice how slowly the comet moves when it is far from the Sun and how fast when it is near.

We have tried to give some sense of the evolution of scientific discovery, and present the evidence for and against those cometary theories that have been rejected, those that are currently fashionable, and some that are neither fashionable nor rejected. We hope we have indicated clearly which is which. Certain concepts and terms are presented more than once for clarity and accessibility by the general reader. This is not by any means a mathematical book, but all modern science is quantitative and so we have included a little arithmetic here and there. When we want to know how far away something is, we could measure in miles or versts or leagues. Nature is just the same no matter what measuring rods we use. But because it is the scientific convention, because the vast majority of nations on the planet have adopted it, and because it is simpler, we will here employ the metric system. A meter is about a yard long. A thousand of them make a kilometer, which is about 60 percent of a mile. If we imagine a meter divided into a hundred equal parts, each part is a centimeter. A little over two and a half centimeters make an inch. A micron is a millionth of a meter, much too small to see. Ten thousand atoms, shoulder to shoulder, make a chain one micron long.

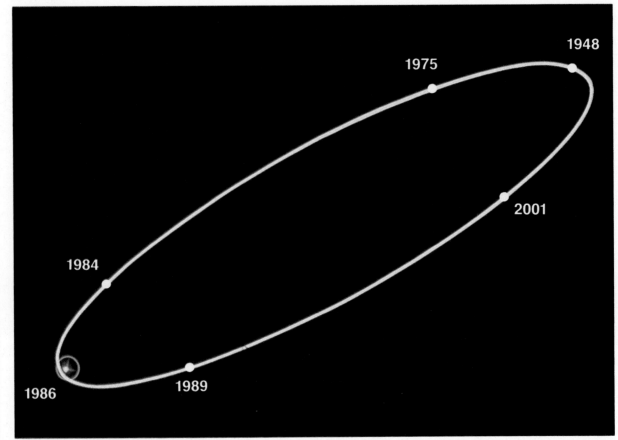

The orbit of Halley's Comet, shown in white. The blue circle represents the Earth's orbit around the Sun. Halley's Comet spends most of its time in the outer solar system sweeping inside the orbit of the Earth once every 76 years. It was last at aphelion, its farthest point from the Sun, in 1948. Its most recent approach to the Sun (perihelion) was in 1985/86. The comet moves very slowly when it is far from the Sun and much more quickly when it is closer, as the dates on this diagram indicate. Diagram and flip movie beginning on page 1 by Jon Lomberg/BPS/Maren Leyla Cooke.

Your fingernail is about ten thousand microns wide, or around one centimeter.

For reasons of length and readability, we have included the names of only a tiny fraction of the cometary scientists alive today. They have transformed the subject into one of the most exciting areas of modern science. Far more of them exist than the total number of cometary scientists who have lived in all previous generations. We beg the indulgence of those whose names we have omitted. The interested reader can follow the trail of original scientific papers to find the names of these specialists through the bibliography in the back of the book. In that bibliography, we have included both popular books on astronomy and representative or especially interesting works from the scientific literature.

We are indebted to the world astronomical community for their generosity to us throughout the preparation of this book. Joseph Veverka of Cornell University, a moving force behind U.S. plans for spacecraft missions to the asteroids and the comets, served as chief technical consultant and reviewer, and provided many astronomical images for the book. We are grateful to Mary Roth, John Kaprielian, and Margaret Dermott for facilitating our acquisition of these photographs. Mark Washburn fulfilled several research assignments with his customary diligence and excellence.

Several colleagues were kind enough to read earlier drafts of this book and provide detailed and valuable comments: Martha Hanner, Joseph Marcus, Steven Soter, Paul Weissman and Donald K. Yeomans, who also calculated the dates of the future returns of Halley's Comet. Others who permitted us to tap their expertise include John C. Brandt, Donald Brownlee, Stephen Jay Gould, Brian Marsden, Richard Muller, Marcia Neugebauer, Ray Newburn, Zdenek Sekanina, J. John Sepkowski, Jr., Eugene M. Shoemaker, Reid Thompson, and Fred L. Whipple. We are deeply grateful to all of them.

One of the most pleasant experiences in writing this book was our encounter with Ruth S. Freitag, Senior Science Specialist of the Science and Technology Division of the Library of Congress. Ms. Freitag has recently published a bibliography of Halley's Comet with over 3,200 citations. The subject will deserve a revised edition soon. Ms. Freitag's knowledge, enthusiasm for her subject, and willingness to share a treasury of comet illustrations speaks well of the nation's library.

The visual content of *Comet* is largely due to Jon Lomberg, an artist whose dreams are informed by science. His diagrams—produced with Simon Bell and Jason LeBel of Bell Production Services, Toronto, Canada—teach painlessly and with elegance. In addition to reviewing the manuscript, Jon also coordinated the efforts of other artists in producing some forty specially commissioned paintings for this book. We are proud to display herein the work of some of the planet's finest astronomical artists:

Michael Carroll	Pamela Lee
Don Davis	Jon Lomberg
Don Dixon	Anne Norcia
William K. Hartmann	Kim Poor
Kazuaki Iwasaki	Rick Sternbach

Our chapter on the life of Edmond Halley is illustrated with a number of images that have come to us through the kindness of Michael Thomas of A.M. Heath, London. We also thank Maren Leyla Cooke and Takako Suzuki for prehistoric animals and ancient calligraphy, respectively, and Donald K. Yeomans for giving us access to his personal collection of cometary pictures.

The manuscript was expertly guided through its many incarnations by Shirley Arden, whom we acknowledge in the dedication. We also thank Pandora Peabody and Maruja Farge, who in different ways made fundamental contributions to this book.

Patricia Parker, professor of comparative literature, University of Toronto, introduced us to the possibilities of a concordance of the world's cometary literature, launching us on a heady voyage. We have benefited from the remarkable breadth and depth of the scholarly community at Cornell University. Patricia Gill of the Department of English provided us with original translations of numerous ancient comet references and literary allusions and assembled a corps of talented researchers and translators, including Ann Bishop, Milad Doueihi, Michael D. Layne, Jim LeBlanc, Gina Psaki, Heather Smith, Karen Swenson, and Xie Yong.

We thank Howard Kaminsky, Jason Epstein, Robert Aulicino, Nancy Inglis, and Ellen Vanook at Random House for their efficient production; Derek Johns for helpful line-editing and many courtesies; Scott Meredith, Jack Scovil, Jonathan Silverman, Bill Haas, and others at the Scott Meredith Literary Agency for services that went beyond the usual norms in making *Comet* possible. Our thanks to Dorion Sagan for his support. We are also grateful to Kel Arden, David Aylward, Daniel Boorstin, Frank Bristow, Brian Dias, George Finlay, Andrew Fraknoi, Louis Friedman, A.L. Gabriel, Irving Gruber, Annie Guehenno, Jean-Marie Guehenno, Theodore Hesburgh, P.D. Hingley, Michel-Henri Lepaute, Bob Marcoux, Jerred Metz, Nancy Palmer, David Pepper, N. W. Pirie, George Porter, Roald Sagdeev, Alan Stahl, Andy Su, Peter Waller, Jean Wilson, Eleanor York, and Robert Zend.

This book, and our knowledge of the solar system, is indebted to the openness and relative freedom from secrecy of the National Aeronautics and Space Administration of the United States of America and, increasingly, its counterpart institutes in the Academy of Sciences of the Union of Soviet Socialist Republics.

The thirtieth recorded apparition of Halley's Comet brings us face to face with the great question of our time. This return is the first since we have become a spacefaring civilization, and the first since we have devised the means to destroy ourselves. We recall those eons during which there were no beings on Earth to marvel at the comets; we hope, at least until the Sun dies, that will never be the case again.

You lift your children onto your shoulders that they may better see a comet and, in so doing, join a chain of generations that stretches back far beyond the reach of written memory. There is no cause more important than protecting that ancient and most precious continuity.

—Carl Sagan and Ann Druyan
Ithaca, New York
August 6, 1985

PART I
THE NATURE OF COMETS

The Earth as seen from the icy surface of a comet—clearly a member of a shower of comets that has entered the inner solar system at some time in the distant past. Painting by Jon Lomberg.

Chapter I

ASTRIDE THE COMET

How vast is creation! I see the planets rise and the stars hurry by, carried along with their light! What, then, is this hand which propels them? The sky broadens the more I ascend. Worlds revolve around me. And I am the center of this restless creation.

Oh, how great is my spirit! I feel superior to that miserable world lost in the immeasurable distance beneath me; planets frolic about me—comets pass by casting forth their fiery tails, and centuries hence they will return, still running like horses on the field of space. How I am soothed by this immensity! Yes, this is indeed made for me; the infinite surrounds me on all sides. I am devouring it with ease.

—Gustave Flaubert,
Smarh, 1839
Translated by Jim LeBlanc

THESE ARE THE SNOWS OF YESTERYEAR, the pristine remnants of the origin of the solar system, waiting frozen in the interstellar dark. Out here trillions of orbiting snowbanks and icebergs are stored, gently suspended about the Sun. They cruise no faster than a small propeller-driven aircraft would, buzzing through the blue skies of far-off Earth. The slowness of their motion just balances the gravity of the distant Sun, and, poised between feeble contending forces, they take millions of years to complete one orbit around that yellow point of light. Out here you are a third of the way to the nearest star. Or rather, to the next nearest star: In the depth and utter blackness of the dark sky around you, it is entirely clear that the Sun is one of the stars. It is not even the brightest star in the sky. Sirius is brighter, and Canopus. If there are planets circling the star called the Sun, there is no hint of them from this remote vantage point.

These trillions of floating icebergs fill an immense volume of space; the nearest one is three billion kilometers away from you, about the distance of the Earth from Uranus. There are many icebergs, but the space they fill, a thick shell surrounding the Sun, is incomprehensibly vast. Most of them have been out here since the solar system began, quarantined from whatever mischief may be going on down there, in that alien and hostile region bordering the Sun.

Beyond the occasional soft ping of a cosmic ray from some collapsed star at the other end of the Milky Way, hardly anything ever happens here. It is very peaceful. But something *has* happened, a gravitational intrusion, not by the Sun or its possible planets, but by another star. It was slow in coming, and at its closest it was never very near. You can see it over there, glowing faintly red, much dimmer than the Sun. This cloud of icebergs has been carried with the Sun on its motions through the Milky Way Galaxy. But other stars have their own characteristic motions, and sometimes by accident approach us. So on occasion, as now, there is a little gravitational rumbling, and the cloud trembles.

Since your iceberg is bound so weakly to the Sun, even a little push or tug is enough to throw it onto some new trajectory. The neighboring icebergs—much too small and distant for you to see directly—have been similarly affected, and are now hurrying off in many directions. Some have been shaken loose from the gravitational shackles that had bound them to the Sun, and are now liberated from their ancient servitude, embarking on odysseys into the vast spaces between the stars. But for your iceberg there is a different destiny working itself out: You have been tugged in such a way that you are now falling, slowly at first, but with

gradually increasing speed—down, down, down toward the point of light about which this vast collection of little worlds slowly revolves.

Imagine that you are as patient and long-lived as the iceberg on which you are standing, that you have adequate life expectancy and life-support equipment for a journey of a few million years. You are falling toward the bright yellow star. Your worldlet and its brethren have been given a name. Comets, they are called. Your comet is an emissary from the kingdom of ice to the infernal realm near the Sun.

Out here a comet is only an iceberg. Later on, the iceberg will be just one part of the comet, called the nucleus. A typical cometary nucleus is a few kilometers across. Its surface area is the size of a small city. If you were standing on it you would see the smoothly curving contours of gracefully sculpted hillocks built of dark, reddish-brown ice. There is no air on this small world, nothing liquid, and—apart from you yourself—nothing alive, at least so far as you can see. You can, over the following millions of years, explore every corner, every mountain, every crevice. With the skies perfectly clear, and with no particularly urgent tasks before you, you can also spend a little time studying the magnificent array of unwinking bright stars that surrounds you.

Your footprints are deep, because the snow beneath your feet is weak. In a few places there are patches of ground so fragile that were you to walk upon them imprudently you would fall through—as in the legendary quicksand of Earth—onto deep shadowed ice, perhaps meters below. Your fall would be slow, though, almost languorous, because the acceleration caused by gravity, the downward force you feel here on the comet, is only a few thousandths of a percent of the familiar 1 g of Earth.

On more solid ground, the low gravity might tempt you into unprecedented athletic feats. But you must be careful. If you so much as stride purposefully, you walk off the comet altogether. With only a little effort, from a standing position you leap thirty kilometers into space, taking almost a week to reach the peak of your trajectory. There, gently tumbling, you have a comprehensive view of the comet slowly rotating beneath you, its axis by accident almost pointing toward the Sun. You can make out its lumpy shape; the comet is far from a perfect sphere. Perhaps you worry that you have jumped too high, that you will not fall back to the comet, that you will drift alone through space forever. But no, you see that your outbound velocity is gradually diminishing, and eventually, ten or twelve days after making this modest exertion, you tumble lightly back onto the somber snows. On this world you are dangerously strong.

Since it is hard to take a step without launching yourself

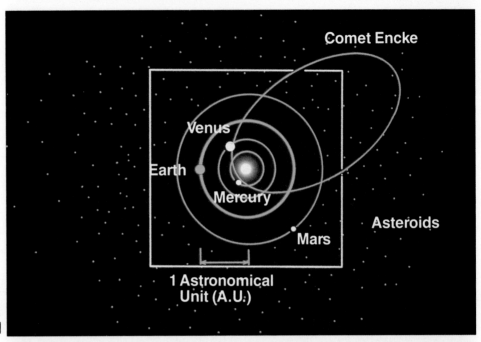

**INNER
SOLAR SYSTEM**

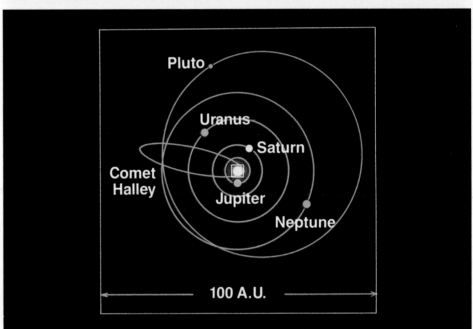

**OUTER
SOLAR SYSTEM**

The scale of the solar system.

Upper left: The inner solar system, with the Sun in the center and, shown as four concentric blue circles, the orbits of the terrestrial planets, Mercury, Venus, Earth (in blue) and Mars. Beyond the orbit of Mars is a cloud of small asteroids, shown here as dots. Also displayed is the elliptical orbit of Comet Encke. The Earth's distance from the Sun—150 million kilometers or 93 million miles—is called 1 Astronomical Unit (A.U.) and is shown in the scale.

Lower left: The boundaries of the planetary part of the solar system, as known today. The yellow boundary of the previous figure is embedded in this diagram as the small square in the center. Shown are the concentric orbits of the jovian planets, Jupiter (yellow), Saturn, Uranus (green), Neptune, as well as the smaller planet Pluto—most of the time the outermost planet, but in the late twentieth century

1877

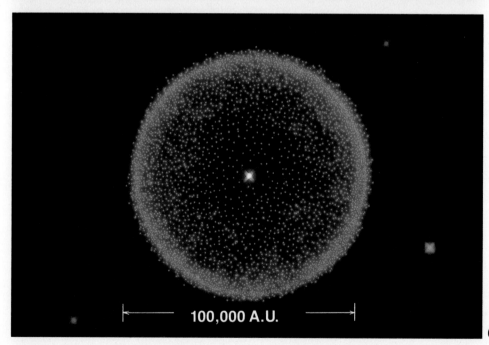

Comet Mrkos

1000 A.U.

INNER EDGES OF OORT CLOUD

100,000 A.U.

OORT CLOUD

slightly closer to the Sun than Neptune. The highly elliptical orbit is that of Halley's Comet, which at aphelion takes it beyond the orbit of Neptune. The scale of this diagram is 100 Astronomical Units.

Upper right: The inner edges of the Oort Cloud. The planetary part of the solar system is contained in the small yellow square at center, and the highly eccentric orbit of a long-period comet is shown in blue. The dots represent the inner edge of the cloud of comets, which perhaps begins 1,000 A.U. to the Sun, or even closer.

Lower right: The solar system on its grandest scale. The previous three figures are subsumed in a dot too small to see at the center of this picture, where the Sun and the planets are located. Surrounding the Sun out to a distance of 100,000 Astronomical Units is a great spherical assemblage of comets called the Oort Cloud. It reaches halfway to the nearest stars. Diagrams by Jon Lomberg/BPS.

on a small parabolic trajectory, team sports would be played in agonizing slow motion, the cluster of players rising and spinning in the space surrounding the comet like a swarm of gnats sizing up a grapefruit. A game of baseball would take years to complete—which is just as well, since you have a million years or so to idle away. But the ground rules would be unorthodox.

You pack the snow into an odd dark snowball and easily fling it off the comet, never to return again. With a flick of the wrist, without even engaging your arm in the throwing motion, you have launched a new comet on its own long, falling trajectory into the inner solar system. High above the equator of your comet, you can lightly pitch a snowball so that it hovers forever above the same point on the surface. You can make arrays of objects stationary in space, vast, apparently motionless three-dimensional assemblages, poised above the cometary surface.

As the millennia pass, you cannot help but notice that the yellow star is gradually growing more intense, until it has become by far the brightest star in the sky. The early phase of your voyage has been tedious, even if you are endowed with heroic patience, and in several million years hardly anything has happened. But you can at least see your surroundings more clearly now. The icy ground beneath you has hardly changed at all. The journey has been so long that you have been able to detect variations in the positions and, you think, even the brightness of many nearby stars. Your world is moving faster now, but otherwise everything is still, silent, cold, dark, changeless.

The comet eventually begins crossing the orbits of other kinds of objects, much larger bodies that are also bound in gravitational thrall to that beckoning point of light. As you pass close by them, you career perceptibly. Their gravities are so large that they retain massive atmospheres. Your comet, by contrast, is so insubstantial that any puff of gas released escapes almost instantly to space. Accompanying the giant gas planets with their multicolored clouds is a retinue of smaller, airless worlds, some of them made of ice—much more kin to the comets than the huge ball of hydrogen that fills your sky.

You can feel the warmth of the Sun increasing. The comet feels it too. Little patches of snow are becoming agitated, frothy, unstable. Grains of dust are being levitated over the patches. Considering the feeble gravity, it is no surprise that even gentle puffs of gas send grains of ice and dust swirling skyward. A powerful jet gushes up from the ground and a fountain of fine particles is launched far above you. The ice crystals sparkle prettily in the sunlight. After a while the ground becomes covered with a light snow. As you

1877

plunge onward, closer to the Sun, its disk now easily visible, such blowoffs become more frequent. While on one of your excursions aloft, you chance to see an active jet, a geyser pouring out of the ground. You give it wide berth. But it reminds you of the instability—the literal volatility—of this tiny world.

Far out in space, the vanguards and outriders of the columns of crystals are being blown back by some invisible influence. Eventually the cometary nucleus on which you have been riding is enveloped in a cloud of dust particles, ice crystals and gas, and the material being blown back behind you slowly forms an immense but graceful tail. If you stand on solid ground, far from the kinds of unstable ices that produce the geysers, you can still see a fairly clear sky, and track your motion by the stars. When the big jets go off, you can feel the ground shift. Here and there the ice has sheared or cracked or fallen, revealing intricately stratified layers of various colors and darkness—a historical record of the building of the comet from interstellar debris billions of years ago. By looking at the stars, you can tell that your worldlet is darting a little, rebounding in the opposite direction every time a new gusher erupts. The fountains of fine particles cast diffuse shadows on the ground, and there are now a sufficient number of them—most still of modest dimensions—that the field of darkish ice has taken on a mottled, dappled appearance.

The fine, icy grains evaporate in a few moments when heated by the increasingly fearsome sunlight, and only the dark grains that they contain will persist as solids. The substance of the comet is being converted into gas before your eyes. And the gas, illuminated by sunlight, is glowing eerily. You realize that there is now not just one tail, but several. There are straight blue tails of gas, and curving yellow tails of dust. No matter in what direction the jet happens to gush at the beginning, the unseen hand carries it away from the Sun. As the jets turn on and off, and the streamers above you curve because of the rotation of the comet, a rococo skywriting takes form. But everything aloft is relentlessly redirected by the invisible forces, the pressure of light and the wind from the Sun. The solar wind seems intermittent. So the gradually burgeoning tails form, merge, separate, and dissipate, and knots of higher brightness abruptly accelerate and then decelerate leeward from the Sun. And windward, sheets of gas and fine particles form complex and exquisite veils that change their aspect in the twinkling of an eye. This is a kind of polar fairyland, and its beauty momentarily distracts you from how dangerous it has become.

Because of the evaporation of so much ice, the ground near the exhausted gushers is friable, delicate, fragile, often no more than a matrix of fine particles stuck together billions of years ago. Soon hills of snow that had stolidly resisted the importuning sunlight for eons show signs of stirring. There is an internal motion. The ground buckles. Tentative puffs of gas are released to space, then many geysers simultaneously erupt, and you know that nowhere on the entire surface of this cometary nucleus is there a safe refuge. The comet has awakened from a four-billion-year-long trance into a wild and manic frenzy.

Later, after you pass the Sun and retreat into the interstellar night, the comet will lose its tail, and settle down. Its orbit will one day carry it back again to the inner solar system. Perhaps in some future pass by the Sun, millions of years hence, the surface will be safer because all the outer layers of ice will already have been vaporized by the heat. Only dusty and rocky stuff will be left. After many passages, a comet becomes less active, produces fewer jets, generates a less spectacular tail. As they get older, they settle down. But you are aboard a new comet, and swirling geysers are spraying the skies with dust. Maiden voyages are always the most dangerous.

You are venturing still closer to the Sun, and although the sky above you is overcast, the temperature is rising. But the surrounding haze of fine, bright particles that diffuses the sunlight also reflects it back to space, and evaporating the ice uses energy that otherwise would go into heat. If not for this protection the cometary nucleus and you yourself might become dangerously hot.

You are rounding the Sun now, racing, hurtling through this treacherous regime. You have never moved this fast before. The ground is creaking and straining, new fountains are violently erupting. You take refuge at the shadowed base of a hillock of ice whose sunward side is crumbling, evaporating, and shooting pieces of itself out into space. But eventually the activity subsides, and the skies partially clear. Formerly the Sun was ahead of you; now it is behind, on the other side of the nucleus.

Through a break in the surrounding nebulosity you realize that you are passing close to a small, blue world with white clouds and a single battered moon. It is the Earth in an unknown epoch. There may be beings there who will look up and see this apparition in their skies, who will note the great blue and yellow tails streaming away from the Sun, the complex pinwheel-shaped patterns of fine-grained material being jetted off into space, and they will wonder what it means. Some of them might even wonder what it is.

It is surprising how closely you are passing, and it occurs to you that sooner or later some comet is going to run smack into this little planet. The Earth would survive such an

1878

impact, of course—although there would doubtless be minor attendant changes, some species of life having failed, perhaps, and others newly promoted. But the comet would not survive. It would fall deeper and deeper into that atmosphere, large fragments, whole hills of the nucleus separating off, flames licking through the crevices into its hidden interior. Perhaps enough of the comet would survive to make a huge explosion, generating a large hole in the ground down there, and spraying up a cloud of surface dust. But of the comet itself, all the ices would have vaporized. All that would be left would be a sprinkling of fine, dark grains scattered like birdseed or buckshot upon this alien land.

But, you reassure yourself, running into a world is an unlikely event. On this swing past the Sun, at any rate, you will not collide with anything larger than occasional motes of interplanetary dust, the remnants of past comets that have spent their substance dashing through the realm of fire. With a last glance at the blue world, you silently wish its inhabitants well. Perhaps they will consider their skies comparatively drab and cheerless once your comet has departed. For yourself, you are relieved to be on the return trajectory, out of the deadly heat and light, and heading back to the placid cold and dark, where, except for an occasional unlucky jostling by a passing star, comets can live forever.

The long and lovely tails now precede you on your journey; the wind from the Sun is at your back.

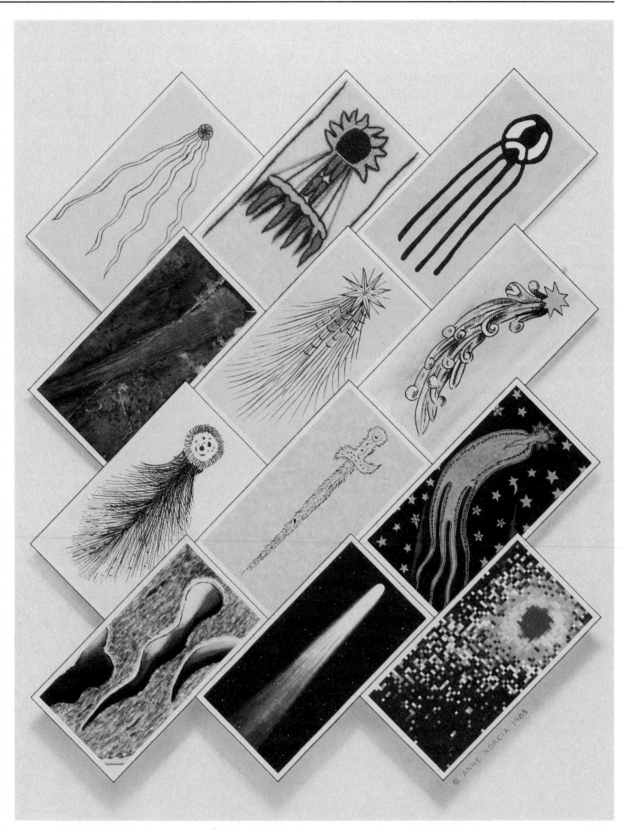

The comet as depicted in a variety of different ages and cultures. Painting by Anne Norcia. For a detailed key see page 378.

Chapter II

PORTENT

If at your coming princes disappear,
Comets! come every day—and stay a year.

—Samuel Johnson,
Letter to Mrs. Thrale,
October 6, 1783

FOR A MILLION YEARS OR MORE, THERE HAVE BEEN human observers to wonder at the grandeur of the comets that intermittently grace the skies of Earth. Memorable comets visible to the naked eye have appeared on average once a decade, a few times a lifetime. On a hundred thousand occasions during the tenure of humans on Earth our ancestors must have seen a diffuse streak of light, brightening night after night, sometimes outshining even the brightest stars, and then, weeks or months later, slowly fading from view. A hundred thousand apparitions. And yet we have not a single concrete recollection of any comet earlier than the last three or four millennia. In this sense, as in many others, we are a species of amnesiacs. We are estranged from our past. We can only guess what our ancestors thought when the habitual serenity of the heavens was so spectacularly interrupted.

For most of that time humans enjoyed a far more intimate relationship with the sky than we do now. We may know much more than they did about what is actually up there, but they were much more a part of it. They slept among the stars. They looked to the sky to tell them when to camp and when to move on, when to expect the migratory game and the rains and the bitter cold. Appropriately, they watched the sky as if their lives depended on it, but also because it was a puzzle whose complex beauty moved them. They invested the heavens with hypotheses, explanations, and metaphors that we call myth. The apparition of a bright comet is a recurring occasion for marvel and speculation, and thus the comets prodded us along, a little at least, as we were finding our way, along the road to consciousness.

Comets were a kind of psychological projective test—something wholly unfamiliar that you must describe in ordinary language. The Tshi people of Zaire call comets "hair stars," and the word comet—the same in many modern languages—comes from the Greek word for hair. A comet suggests flowing tresses. To the Chinese, comets were "broom stars," and much else. In other cultures, they are "tail stars" or "stars with long feathers." Scientists today still describe comets as having "tails." To the Tonga, comets are "stars of dust," which is much closer to the truth. The Aztecs saw them as "smoking stars." Among the Bantu-Kavirondo, all cometary apparitions are returns of one and the same comet: "There is but one comet, Awori, the feared one, with his pipe."

It was hard not to take the comets personally. Nearly everybody did. This tendency to construe each apparition as a telegram to the locals from the gods permeates virtually every cometary record until the sixteenth century. Rarely have so many diverse cultures, all over the planet, agreed so

The tails of many comets suggest flowing hair. This is a drawing of Donati's Comet on September 29, 1858, as seen through the telescope by G. P. Bond, Harvard College Observatory.

1880

well. In the history of the world, more societies have advocated incest or infanticide than have taught that comets were benign, or even neutral. Everywhere on Earth, with only a few exceptions, comets were harbingers of unwanted change, ill fortune, evil. It was common knowledge.

In their myths, the tribal peoples of Africa may have preserved something of our original perceptions of comets. To the Masai of East Africa, a comet meant famine; to the Zulu of South Africa, war; to the Eghap of Nigeria, pestilence; to the Djaga of Zaire, specifically smallpox; and to their neighbors, the Luba, the death of a leader. The !Kung of the Upper Omuramba in what is now Namibia were alone in their optimism. They saw the comet as a guarantee of good times ahead. This is such an unusually cheerful interpretation that you might well ask who the !Kung are. (The ! stands for a click you make by touching your tongue to the roof of your mouth at the same instant you say the K. It takes some practice.) They are hunter-gatherers, with a rich culture that is closer to the long-term human norm than almost any other culture today. So you might wonder whether the !Kung knew something about nature—or themselves—that the rest of us have forgotten.

There is an overwhelming sadness to the literature on comets. With melancholy consistency we discover that disaster has always been a commonplace; that any comet at any time viewed from anywhere on Earth is assured of some tragedy for which it can be held accountable. This connection of comets and misfortune is made in the earliest surviving reference to a comet,* a single Chinese sentence from the fifteenth century B.C.:

> When Chieh executed his faithful counselors, a comet appeared.

These words, associating official perfidy and murder with the early stirrings of astronomy, were written two centuries before the birth of Moses. Three hundred years later, another writer noted,

> When King Wu-wang waged a punitive war against King Chou, a comet appeared with its tail pointing toward the people of Yin.

---

*And the most recent. Thirty-five hundred years later, comets continue to be linked with catastrophe. In Chapters 15 and 16 we discuss the current scientific debate over whether comets are responsible for the extinction of the dinosaurs and most other species. Even "apparition"—the word we use today to describe the presence of a comet in our skies—is an echo, with ominous and supernatural overtones, of our ancient belief about comets.

Japanese comet calligraphy by Takako Suzuki.

A Japanese tsuba, or sword guard, from the seventeeth century, illustrating what may be a comet over the Sumiyoshi Shrine. (The putative comet can be seen just below the crescent moon at top.) Courtesy David Pepper/Okamé Antiques, Toronto.

The Yinnites were in trouble.

If there was unanimity that comets caused disasters, there was disagreement as to what to do once you saw one. In Master Tso Ch'iu Miu's *Enlargement of the Spring and Autumn Annals*, written between 400 and 250 B.C., there is an entry entitled, "Yen Tsu Argues Against the Use of Prayer in Averting Disaster from Comets":

In the year [516 B.C.] a comet appeared in the kingdom of Ch'i. The king of Ch'i wished to send his ministers to pray to heaven. Yen Tsu dissuaded the king, saying, "It's useless. You're only fooling yourself. Whether heaven will give you a disaster or good fortune is set; it won't change. How can you expect prayer to change anything? A comet is like a broom: It signals the sweeping away of evil.* If you have not done anything evil, why do you need to pray? If you have done something evil, praying won't avert disaster. The work of the Ministers of Prayer won't change fate."

The king was delighted to hear this, and ordered the praying stopped.†

A survey of the early history of cometary observation gives the impression that for almost a thousand years everybody except the Chinese went to bed early. The Chinese logged at least 338 separate apparitions from roughly 1400 B.C. to 100 A.D. Since 240 B.C. they have missed a return of Halley's Comet only once, in 164 B.C. Their neighbors, the Koreans and the Japanese, made valuable but far less frequent observations. In the West, nothing approaching systematic observation of comets was in evidence until the fifteenth century.

In the 1970's, an excavation of the "Number Three Tomb" at Mawangdui, near Changsha, revealed the most impressive example of Chinese priority in this field—an illustrated textbook of cometary forms, painted on silk. Part of a larger work concerning clouds, mirages, halos, and rainbows, it was compiled around 300 B.C. Twenty-nine comets are displayed, classified by their appearance and by the particular brand of mayhem each foretells. Eighteen of the thirty-five different names known in Chinese for comets are given here. The four-tailed comet signifies "disease in the world," the three-tailed one, "calamity in the state." A comet with two tails that curve to the right

---

*This is a pun. One of the many Chinese ideograms for comet is "broom star."
†Translated by Heather Smith and Xie Yong.

Small Arbiter of Human Destiny

Chinese comet
calligraphy.
Translation:
Broom Star.
Calligraphy by
Takako Suzuki.

1880

Autumn orchids, luxurious jungle
Spread life below the hall.
Green leaves, white blossoms
A rich and fragrant scent overtakes you.
From every person comes lovely children.
Why then, my Lord, such bitter sorrow?

Autumn orchids, fresh and lush
Green leaves, purple stem.
The hall is filled with lovely people.
Suddenly alone with me, a meaningful glance.

He came without a word, went without saying goodbye.
Riding the whirlwind, carrying the cloud banner.
Grief beyond grief are life's separations.
Joy beyond joy are new friends.

Lotus garments, sweet basil belt
Suddenly came, hurriedly left.
His evening lodgings, the gods' frontier.
You wait for whom at the clouds' edge?

With your lady, bathe in the Pool of Union.
Dry her hair under the Sun.
I search the sky for my lovely one—why is he not yet here?
Face to the wind, indistinct, I lift up my voice in song.

Peacock-feather canopy, kingfisher-blue flag
Climb up to the Ninth Heaven, soothe the comet.
Grasping his long sword, he protects and nurtures the young.
My Lord alone is fit to bring justice to mankind.

> —Ch'ü Yüan (340–278 B.C.)
> Translated by Heather Smith
> and Xie Yong

Ch'ü Yüan, statesman and one of China's most beloved ancient poets,
committed suicide while in exile by drowning himself in a river. Every May
5, people in China still throw specially prepared rice into the rivers in a
symbolic placatory gesture to prevent the fish from devouring his body. In
this poem, the comet is both a metaphor for the lost lover and a link to the
wise and compassionate god. Many of the seemingly irrelevant images—
such as "cloud banner" and "blue flag"—are in reality allusions to the
numerous Chinese names for comets.

As late as 1528, European perceptions of comets had distinct hallucinatory elements. In this woodcut of the comet of that year, a mélange of decapitated heads and miscellaneous implements of warfare are depicted. The picture was based on the description of this comet by Ambrose Paré. After Amédée Guillemin, *The World of Comets*, Paris, 1877.

But after the conquest of Troy and the annihilation of its descendants,… overwhelmed by pain she separated from her sisters and settled in the circle named arctic, and over long periods she would be seen lamenting, her hair streaming. That brought her the name of comet.

—Hyginus, *De Astronomia*, ca. 35 B.C.
Translated from the Latin by Milad Doueihi

promises a "small war," although at least "the corn will be plentiful."

How long would it take to assemble a catalogue of 29 distinct cometary forms? With 338 separate sightings recorded over 3,000 years in the surviving Chinese annals, the average discovery rate is about one bright naked-eye comet a decade—not too far from current values. If each of the 29 forms appears equally often, you would have to wait $29 \times 10 = 290$ years to see them all. But some cometary forms are much rarer than others. Thus, if every form depicted corresponds to a different comet, the Mawangdui atlas must have drawn on an earlier continuous tradition of systematic observation which precedes it by many centuries, possibly millennia. Accordingly, this splendid tradition of recording cometary forms must date back to 1500 B.C. or earlier. The earliest written and the earliest graphic representations of comets thus trace back to, or at least through, the same epoch. Perhaps there was an out-of-the-ordinary comet then that commanded their attention.

Our admiration for the Mawangdui silk is further enhanced when we consider the depictions themselves, which are at least roughly consistent, in a few bold strokes, with modern photographs of the comets. These observers drew what they saw. We have only to compare these images with a European woodcut of the Comet of 1528 (this page) to appreciate their sobriety: no dragons; no devils; no implements of torture. Just comets.

In surveying the long-standing Chinese absorption with comets, we are reminded of another area in which the Chinese led the way, the invention of fireworks. Might the ancient Chinese have devised skyrockets to adorn the heavens during the long and tedious intervals between comets? Even if rockets and comets were not connected then, they certainly are today (Chapters 6 and 18).

The ancient Chinese assembled a large, accurate, and detailed body of data on comets. Their catalogues list, for hundreds of apparitions, many of the following pieces of information—a date, the kind of comet, the constellation in which it was first seen, its subsequent motion, its color and apparent length, and how long before it disappeared. Sometimes day-by-day changes in the length of the comet's tail are recorded. But for all this, they never had an inkling of what comets really were. This was to be wholly an achievement of the West, although it was a long time in coming. Western cometary astronomy prior to the Renaissance is a chronicle of occasional episodes of lucidity—especially in Ionia, Athens, and Rome—punctuating a far longer and more widespread gloom of ignorance, superstition, and delusion.

Record of the World's Change

Comets are vile stars. Every time they appear in the south, something happens to wipe out the old and establish the new. Also, when comets appear, whales die. In Sung, Ch'i and later Ch'in times, when a comet appeared in the Constellation of the Big Dipper, all soldiers died in chaos....

When a comet appears in the North Star, the emperor is replaced. If it appears in the end of the Big Dipper, everywhere there are uprisings and war continues for several years. If it appears in the bowl of the Dipper, a prince controls the emperor. Gold and gems become worthless. Another explanation: Scoundrels harm nobles. Some leaders appear, causing disturbances. Ministers conspire to rebel against the emperor....

When a comet travels north but points south the country has a major calamity. Western neighbors invade and later there are floods. When a comet travels east and points west, there are uprisings in the east.

...When a comet appears in the Constellation Virgo, some places are flooded and there is severe famine. People eat each other....If the comet appears in the Constellation Scorpio, there are uprisings, and the emperor in his palace has many worries. The price of rice goes up. People migrate. There is a plague of locusts.

...When a comet appears in the Constellation Andromeda, there are floods and migrations of people. Many rise up and the country is divided by civil war. When a comet appears in the Constellation Pisces there is first drought and later flooding. Rice is expensive. Domesticated animals die and an epidemic strikes the army.

When a comet travels into the Constellation Taurus, in the middle of the double month,* blood is shed...dead bodies lie on the ground. Within three years the emperor dies and the country is in chaos. When a comet appears in Orion there are major uprisings. Princes and ministers conspire to become emperor. The emperor has many worries. Everywhere there is disaster by war....

When a comet appears in the Constellation Hydra, there is war and some conspire to overthrow the emperor. Fish and salt are expensive. The emperor dies. Rice also becomes expensive. There is no emperor in the country. The people hate life and don't even want to speak of it.

—*Record of the World's Change*
Li Ch'un Feng, 602–667 A.D.
Translated by Heather Smith and Xie Yong

*The Chinese calendar at this time was a lunar one, but in every year there was a "double month," to adjust to solar time.

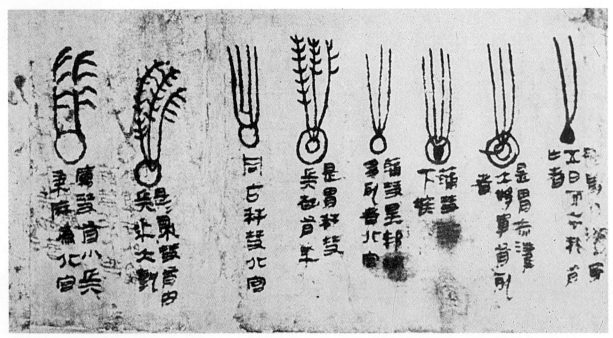

A portion of the world's first cometary atlas, the Mawangdui silk, ca. 300 B.C. Wen wu. "Ma Wang Tui po shu 'T'ien wen ch'i hsiang tsa chan' nei jung chien shu" and "Ma Wang Tui Han ts'ao po shu chung to hui hsing t'u." Beijing: Wen wu ch'u pan she. 1978, Volume 2, pp. 1–9.

The earliest unambiguous references to comets in the West come from what is today Iraq. The few surviving Babylonian fragments remind us of their African and Chinese counterparts. Consider this, from the time of Nebuchadnezzar I, in the twelfth century B.C.:

> When a comet reaches the path of the Sun, Gan-ba will be diminished; an uproar will happen twice...

The diminishment of Gan-ba is bad news, you can be sure. Occasionally, as in this notice from the same time and place, the auspices are favorable:

> When a star shines and its brilliance is as bright as the light of day, [when] in its shining it takes a tail like a scorpion, it is a fortunate omen, not for the master of the house, but for the whole land.

A comet bright enough to be seen in the daytime sky must have come very close to the Earth, or the Sun.

The confidence with which these ancient astrological pronouncements are made is striking. There is no hedging of bets, no ambiguity, and no curiosity. We never find a confrontation of two contending hypotheses, much less an appeal to observations to decide the issue. Science had not yet been invented.

In the works of Diodorus of Sicily (ca. 60–21 B.C.) and Lucius Annaeus Seneca of Rome (ca. 4 B.C.–65 A.D.) there is indirect evidence—or perhaps it is only hearsay—that the

Egyptians and Babylonians had devised some scientific understanding of the comets. Diodorus wrote,

> As a result of their long observations, they [the Egyptians] have prior knowledge of earthquakes and floods, of the rising of comets, and of all things which the ordinary man looks upon as beyond all finding out.

The annual times of flooding of the Nile Valley were well-known to the ancient Egyptians. From the odd behavior of animals, it is possible—as the modern Chinese have demonstrated—to predict an earthquake early enough to save many lives. But correctly foretelling the apparition of a comet is much more difficult. Maybe someone made a lucky guess.

Seneca reports an opinion that the Babylonians believed comets were bodies something like planets. There is no elaboration. We do know that the Egyptians and Babylonians made seminal contributions to mathematics. However, it was in Greece in the fifth century B.C. that curiosity turned away from the supernatural, and first found its world-changing means of expression: science.

Everything we know of the inventors of this new way of thinking comes to us secondhand. Democritus (born around 460 B.C. and believed to have lived to a very old age) wrote at least seventy works, all of which have been destroyed or lost. Our knowledge of Democritus comes mostly from Aristotle (384–322 B.C.), who held him in high regard and disagreed with virtually everything he had to say. We are told that Democritus believed comets to be produced when one "star" passed near to another. Democritus may have properly distinguished star from planet, but even if not, he was on the right track: Comets, he was saying, are celestial bodies and arise by natural processes. So far as we know, no one had ever proposed so outlandish an idea before.

Aristotle believed he could disprove this hypothesis by noting that in his time Jupiter had come close to a star in the constellation Gemini, and had not produced a comet. But Aristotle did not know that the star was light-years behind the planet, and that only from our perspective were they passing "near" one another. In this matter, as in most others, Democritus seems to have been the better scientist. But Aristotle's argument appealed to observation, not myth or conventional wisdom. The debate was scientific. This also seems to be the earliest mention of the possibility that comets are disgorged by Jupiter; the idea has had nine lives at least, and was not so long ago a cause of vigorous controversy both in the United States and in the Soviet Union.

Aristotle had other reasons for believing that comets could not live among the planets, reasons again based partly on observation. He framed a supportable scientific hypothesis, which went something like this: The zodiac is the succession of constellations, many named after animals, through which over the months and years the planets, the Sun, and the Moon move. (During daylight, you can't see what constellation the Sun is in, of course, but with a star map near twilight or dawn you can tell.) The zodiac runs all the way around the sky at an angle to the horizon. For all our ancestors knew, the planets, the Moon, and the Sun might, over the course of a lifetime, wander through every constellation in the sky. Since this does not happen, all the planets must lie very nearly in the same plane. In contrast, comets are observed to travel sometimes within, but sometimes well outside the zodiac. Furthermore, unlike planets, comets change their forms in a few days, before the eyes of the observers. Thus comets could have nothing in common with planets. They must be sublunary, beneath the Moon—that is, within the Earth's atmosphere. (Aristotle thought the Moon represented the farthest reach of the atmosphere.) The conclusion was clear: Comets were a form of weather. Although there was some debate early on, this view held sway for two thousand years.

All of Aristotle's astronomy was predicated on his deep conviction that the heavens were "free from disturbance, change and external influence..." He believed that the Earth was absolutely stationary in space—as if nailed down. The heavens, on the other hand, were whipping around the Earth at the brisk rate of a rotation a day. The bottom of the atmosphere clearly is stationary with the Earth. But the top of the atmosphere must share the sky's rotation. Now, imagine an exhalation of hot, dry gas from the Earth—perhaps through a fissure, a crevice, or a volcano. That gas will rise and, when reaching the sky, be heated by the Sun and, he thought, burst into flame. But since the burning gas has reached the realm of the heavens, it must now move with the stars and the planets. This was Aristotle's explanation of the comets. And—given the limitations of the science of his day—it was far from foolish.

He taught that the aurora borealis and even shooting stars were examples of the same sort—exhalations from the interior of the Earth rising to the stars. Comets survive, he taught, until all the gas had burst into flame. New comets were due to new exhalations. There was thus a balance or steady state between the production and the destruction of visible comets, an idea still central to understanding them. Aristotle held that there are so few comets because most of the flammable vapors outgassed by the Earth were otherwise employed—in producing the continuous band of fire called the Milky Way. In contrast, Democritus had concluded that

the Milky Way is composed of an enormous number of stars, so far away that we cannot see them individually—exactly the right answer.

Underlying Aristotle's scientific arguments was a quasi-religious doctrine: He was forced to concoct a terrestrial source for the comets, since he had locked them out of his changeless skies, decreeing that no new celestial bodies may be born, and no old ones die. His insistence on the immutability of the heavens was the most influential error in the history of astronomy, contributing to a detour from reality that lasted nearly two millennia. But Aristotle cannot be held fully accountable for the credulous acceptance accorded his opinions by succeeding generations.

Seneca, born in Cordoba, Spain, to a wealthy and famous family, was a contemporary of Jesus, although the two never met. His brother, though, was an acquaintance of Saint Paul. As a young man, Seneca came to Rome, where he studied grammar, rhetoric, law, and philosophy. He enjoyed a substantial reputation as writer and orator until the year 41, when he was banished to Corsica for sleeping with Caligula's sister. Considering this emperor's predilection for cruelty, the punishment was mild. Seneca spent his years of exile writing and studying philosophy and natural science; astronomy was among his pursuits.

In the year 49, he was recalled to Rome to teach. As a tutor, his success was mixed, his only pupil being the future emperor, Nero. When at the age of seventeen Nero was raised to the purple, Seneca became political advisor to the emperor and minister of state. For the next eight years, Seneca and Sextus Afranius Burrus, the commander of the Praetorian Guard, ran the Roman Empire. By all accounts they performed well, fostering fiscal and judicial reforms and somewhat easing the lot of the slaves. But Nero grew more tyrannical, Burrus died—perhaps of foul play—and Seneca's political power withered. He withdrew from public life, and wrote some of his most celebrated works until, in the year 65, he received an imperial command to commit suicide for his alleged involvement in a conspiracy against the Empire. He died with courage and composure.

Seneca left works on many subjects, but it is his *Natural Questions*, written during the last years of his life, that concerns us here. The seventh "book" is entitled "Comets," and Seneca does the subject considerable justice, taking on Aristotle with some success. He argues that comets could not be atmospheric disturbances: they move with stately regularity and are not dissipated when the wind is blowing. Thus, "I do not think that a comet is just a sudden fire, but that it is among the eternal works of nature." In refuting

A bronze bust of Seneca from the Hellenistic period, now in the National Museum, Naples. The Bettmann Archive.

This one description ought to be generally agreed upon: [A comet is] an unusual star of strange appearance ...seen trailing fire which streams around it.

—Seneca,
Natural Questions,
Book 7, "Comets"

Aristotle's argument that comets cannot be planets since they are not restricted to the zodiac, Seneca asks, faintly echoing Job:

> Who places one boundary for planets? Who confines divine things in a narrow space? [The planets]...have orbits that are different from one another. Why, then, should there not be other stars which have entered on their own route far removed from them? What reason is there that in some part or other of the sky there should not be a passageway?

To the objection that stars can be seen through comets, so comets must be incorporeal and cloudlike, Seneca replies correctly that the transparency applies only to the tail, not necessarily to the head.

One of the most fascinating passages is an exposition and critique of the views of one Apollonius of Myndos, an otherwise unknown Greek scholar from the fourth century B.C.:

> ...Many comets are planets...a celestial body on its own, like the Sun and the Moon. It has a distinct shape...not limited to a disk, but extended and elongated lengthwise....A comet cuts through the upper regions of the universe and then finally becomes visible when it reaches the lowest point of its orbit.... Comets are many and various, different in size, unlike in color....Some are bloody, menacing—they carry before then the omen of bloodshed to come. Others diminish and increase their light, just as the other stars [planets] do which are brighter and larger when they descend because they are seen from a closer position, and are smaller and dimmer when they recede because they are withdrawing far away.

Except for the business of comets as omens, Apollonius' views—which Seneca shares and extends—seem astonishingly modern.

In "Comets," Seneca's style is so direct that you have no difficulty hearing his voice speaking inside your head. "We can only investigate these things," he tells you, "and grope in the dark with hypotheses, not with the assurance of discovering the truth, and yet not without hope." He continues,

> Many things that are unknown to us the people of a coming age will know. Many discoveries are reserved for ages still to come....Nature does not reveal her mysteries once and for all. We believe we are her initiates, but we are only hanging around the forecourt.

And the loiterers in the temple forecourt were dwindling. Seneca laments what he saw in his time as a waning of "interest in philosophy. Accordingly, so little is found out from these subjects which the ancients left partially investigated, that many things which were discovered are being forgotten." For good reason, he surmised a growing intellectual lassitude in his world. His reliance on reason, his willingness to decide between alternatives on the basis of objective evidence, stand in marked contrast to the approach of the next generation—of Lucan (39–65), for example, Seneca's own nephew, who wrote: "The heavens appeared on fire, flaming torches traversed in all directions the depths of space; a comet, that fearful star which overthrows the powers of the Earth, showed its horrid hair." Or consider the view of the naturalist Pliny the Elder (23/24–79):

> A fearful star is the comet, and not easily appeased, as appeared in the late civil troubles when Octavius was Consul; a second time by the...war of Pompey and Caesar; and, in our own time, when, Claudius Caesar having been poisoned, the empire was left to Domitian, in whose reign there appeared a blazing comet.

Lucan and Pliny were the wave of the future.

Josephus mentions in his *History of the Jews* that a "sword" hung over Jerusalem for a whole year, foretelling the destruction of the city in the reign of the Emperor Vespasian. This is probably a reference to the apparition in the year 66 of Halley's Comet. But how could a natural object hang over a city for a year? The Earth turns. A comet would rise and set with the stars, and surely cannot be described as "hanging" over any place on the Earth's surface. Meteors streak and vanish in an instant, planets cannot look like a sword, the aurora borealis is too far north, and even an artificial satellite cannot hover over nonequatorial Jerusalem. Barring a miracle, Josephus' hanging sword—unrecorded by any other chronicler of the time—must, along with other matters in his chronicle, be considered skeptically.

As classical learning dwindled over the centuries, the triumph of superstition was symbolized by the deaths of two emperors. The comet of the year 79 was said to portend the death of the Emperor Vespasian, who had already found himself involved with cometary portent. Although a shameless contriver of divine omens to justify and authenticate his seizure of the empire, Vespasian in this case expressed a healthy skepticism. "That hairy star does not portend evil to me," he said. "It menaces, rather, the King of the Parthi-

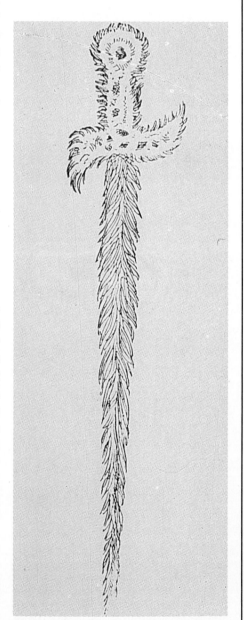

Representation of a comet as a sword; the figure was drawn, after the description in Pliny, but in a much later time. From Hevelius' *Cometographia*, 1668.

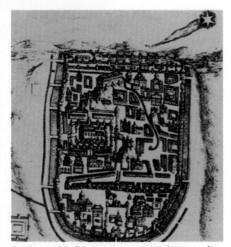

Seventeenth-century woodcut of a comet hanging over Jerusalem according to the account of Josephus. From *Theatrum Cometicum* by Stanislau Lubienietski, Amsterdam, 1668. From the collection of D. K. Yeomans.

ans," his longtime adversary. "He is a hairy man," Vespasian explained, "I am bald." His skepticism did not save him, though, and he died the same year. By the early Middle Ages, the association of comets with the deaths of princes was so deeply ingrained that when a ruler died and nothing untoward was seen—as was the case at the death of Charlemagne in 814—a fiery apparition, otherwise invisible, was generally acknowledged to be blazing in the skies: the Emperor's new comet.

You steep yourself in the cometary discourse of the whole of the Middle Ages, and you cannot find anyone within shouting distance of Democritus, Apollonius, Aristotle, or Seneca. The medieval treatises are full of divination and portent, omen and blood, mysticism and superstition, with hardly ever, even as a passing thought, the view that comets might be just a part of nature, and not a warning to the wayward. Even the historian Isidor, Bishop of Seville (602–636), who denounced astrology and astrologers, believed that comets presaged "revelations, wars, and pestilence." For more than a millennium there was no dissenting opinion, and even the occasional attempt at purely factual statements—as the announcement by the Venerable Bede (673–735) that comets are never seen in the western sky—were often wholly in error on straightforward matters of observation.

So, when the Renaissance and the Enlightenment finally came, a new breed of scholars arose who were predisposed to hold the Church responsible for superstition and ignorance —about comets, and many other matters. Callixtus III was a Spanish-born Borgia, untainted by scandal, who was made Pope in 1455 at the age of seventy-seven. It was the obses-

A depiction of what would later be called Halley's Comet over the city of Nuremberg in 684. The woodcut was made well after the event. From *The Nuremberg Chronicle* (1493). Courtesy Rare Books and Manuscripts Division, The New York Public Library, Astor, Lenox and Tilden Foundations.

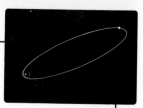

sion of the end of his life to recover Constantinople from the Turks, the adversary ideologies of the time being as fervent and narrow as the still deadlier struggle in our own day between capitalism and communism. Each side confidently called upon the One True God, and publicly proclaimed the inevitability of victory.

In 1456, a great comet appeared in European as in Chinese skies. Once detected, the comet seems to have struck real terror in beleaguered Christendom. We know today, as Callixtus and his contemporaries did not, that what they were seeing was the periodic return of Halley's Comet. Callixtus was convinced, so the story goes, that the visitor was an evil omen, somehow allied with the Turkish cause; accordingly, he excommunicated it, ordering this heartfelt plea inserted into the *Ave Maria*: "From the Devil, the Turk, and the Comet, Good Lord, deliver us."

According to many accounts, Callixtus sent 40,000 defenders to Belgrade, a city under Christian control besieged by the Turks, where, on August 6, 1456, with Halley's Comet hanging overhead, a great battle was fought which lasted two days. A later historian described the battle in these words: "The Franciscans, unarmed, crucifix in hand, were in the front rank, invoking the papal exorcism against the comet, and turning upon the enemy that heavenly wrath of which none in those times dared doubt." The armies of Mohammed II were repulsed, and both the comet and the Turks retreated. (The Turks composed memorable images of the comet, of which a particularly lovely example can be seen on page 12, second from the bottom on right.) Constantinople was never recovered by the Christians.

One astronomical writer after another has quoted this account of the evenhanded anathematization of Turk and comet, and yet there is no record of such a prayer or curse in the Vatican archives, and no evidence that Callixtus excommunicated the comet, its previous adherence to the tenets of Christianity being at any rate in doubt. On June 29 a papal bull was issued ordaining public prayers for the success of the Crusade; there is not a word in it about the comet, which was invisible to the naked eye by the middle of July; the decisive victory over the Turks was a few weeks later.

As far as can be determined, the story most astronomical writers relate about Callixtus and the comet traces back to *The System of the World*, by Pierre Simon, the Marquis de Laplace, whom we will meet again in these pages. A brilliant scholar who left an enduring mark on the history of physics and astronomy, he was also a partisan of the French Revolution and its rationalistic underpinnings. His *System of the World* was published "in Year IV of the Republic"

The Comet of 1066 as depicted on the Bayeux Tapestry, which was completed soon after this apparition. King Harold of England, about to be overthrown by the Norman invaders, wonders whether to heed the portent. Crowds of onlookers (*left*) admire the comet. Courtesy International Halley Watch.

Montezuma II of Mexico is distressed to see an omen in his sky. When Hernán Cortés appeared in 1519, Montezuma regarded his arrival as the fulfillment of dire cometary prophecy. This image, from *Los Tlacuilos de Fray Diego Durán* (México, Cartón y Papel de México, 1975), was painted sixty years afterwards. Courtesy Ruth S. Freitag, Library of Congress.

(1796). Because the Church had been intimately allied with the brutal regime of the Bourbon kings, it is possible that Laplace was disinclined to be overly generous to Callixtus III. But Laplace did not make the story up, and the confusion seems to trace back to the 1475 work *Lives of the Popes* by the Vatican librarian, Platina. The image, however spurious, of a Pope solemnly excommunicating a comet was deemed at least consistent with the prevailing opinions on comets, both secular and ecclesiastic.

Far from the Vatican, and a generation later: In Tenochtitlán, the Aztec emperor, Montezuma II (1466–1520), was awaiting the great white-bearded god, Quetzalcoatl, whom prophecy had decreed would return to Mexico and reclaim his empire. When two brilliant comets arrived in rapid succession, and appeared to rendezvous in the sky, Montezuma took them as a sure omen that Quetzalcoatl was on his way, that the Aztec empire was his no longer. Withdrawn and disconsolate, he took every fire, storm, and oddity of nature as a further portent. The master of the greatest empire in the Western Hemisphere was reduced to immobility by two comets and a prophecy. And so, in 1519, when the white-bearded conquistador Hernán Cortés arrived out of the eastern sea with an expeditionary force of 600 men and a few horses, Montezuma did not have to be much persuaded. He

1885

handed the empire back to Quetzalcoatl. The Aztecs were helpless against Cortés' small band for more reasons than one. But the conquest and plunder of Mexico, and the annihilation of the Aztec civilization, were in some significant measure due to a fatalistic dread of comets.

Across the Atlantic, the Protestant Reformation was soon underway. The contending sects, divided on many theological issues, were in perfect harmony on the matter of comets. On this issue at least, the movers and shakers of the Reformation saw eye to eye with Montezuma II. Martin Luther (1483–1546) spoke for all of them when, in an Advent sermon, he preached: "The heathen writes that the Comet may arise from natural causes; but God creates not one that does not foretoken a sure calamity." No hedging of bets here. Andreas Celichius, the influential Lutheran bishop of Altmark, in 1578 described comets as

the thick smoke of human sins, rising every day, every hour, every moment, full of stench and horror before the face of God, and becoming gradually so thick as to form a comet, with curled and plaited tresses, which at last is kindled by the hot and fiery anger of the Supreme Heavenly Judge.

This view of comets as congealed sin ignited by the Wrath of Heaven shows how many steps backward had been taken since the time of Aristotle (to whom Celichius is clearly indebted). A retort was provided by Andreas Dudith in the following year: "If comets were caused by the sins of mortals, they would never be absent from the sky."

From all this, you might have never guessed in the middle of the sixteenth century that a revolution in human understanding of comets, and much else, was about to happen. In Denmark, in the very year that Luther died, there was born just such a "heathen" as Luther had denounced. His name was Tycho Brahe. With his artificial nose of gold or brass, his entourage of dwarfs, his legendary drinking parties, and his palatial island observatory, he was not your typical astronomer.

In Tycho's time, the unimpeachable authority on comets was still Aristotle. His doctrine that comets were confined to the Earth's atmosphere because the heavens were fixed and changeless was a cornerstone of the sixteenth-century model of the universe, endorsed by secular and religious authorities alike. This was not a controversial issue. All knowledgeable experts agreed with Aristotle. The first real doubts were raised on a night in 1572 when Tycho looked up at the constellation Cassiopeia and saw a star "brighter than Venus," where no star had been before. The new star was far beyond the Earth's atmosphere, and is even now known as

A Dutch coin bearing the date November 14, 1577, and depicting the Great Comet of that year flying above the clouds and the landscape of the Earth. In English, the Latin inscription reads, "The Star of Offended Divinity." Courtesy American Numismatic Society.

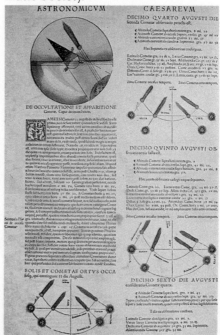

The 1531 apparition of Halley's Comet, as depicted in the *Astronomicum Caesareum* of Apianus (1540). Note that the tail of the comet is correctly shown pointing away from the Sun. Courtesy Ruth S. Freitag, Library of Congress.

If a comet were something in the Earth's atmosphere, as Aristotle maintained, then two observations from widely separated points would see it against quite different background stars *(opposite)*. In this diagram the blue cones represent the lines of sight of the two observers. But if the comet is very far away *(above)*, the widely separated observers see the comet against almost the same stellar background. Simultaneous observations can therefore tell us the distance from the Earth to the comet. Diagrams by Jon Lomberg/BPS.

Tycho's Supernova—*nova* is the Latin word for new. Evidently, and to nearly everyone's surprise, the heavens were not immutable. Aristotle and the Church had been wrong. The supernova of 1572 was a reveille for the astronomers of Europe and, soon, for the culture of the world.

Five years later, a great comet blazed across the skies of Europe to overturn decisively Aristotle's now-tottering world view. Because the Comet of 1577 was visible for an extended period, Tycho and his colleagues were able to share information and test each other's hypotheses. The supernova of 1572 had prompted Tycho to approach the comet as if it were an astronomical body, rather than an atmospheric disturbance.

If you look up at a comet you see it against a backdrop of more distant stars. As time goes on, it may move from constellation to constellation (page 33, top left), but in any given few-day period it seems frozen within its constellation, and rises and sets with the stars. Tycho asked himself how the comet would look if it were merely an atmospheric disturbance and close to the Earth, and also how it would look if it were a body like the planets or the stars, far from the Earth. He was able to hold two contradictory ideas in his head at the same time. Place your finger in front of your nose and alternately wink your left and right eyes. You will see the finger seem to move against the more distant landscape. Now move the finger to arm's length and wink again.

1886

An indication on the coinage of changing attitudes about comets. Shown here is a coin issued by the Dutch complaining about the Danes for "The Vain Siege of Hamburg," in 1686. The Latin legend above the comet reads "Not All That Terrifies Harms"—a reference both to the king of Denmark and to the comet. When the armies of neighboring cities came in aid of Hamburg, the Danish king withdrew. Courtesy American Numismatic Society.

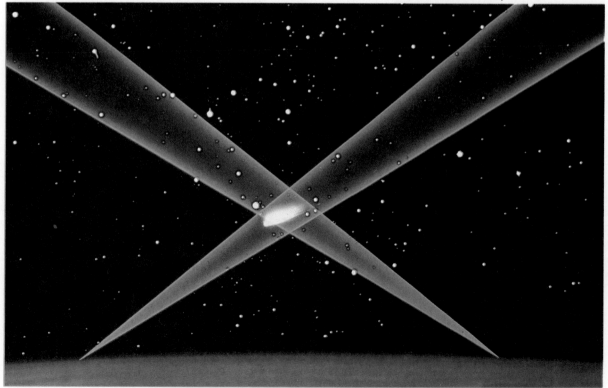

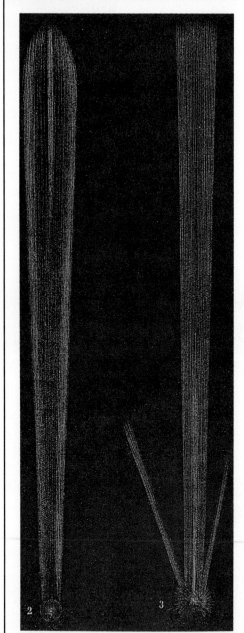

Tycho's and Newton's Comets. The Great Comet of 1577 *(left)*, as seen by Cornelius Gemma, and the Great Comet of 1680 *(right)*, as seen by J. C. Strum. From Amédée Guillemin, *The Heavens*, Paris, 1868.

The finger still moves, but less so. The apparent motion is called parallax; it is merely the change in perspective from left eye to right. The closer the finger is to you, the more parallax or apparent motion against the background; the further the finger is from you, the smaller is the parallax.

Tycho realized that this same principle can be applied to a comet—provided you could observe it from two widely separated observatories. If the comet is close to the Earth, the perspective will change greatly between the two observatories, and each observer will see the comet in front of a very different constellation. But if the comet is far from the Earth, then both observatories will see the comet in front of the same constellation. Through parallax, it is possible to measure the distance of a comet from the Earth. Telescopes are not needed—only sighting markers and graduated circles with angular degrees marked off on them. Tycho lived in the last generation before the invention of the astronomical telescope, and his measurements could have been done—at least crudely—at any time in the previous few thousand years, had anyone thought to make the measurement. (Indeed, error-ridden parallax measurements had been attempted more than a century earlier, at the 1456 apparition of Comet Halley.)

Tycho was not the only one to make these measurements and do these calculations on the Comet of 1577, and some of his contemporaries—perhaps still in the thrall of Aristotelian thinking—got the wrong answer, and deduced that the comet was in the Earth's atmosphere. Tycho's meticulous measurements and calculations, however, have stood the test of time. If the comet had been within the Earth's atmosphere, a sizable parallax would have been detected. Tycho was able to find no significant parallax. With his precision of measurement, the Comet of 1577 had to be much further from the Earth than is the Moon. The comet must therefore be somewhere up there among the planets and the stars. Combining international cooperation, elementary mathematics, and simple observations, Tycho found that the conventional wisdom had been dead wrong for two millennia.

If previous generations had known how distant the comets really are, perhaps they would have been less frightened by them. Tycho liberated the comets from the narrow circumterrestrial confines into which they had been stuffed by Aristotle and allowed them to soar into space. Now science was free to do the same. The upheavals that mystics had always associated with comets was finally justified by the Comet of 1577.

In the vast literature of prophecy inspired by the comets up to this time, there is only one that commands our admi-

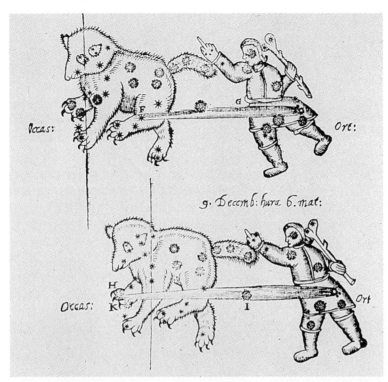

One of the first drawings of a comet as seen through the telescope. Shown are two different sightings of the Comet of 1618 on consecutive nights. The comet is moving through the constellations of the Big Bear (or Big Dipper) and Boötes. Drawn at the telescope by J. B. Cysat. From *The Mathematica Astronomica Ingolstaat*, 1619. From the collection of D. K. Yeomans.

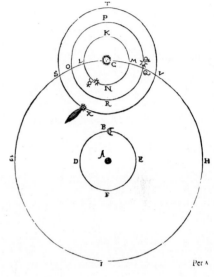

Tycho Brahe's own drawing of the Comet of 1577 (at position X). In Tycho's mistaken scheme of the solar system, the comet went around the Sun, but the Sun went around the Earth (A). From Tycho Brahe's *De Mundi Aetherei*, 1603. From the collection of D. K. Yeomans.

ration. The prediction was made not by a soothsayer or a priest, but by a scientist. It was Seneca:

> Why, then, are we surprised that comets, such a rare spectacle in the universe, are not yet grasped by fixed laws and that their beginning and end are not known, when their return is at vast intervals?...The time will come when diligent research over very long periods will bring to light things which now lie hidden.... Someday there will be a man who will show in what regions comets have their orbit, why they travel so remote from other celestial bodies, how large they are and what sort they are.

Tycho set astronomy back on course, but the man thus prophesied was Edmond Halley.

Who vagrant transitory comets sees,
  Wonders because they're rare; but a
  new star
Whose motion with the firmament
  agrees,
  Is miracle; for there no new things
  are.

—John Donne,
*To the Countess of
Huntingdon,* 1633

Donne's astronomy was 56 years out of date; after Tycho, every comet as well as every supernova was a disproof of the Aristotelian view that in the heavens "there no new things are."

Edmond Halley, age thirty. Painting by Thomas Murray. Courtesy The Royal Society of London.

Chapter III

HALLEY

[Mr. Halley possessed] the qualifications necessary to obtain him the love of his equals. In the first place, he loved them; naturally of an ardent and glowing temper, he appeared animated in their presence with a generous warmth which the pleasure alone of seeing them seemed to inspire; he was open and punctual in his dealings; candid in his judgment; uniform and blameless in his manners, sweet and affable, always ready to communicate....

—Jean Jacques d'Ortous de Mairan,
"Elegy for Mr. Halley,"
Memoires de l'Académie Royale des Sciences, Paris, 1742

WHEN WE THINK OF EDMOND HALLEY, IF WE THINK of him at all, it tends to be solely in connection with his namesake—humanity's favorite comet. The comet becomes a kind of mnemonic device, operating at roughly seventy-five-year intervals, that chides us to remember him. To most of us, Halley is like the athlete who makes his way into the record books on the basis of one extraordinary season, or even one memorable play. We consult the records, expecting to find a journeyman who like countless others has contributed a brick or two to the edifice of science. Instead, we find a master builder.

The date of his birth is uncertain; Halley believed it to be October 29, 1656. He began life in the Borough of Hackney, then a rural community outside London, but since subsumed by the spread of the city. Although we have not a single anecdote, even apocryphal, that indicates the texture of his childhood, we do know that it was then that he first dreamt what he would later become. "From my tenderest years I gave myself over to the consideration of Astronomy," he recalled while still quite young. "[It gave me] so great pleasure as is impossible to explain to anyone who has not experienced it." Halley's sense of science, not as livelihood, but as rapture, was never to leave him. In his boyhood, two comets were seen, one, in 1664, popularly associated with the Great Plague of London, the other, in 1665, connected with the Great Fire. Although no record exists of Halley witnessing these visitors, a young person of his inclination and ability must have been influenced by the comets of 1664 and 1665, so steeped were they in portent, calamity and disaster.

His father, also named Edmond, was a businessman, a soap boiler and salter who owned lucrative properties in London. In 1666 the Great Fire devoured his real estate holdings, but his other businesses flourished. The recent horrors of the bubonic plague had instilled in Londoners a regard for personal hygiene. Soap manufacture became a growth industry. Moreover, the expanding British navy had a chronic need for salted meat to sustain its sailors on long voyages. The elder Halley, his businesses flourishing, was happy to use his new wealth to see his son's evident new promise realized.

He sent him to St. Paul's, one of the best schools in England, where Edmond did brilliantly. In 1671, he was elected Captain of the school—a mark of that rarity, an excellent student who is popular with his classmates. We know nothing of Halley's mother other than that her name was Anne Robinson and that she died on October 24, 1672, nine months before her son's departure for Queen's College, Oxford. Additional evidence of his father's largesse is pro-

A contemporary horoscope of Edmond Halley, who had no use for astrology. The original is in the Bodleian Library, Oxford. Courtesy Joseph Veverka.

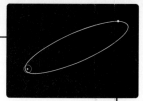

vided by the quantity and quality of the astronomical instruments that Halley took along to college—among them a twenty-four-foot-long telescope, which he immediately put to good use.

We know this because on March 10, 1675, the eighteen-year-old Edmond Halley had the audacity to write to John Flamsteed, England's first Astronomer Royal, informing him that the authoritative published tables on the positions of Jupiter and Saturn were in error. The young man had also found errors in the star positions published by the incomparable Tycho Brahe. The tone of Halley's letter is not that of the young cowboy come to challenge the legendary gunslinger; but much more the youthful enthusiast, full of admiration for those who have come before him, anxious to join their club—and even more anxious to discover the true nature of the universe. We do not know exactly what Flamsteed's response was, but it must have been positive, because the next year he helped Halley publish his first scientific report, or "paper." It appeared in *Philosophical Transactions*, the journal of the Royal Society of London—then, as now, the leading scientific organization in Britain—and was titled "A Direct and Geometrical Method of finding the Aphelia, Eccentricities, and Proportions of the Primary Planets, without supposing equality in angular motion."

What was this paper about? Since the work of Tycho's student, Johannes Kepler, it had been known that each planet moves along a path called an ellipse, a kind of stretched-out circle (see page xii). The eccentricity is a measure of how stretched out the ellipse is; an ellipse with zero eccentricity is a circle, and an ellipse with an eccentricity of 1 or greater is not even a closed curve, but rather a parabola or hyperbola (see page 38). The eccentricity of the Earth's orbit is 0.017—to the naked eye, indistinguishable from a circle. Mercury's orbital eccentricity, by contrast, is 0.21—perceptibly elongated. Many years later, it would be one of Halley's triumphs to determine that comets move in elliptical orbits, one of the keys to their origin.

In a roughly circular orbit such as the Earth's, we are always pretty much at the same distance from the Sun. But in a highly elliptical orbit, the distance to the Sun varies depending on where the moving object is in its orbit. The nearest point to the Sun, when the planet or comet is moving fastest, is called the perihelion (plural, perihelia). As the object sweeps around the Sun, it is said to be making perihelion passage. The phrase has a nice ring to it, as if it were something experienced on steamships of the old P&O line. The far point in the orbit is called aphelion. The more elliptical the orbit is, the bigger the difference between aphelion and perihelion. It is perfectly possible for a comet to have its

The conic sections. When a cone is sliced or sectioned at various angles, several different shapes or curves are produced which collectively are called conic sections. Here the exterior of the cone is shown in gray and three of the conic sections in blue. If the four pieces shown here were put together, the cone would be reassembled. The top of the cone is called the apex, and the flat bottom on which it stands is called the base. If you make a cut parallel to the base, you produce a circle. If you cut at an angle, you produce a stretched-out circle, a little like an oval, called an ellipse. If you slice perpendicular to the base, the surface you expose is a hyperbola; unlike the circle and the ellipse, the hyperbola does not curve back upon itself. There is one other curve, not shown here, which lies at the transition between an ellipse and a hyperbola; it is called a parabola. These pretty conic sections were first described by Apollonius of Perga in the second half of the third century B.C. It is astonishing that planets and comets know to move about the Sun precisely along such paths; and that the trajectory of a rock thrown up into the air follows a parabola. So does a ballistic missile. Newton showed that the inverse square law of gravitation forces bodies to move through space along conic sections. Such connections between seemingly abstract mathematics and the way the world actually works characterize the major discoveries which have molded modern science. Diagram by Jon Lomberg and Jason LeBel/BPS.

perihelion near the Earth, and its aphelion far beyond the most distant known planet. But this was before Halley began studying the comets. In his first paper, Halley proposed a new and more accurate method of calculating the orbits of the planets.

The paper had to be rewritten many times, in part due to Halley's inexperience, but also because the Bishop of Salisbury had published a contrary view, and so, Halley was told, might construe the paper as a personal insult. Halley had no intention to offend, and made the suggested changes happily. This was not the last time that Halley's pursuit of science would offend ecclesiastical sensitivities.

The same year that Halley's paper on orbits brought him to the attention of the world astronomical community, he chose to leave Oxford without taking his degree, and to travel to the distant island of St. Helena, west of Africa, to make the first map of the southern skies. The constellations you can see from high northern latitudes are almost completely different, of course, from those you see from close to the South Pole. At the equator you see all of the northern and all of the southern constellations. St. Helena was then the southernmost outpost of the British Empire and a good vantage point from which to observe not only the southern skies, but also some of the charted stars of the Northern Hemisphere—important because, if Halley's proposed map

1889

of the southern stars were to be useful to European astronomers, it would have to share some points of reference with star positions that had been previously determined. Besides its latitude, St. Helena had something else to recommend it; by all accounts, the weather was invariably clear, a matter of critical concern for astronomical observations.

Halley's new friends at the Royal Society wrote a letter to the government in support of the proposed expedition, which soon found favor with King Charles II. But the king was only the titular monarch of far-off St. Helena; in reality, the little island was a fiefdom of the powerful East India Company. Charles was, however, not without influence, and after he wrote to the company's directors, they volunteered to provide passage for Halley and a scientific companion. Halley's father, "willing to gratify [Edmond's] Curiosity," provided him with a huge allowance that more than covered the cost of the best observing equipment then available and such other expenses as he might incur.

In November 1676, Halley set sail on the *Unity*, beginning a journey that would take three months and cover nearly ten thousand kilometers of ocean. His destination was an island so remote that 130 years later, the British would deem it the only prison secure enough to contain the captive Emperor Napoleon. For almost two centuries European mariners had been sailing the southern seas, mapping every coastline they could lay their eyes on, and not one of them had accurately mapped the very different constellations above them. Halley's self-imposed assignment was to bring back half the sky. He had just turned twenty-one.

Traveler's tales to the contrary, the weather on St. Helena was rotten. Halley would wait for weeks for an hour's window on the stars. *English* weather was better than this. At least in England, in inclement weather, there were other things to do. Halley had exiled himself to little more than a rock in the middle of a vast ocean. And he had more than clouds and boredom to worry about; the governor of St. Helena was a certified madman who quickly grew to hate Edmond Halley. Eventually the governor's behavior became so bizarre that he was recalled and dismissed, but not before Halley himself was ready to return. Despite the difficulties, his trying year on St. Helena was well-spent. He came home with the first map of the southern skies and much more. He had discovered stars and nebulae that European astronomers had not known. In St. Helena he had observed a transit of the planet Mercury across the face of the Sun, later significant in the determination of the Sun's distance from the Earth. He confirmed that there was no Pole Star in the southern skies, and that accurate timekeeping on St. Helena with a pendulum clock calibrated in England was

Winchester Street, London, where both Edmond Halley, the astronomer, and his father lived. As a young man, Halley made astronomical observations from these rooftops. This early nineteenth-century print shows the street very little changed from Halley's time. Ann Ronan Picture Library.

The frontispiece of Halley's Chart of the Southern Skies. The dots representing the stars are connected into fanciful and mythological patterns, the constellations. Around the periphery are constellations known to Northern Hemisphere astronomers, while toward the center of the chart are the constellations that can be seen only in the southern sky. Ann Ronan Picture Library.

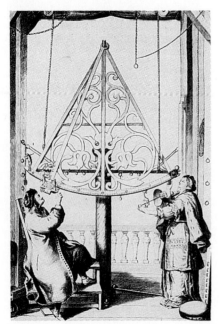

A sextant for measuring the positions of the stars, being attended to by Johannes Hevelius and his wife. This was the sort of technology that Halley came to Danzig to check out. Courtesy Royal Astronomical Society.

impossible unless you shortened the pendulum. (Although Halley didn't know it at the time, this is because the spin of the Earth creates a small centrifugal force that slightly counteracts gravity at the equator, but weakens with increasing latitude.)

His catalogue of the southern skies was presented to the Royal Society by Robert Hooke, a neurotic polymath who was the first person to see Jupiter's Great Red Spot through the telescope, and the first person to see a living cell through the microscope. (He also was the first to use the word "cell" in its biological context.) Hooke made lasting contributions to physics, astronomy, biology, and engineering. The Fellows of the Royal Society were quick to appreciate Halley's achievement, but as far as Oxford was concerned, Halley was just another dropout. They would not let him return to take his degree because he had departed for St. Helena before fulfilling his residency requirements—so serious a breach of regulations that nothing short of a royal decree could put the matter right. He appealed once more to Charles II, and once more Charles signed a letter on his behalf, this time requesting that he be granted the degree of Master of Arts "without any condition of performing any previous or subsequent conditions of the same." The Vice-Chancellor of Oxford complied. At about the same time he was granted his degree, Halley was elected a Fellow of the Royal Society, a considerable distinction for so young a man.

Halley soon turned his attention to an intensifying controversy involving Hooke, Flamsteed, and Johannes Hevelius of the free city of Danzig, the preeminent observational astronomer of the time. Hooke and Flamsteed swore by the recently invented telescopic sight, which, when mounted on a measuring instrument, improved the precision with which the relative position of the stars could be determined. But the older Hevelius rejected the new technology and persisted in using an open sight, without optics, a mechanism as simple as the sight on a rifle. Appalled, Hooke and Flamsteed conducted an increasingly shrill campaign against Hevelius, asserting to everyone who would listen that Hevelius' observations were untrustworthy. It was an unlikely topic for a vitriolic controversy. Hevelius, quite reasonably, began to feel beleaguered. He too was a Fellow of the Royal Society, but he lived in Danzig, far from the pubs and parlors in which Hooke and Flamsteed were pursuing their curious assault. He appealed to the Fellows, writing that every scientist should be permitted to serve astronomy as he believed best; that the results should be tested and judged solely on their merits; that it was unseemly for scientists to disparage the work of a colleague in the absence of supporting evidence. He had been observing with an open sight for more than thirty years, and if it had been good enough for Tycho Brahe it was good enough for him. The Royal Society responded officially by affirming its confidence in Hevelius and challenging Hooke and Flamsteed to document their case, or desist. They would do neither. The parties had reached an impasse.

With his tact, integrity, and genius for observation (and his father's money), there could be no one better suited than Halley to make the journey to test Hevelius' method. Halley sent Hevelius a copy of his southern star catalogue with a request for an invitation to visit. Hevelius agreed, and Halley arrived in Danzig in May of 1679. For ten nights they observed together. Halley was soon convinced that Hevelius, with the obsolete open sight, consistently achieved better results than Flamsteed and Hooke with state-of-the-art equipment. He immediately wrote to Flamsteed, describing his painstaking, systematic efforts to find flaws in Hevelius' observations: "Verily I have seen the same distance repeated several times without any fallacy," he informed his mentor, who was doubtless anxious to learn otherwise, "...so that I dare no more doubt his [Hevelius'] Veracitye."

Despite Halley's testimonial, Flamsteed and Hooke lacked the capacity to apologize, or even to relent. The "open sight" controversy did not die until Hevelius himself died, eight years later. Halley's conduct had been a model of fairness and candor, risking the displeasure of powerful friends in the interest of truth.

✳ ✳ ✳

In 1680, as a great comet appeared in European skies, science was certainly not widely accepted as the favored approach to an understanding of nature. Gibbon mentions, as a perfect example of a world view in transition, an astronomer who "was forced to allow that the tail, though not the head [of the Comet of 1680] was a sign of the wrath of God." Even Gottfried Kirch, the German astronomer who discovered the Great Comet of that year, was convinced of the supernatural nature of comets:

> I have read through many books on comets, heathen and Christian, religious and secular, Lutheran and Catholic, and they all declare comets to be signs of God's wrath....There are some that oppose the belief but they are not very important.

Halley also saw the Great Comet of 1680, but his response was different from Kirch's. On board a ferry in the English

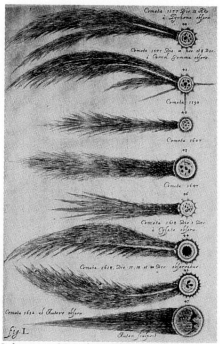

Johannes Hevelius of Danzig was also an observer of comets. Here from his *Cometographia* (1668) are a variety of cometary forms seen between 1577 and 1652. Compare with the Chinese cometary atlas, page 20.

Comet Arend-Roland (1957 III), famous for its sunward spike. Compare with Hevelius' depiction of the Comet of 1590 (second comet from top in figure above). Photographed through the University of Michigan telescope by F. D. Miller, April 24, 1957. Courtesy National Aeronautics and Space Administration.

Photomontage of Saturn, the major Cassini Division in the rings clearly seen, and some of its large moons, four of which were discovered by J. D. Cassini in the seventeenth century. Voyager 1 photographs. Courtesy National Aeronautics and Space Administration.

Channel, somewhere between Dover and Calais, Halley glanced up as the clouds broke and saw something splendid. As soon as he reached France, he hurried to the Paris Observatory to confer with its director.

Jean-Dominique Cassini, discoverer of the major gap in Saturn's rings, as well as four of the planet's moons, received the young author of *A Catalogue of the Southern Stars* with lavish hospitality. Cassini entertained Halley, introduced him to friends and colleagues, gave him unrestricted use of the observatory's equipment and library, and, most significantly, presented Halley with an idea. In a remarkable letter to Hooke dated May 1681 Halley wrote:

> Monsieur Cassini did me the favour to give me his booke of ye Comett Just as I was goeing out of towne; he, besides the Observations thereof, wch. he made till the 18 of March new stile, has given a theory of its Motion wch. is, that this Comet was the same with that that appeared to Tycho Anno 1577, that it performes its revolution in a great Circle including the earth.

Halley recounts the details of three cometary apparitions, adding:

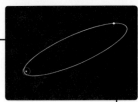

1890

...this is the sume of his Hypothesis and he says it will answer exactly enough to the Motions of the two Comets as likewise to that of Aprill 1665; I know you will with difficulty Embrace this Notion of his, but at the same tyme tis very remarkable that 3 Cometts should soe exactly trace the same path in the Heavens and with the same degrees of velocity.

No one had yet determined the orbit of a comet, but Cassini had noticed that three comets had come from the same part of the sky with similar speeds, and made the daring proposal—unprecedented in the scientific literature,* and explicit only in the folk tradition of the Bantu-Kavirondo people of Africa—that the same comet was returning to Earth in widely separated times.

Halley informs Hooke that his preliminary attempts to sketch out a path for the Comet of 1680, based on Cassini's summary of its apparent movement across the sky, have been unsuccessful; at some future time he hopes to try again. And in the final paragraph, as an aside, Halley lays the groundwork for the science of actuarial statistics. After summarizing the comparative statistics for Paris and London of births, marriages, deaths, and population density—the last facilitated by Halley himself pacing off the dimensions of the city of Paris on foot—he concludes:

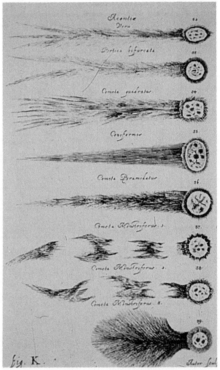

More cometary forms by Hevelius. Note the comets with disrupted tails. The bottom three are called "monstriferous" comets, an echo of the once prevalent cometary mysticism. From the collection of D. K. Yeomans.

...it will from hence follow, supposing it alwaies the same, that one half of mankinde dies unmarried, and that it is nescessary for each married Couple to have 4 Children one with another to keep mankinde at a stand.

It was ten years before he would develop this idea further, and even longer before he returned to the subject of comets. Now he was off to Italy for six months. When he returned home to England, he made his first great discovery.

Her name was Mary Tooke. She was the daughter of the Auditor of the Exchequer. A contemporary memoir describes her as "an agreeable young Gentlewoman; and a Person of real merit." Another tribute portrays her as "a young lady equally amiable for the grace of her person and the qualities of her mind." They were married less than three months after his return from the Continent, at St. James, a church notorious or compassionate—depending on your point of view—for choosing elopements as its specialty.

Some of the Italians called Pythagoreans say that the comet is one of the planets, but that it appears at great intervals of time and only rises a little above the horizon.

—Aristotle,
Meteorology, Book 1, Chapter 6

Although not explicit, this seems also to imply the periodic return of the comets.

*Although some hint of this idea can be found in the writings of Aristotle (who firmly rejected it) and Seneca (who, as we have seen, embraced the suggestion of Apollonius of Myndos that comets move as the planets do).

Halley's observations of the comet that would one day bear his name. On the left-hand page are his observations of September 4, 1682, while the right-hand page contains notes on the parabola, one of several candidate trajectories he considered many years later for the orbits of the comets. The juxtaposition here of comet and parabola may be only a coincidence, although conceivably it reveals a hunch that Halley did not immediately pursue. These pages were reproduced in Arthur Stanley Eddington's article, "Halley's Observations on Halley's Comet, 1682," *Nature*, Volume 83, page 373, 1910. Copyright 1910 Macmillan Journals Limited.

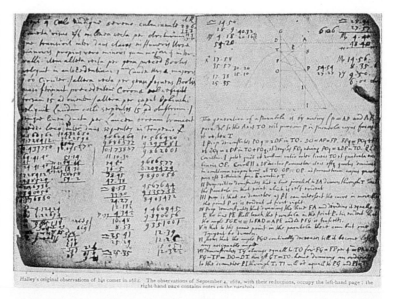

Halley's original observations of his comet in 1682. The observations of September 4, 1682, with their reductions, occupy the left-hand page; the right-hand page contains notes on the parabola.

There does not seem to have been any opposition to the marriage, nor was Mary pregnant. Perhaps they chose St. James because they did not wish to waste another hour. Their marriage and their love were to last until Mary's death, nearly fifty-five years later. The few surviving references to them together make clear a deep and enduring happiness. In 1682, toward the end of their first summer together, Edmond, and possibly, Mary, witnessed another comet, unimpressive compared to the Great Comet of 1680. He made a few notes to record what he saw. It was his only look at the comet that would one day bear his name.

Years after the death of Halley's mother, Edmond Sr. had entered into an unfortunate second marriage. There was talk about his new wife's extravagances, and her apparent disregard for her husband and stepson. The contrast between the marriages of father and son must have pained them both. On the morning of March 5, 1684, the senior Halley complained that his shoes were too tight, and a nephew offered to cut the linings out of the toes. This seemed to help, and Edmond Sr. told his wife that he was going out; he would return by nightfall. A contemporary broadside recounts what happened next:

> When Night came she accordingly expected him, but not returning she was very much concerned at it, and the next day made all possible Enquiry, but after several days not hearing of him, published his Absence in the News Book. From Wednesday the 5th of March, to the 14th of April, notwithstanding all Endeavours, and the strictest Search that could be made, they received no Account where he was, or where he had been. But on Monday last he was found by a River side at Temple-Farm in Strow'd Parish near Rochester on this manner.

A poor Boy walking by the Water-side upon some Occasion spied the body of a Man dead and Stript, with only his Shoes and Stockings on, upon which he presently made a discovery of it to some others, which coming to the knowledge of a Gentleman, who had read the advertisement in the Gazet, he immediately came up to London, and acquainted Mrs Halley with it, withal, telling her, that what he had done, was not for the sake of the Reward, but upon Principles more Honourable and Christian, for as to the money, he desired to make no advantage of it, but that it might be given intirely to the poor Boy; who found him and justly deserved it.

The same nephew who had adjusted Halley's shoes was dispatched by Mrs. Halley to identify the body. It must have been a gruesome task; the face had been obliterated. The broadside continues:

It was concluded by all, that he had not baen in the River ever since he was missing, for if he had, his Body would have been more Corrupted. The Gentleman knew him by his Shooes and Stockings, they being the same Shooes he had cut the Lineing out of, and on one leg he had four Stockings, and on the other three and a Sear-cloath.* The Coroner sat upon him, & the Inquest brought him in Murthere'd.

The death of Edmond Halley, Sr., remains a mystery. No Sherlock Holmes was at hand to sift through the redundant footwear and deduce Halley's whereabouts during the five weeks previous. Two and a half centuries after the fact, Eugene Fairfield MacPike, a distinguished Halley scholar, rendered a verdict of suicide, claiming that "the evidence with which we are supplied appears to point to mental aberration." Perhaps. But the evidence at least as well points to misadventure or murder.

An additional mystery revolves around the younger Halley's response to his father's fate. He is unmentioned in the broadside story, playing no apparent role in the search for his father, the identification of the body, or the coroner's inquest. His whole life is a case study of a massive, almost compulsive curiosity. And yet we find no record that he made any attempt to resolve the mystery of his father's death—the man who had so generously encouraged and supported his intellectual development.

In an unseemly coda to these unhappy events, the "Gentleman" who had brought the second Mrs. Halley news of her husband's corpse sued for nonpayment of the £100

*A cloth imbued with wax, and used as a bandage.

The time it takes for a planet or comet to complete a single orbit of the Sun increases the farther from the Sun it lies—according to a particular law of nature described by the diagonal straight line. The orbital period is given in Earth years and the distance from the Sun is given in Astronomical Units. At lower left, we see that a body 1 A.U. from the Sun takes one year to go around the Sun, the case for the Earth. At a distance of 100 A.U. from the Sun a comet would take a thousand years per orbit, and in the outer parts of the Oort Cloud of comets (Chapter 11), indicated by the blue dots, a comet may take millions of years to go once around the Sun. This relationship, which was central to the work of Newton and Halley, was discovered by Johannes Kepler and is called Kepler's Third Law. Diagram by Jon Lomberg/BPS.

reward, despite his protestations "that he desired to make no advantage of it." The case was heard by a Judge Jeffreys, who was infamous for the frequency and variety of the crimes he committed against the defendants who stood before him— harassment and extortion being the mildest among them. In this particular case, Judge Jeffreys seems to have taken the high road, awarding the "Gentleman" only £20, and ordering Mrs. Halley to pay the remaining £80 to the hapless "poor Boy" who had discovered the body. A decade later, Edmond took his spendthrift stepmother to court in what was described as an effort to defend his inheritance.

* * *

At about the time that his father vanished, Halley had been trying to develop a deeper understanding of planetary motion. Kepler had pointed out that there is a precise proportionality between the period it takes for a planet to go once around the Sun (its year) and the distance of the planet from the Sun (see below). The year on Mercury, close to the Sun, is only 88 Earth days; the year on Saturn, far from the Sun, is almost 30 Earth years. The period of a planet in the outer solar system is longer than an Earth year not only because it has a bigger orbit to traverse, but also because it is moving more slowly. Why? Halley and several

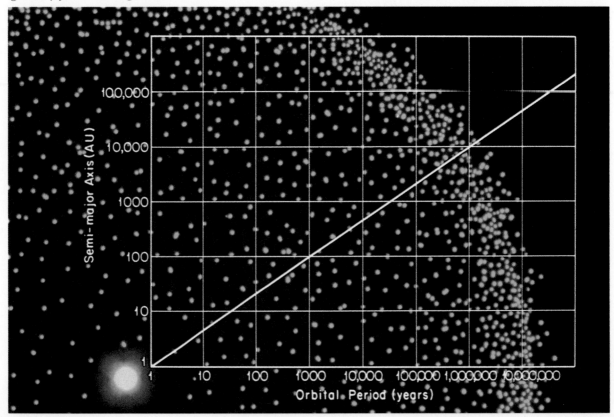

others had a notion that the planets move as they do because of a balance of two forces—one directed outward from the Sun, and provided by a planet's own velocity, and the other directed inward, but provided by a previously undiscovered gravitational force from the Sun. It was clear that the force had to decline with distance, so that far-off planets could move slowly and still balance the force of gravity. But how fast must the gravitational force diminish as you move away from the Sun, in order to account for the observed planetary motion? By intuition—or false analogy with the propagation of light—Halley and his colleagues suggested that gravity was an inverse square law. Move the planet twice as far from the Sun, and the force diminishes to a quarter its original strength; three times further away and the force decreases to one-ninth; and so on. The law of gravity, whatever it might happen to be, dominated the heavens. This problem is central to our understanding of nature.

Halley, Hooke, and Christopher Wren—an astronomer turned to architecture who rebuilt London after the Great Fire—considered the challenge of proving the proposed inverse square law in a coffee house after a meeting of the Royal Society in January 1684. Hooke boasted that he had already done so, but declined to produce the proof. Wren must have had his doubts, because he offered a reward of any book that did not cost more than forty shillings to anyone who could provide a proof before two months had passed. Hooke maintained that he had found the solution, but was delaying showing it to others so all could appreciate the difficulty and significance of his work. Wren was in no danger from this quarter of losing his forty shillings.

Months passed and the challenge remained unanswered, so Halley made up his mind to pay a visit to Trinity College, Cambridge, where there lived a man who might be equal to the task. This scholar had been judged a prodigy, primarily on the basis of his remarkable work on the nature of light and color. But that had been many years before, and he had squandered his genius during most of the interim in the pursuit of alchemical recipes and in strident attacks on the character of Athanasius, a theologian of the early Christian Church who did much to establish the orthodox doctrine of the Trinity. He believed Athanasius guilty of deliberate falsification and historical fraud. Unable to maintain normal relationships with anyone, especially women, this scholar was given to fits of paranoia and depression. Moreover, he was afflicted with an apparent inability to complete anything. Still, he was known to be a brilliant mathematician, and Cambridge was not so far from Halley's home in London. So on an August morning in 1684 he set out to visit Isaac Newton.

Isaac Newton, the greatest British scientist, commemorated on the one-pound note.

Halley's encounter with Newton (1642–1727) was a watershed for Halley, for Newton, for science, and, in ways so numerous as to be incalculable, for the fate of the world. The mathematician Abraham De Moivre took down Newton's version, related many years later, of this momentous encounter:

> The D^r [Halley] asked him what he thought the Curve would be that would be described by the Planets supposing the force of attraction towards the Sun to be reciprocal to the square of their distance from it. S^r [Newton] replied immediately that it would be an Ellipsis, the Doctor struck with joy & amazement asked him how he knew it, why saith he I have calculated it, whereupon Dr Halley asked him for his calculation without any further delay, Sr Isaac looked among his papers but could not find it, but he promised to renew it, & then send it to him....

This promise might well have had a familiar and therefore hollow ring. But where Hooke prevaricated, Newton delivered. In November a copy of Newton's *De motu corporum in gyrum* ("On the Motion of Bodies in Orbit"—scholarly works were still being written in Latin) was hand-carried to Halley. Only nine pages long, it contained the proof that the inverse square law implied all three of Kepler's Laws, as well as the seeds of a broad new science of dynamics. Halley instantly recognized what Newton had accomplished. Rushing back to Cambridge, he extracted from Newton a promise to expand his ideas into a book, and to do it quickly.

Halley's first visit to Newton had awakened the latter from a kind of mystic trance. Now Newton was wide awake to the point of sleeplessness, obsessed with this new challenge, unable to eat or to think of anything else. For the next year and a half he would live as a recluse, in monomaniacal pursuit of gravity and planetary motion.

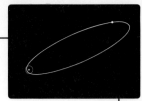

Meanwhile, Halley was back in London, busily trying to get himself demoted. The Royal Society had operated since its inception in the early 1660's as a kind of club for gentlemen with scientific inclinations. But by 1685 the scientific revolution had assumed such dimensions that the voluntary services of its fellows were inadequate to the task at hand. What was needed was a paid full-time secretary, someone who would deal with the growing correspondence, arrange the details of the meetings, and edit the *Philosophical Transactions*. Halley correctly understood that this position would afford him an ideal exposure to everything that was happening in science. He was elected to the job on the second ballot, early in 1686. But his salary from the Royal Society meant that he would have to relinquish his fellowship, sit at the lower end of the table, and be denied the high honor of wearing a wig.

Cheerfully Halley turned his voracious curiosity to the feast at hand, a banquet of geology, geography, biology, medicine, botany, meteorology, mathematics, and of course astronomy. He played a significant role in transforming the Royal Society from a club to the world's central clearinghouse for scientific ideas, and at the same time managed to publish many original papers of his own.

As Newton neared the completion of his masterwork, Halley approached the Royal Society with the proposition that they publish it. Since everyone agreed that the book would be important, the society would under normal circumstances have been only too happy to pay for its publication. However, it had unwisely exhausted all its publication funds on another book. This was the long-awaited *History of Fish*, which unaccountably had failed to find its anticipated audience. So Halley decided that he would pay for the publication of Newton's work out of his own pocket. Oh yes, on the matter of his salary...This, too, would be a problem for the Society. Would Halley mind terribly taking the wages that had cost him his wig in remaindered copies of the *History of Fish?* Halley cheerfully accepted and lugged home seventy-five. By this time he was no longer wealthy, in part because of his stepmother's incursions into his father's estate, but he chose not to make an issue of it.

The level of expectation surrounding Newton's forthcoming work was more than Robert Hooke could bear. He resurrected the old pretense that the inverse square law belonged to him. He would be satisfied with nothing less than an acknowledgment of his priority in the preface of Newton's book. Halley, who had already taken on the responsibilities of agent, editor, publisher, and proofreader for the book, now assumed another role: psychotherapist to the author. Fearing that Newton would hear of Hooke's charges from a

Halley's interests also extended to history and archaeology, as is shown by these papers, published in the *Philosophical Transactions*, on Julius Caesar's conquest of Britain, and on the ancient Near Eastern city of Palmyra. Courtesy The Royal Society of London.

But for him, in all human probability, the work would not have been thought of, nor when thought of written, nor when written printed.

—Augustus De Morgan (1806–1871), on Halley's contribution to Newton's *Principia*

less thoughtful source, Halley wrote him a letter. It begins with expressions of admiration and gratitude, but you can sense Halley's anxiety as he finally gets to the point:

> There is one thing more that I ought to inform you of, viz, that Mr Hook has some pretensions uipon the invention of ye rule of the decrease of Gravity, being reciprocally as the squares of the distances from the Center. He sais you had the notion from him....

Newton's initial reaction was restrained, but the more he thought of it, the more enraged he became. He would withdraw the third volume of the work rather than become enmired in odious controversy with Hooke. But Book III was critical. Richard S. Westfall, in his brilliant biography of Newton, writes of Book III,

> In a word, it proposed a new ideal of a quantitative science, based on the principle of [gravitational] attraction, which would account not only for the gross phenomena of nature, but also for the minor deviations of the gross phenomena from their ideal patterns. Against the background of inherited natural philosophy, this was a conception no less revolutionary than the idea of universal gravitation itself.*

Book III also contained Newton's monumental work on comets. He had painstakingly collected observations of the Comet of 1680, acquired from a wide range of locales, including London, Avignon, Rome, Boston, the island of Jamaica, Padua, Nuremberg, and the banks of the Patuxent River in Maryland. (Even then, worldwide cooperation was essential to understand the comets.) He showed that together they defined a highly eccentric orbit, almost a parabola (see figure on page 38). Newton noted from examination of the history of comets that they are seen much more often in the part of the sky near the Sun than opposite the Sun, and understood this to be evidence that comets generally—not just the Comet of 1680—are in orbit about the Sun, and brighten when they are nearest to it. Tycho had shown that the comets move among the planets; now Newton, with Halley's prodding, had demonstrated that they have the same kinds of orbits (conic sections, page 38) as the planets.

"The comets shine by the Sun's light, which they reflect," Newton wrote. "Their tails...must be due either to the Sun's light reflected by a smoke arising from them, and dis-

*Richard S. Westfall, *Never at Rest*, Cambridge University Press (1980).

1893

persing itself through [space], or to the light of their own heads....The bodies of comets must be hid under their atmospheres."

To Halley, it was unthinkable that these vital contributions might be lost to the world, a casualty of Newton's snit over Hooke's vanity and posturing. He reassured Newton that no one took Hooke's claims seriously and told the story of Wren's wager, back in 1684. He insisted that without Book III, the work would appeal only to mathematicians. He argued that Hooke's demands had been exaggerated by others. He cajoled, he flattered. Finally Newton relented and allowed the publication of the entire work.

And this is how the *Philosophiae Naturalis Principia Mathematica (The Mathematical Principles of Natural Philosophy)*, the central testament of modern science, the keystone of our present understanding of stars, planets, comets, and much more, came to be. In the first edition of the *Principia*, as it has come to be known, printed in July of 1687, a poem of worshipful tribute presented to Newton by Halley appeared as preface. Its last lines read,

> Lend your sweet voice to warble Newton's praise,
> Who searcht out truth thro' all her mystic maze,
> Newton, by every fav'ring muse inspir'd,
> With all Apollo's radiations fir'd;
> Newton, that reach'd th' insuperable line,
> The nice barrier 'twixt human and divine.

The judgment of posterity is not far from Halley's. Among the book's incidental offerings are the invention of the calculus and the theory of interplanetary spaceflight, as well as the central idea behind the intercontinental ballistic missile.

The year after Halley's consummate midwifery delivered Newton of the *Principia*, he and Mary became the parents of two daughters, Katherine and Margaret. At about the same time, Halley became intrigued with Hooke's suggestion that the Flood, as related in the Bible, could be explained by a change in the Earth's poles, the Near East slowly sliding under the equatorial bulge of ocean. Halley was familiar with the observations, made over a period of centuries, of minuscule changes in the latitude of the German city of Nuremberg, and reasoned that since even tiny changes in latitude occur with glacial slowness, then the time between the Creation and the Flood must have been far longer than the period allowed in Genesis, in contradiction to a literal interpretation of the Bible. However, in both the Old Testament and the ancient Babylonian accounts of the Flood, events proceeded swiftly, and recovery took less than a year. So Halley tried to imagine how a quick inundation of the

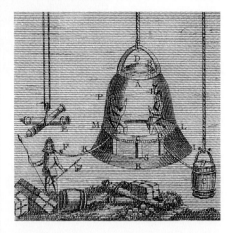

Edmond Halley's diving bell, with two aquanauts aboard. Fresh barrels of air were sent down from a boat above. The explorer on the sea floor is tethered to the diving bell by a breathing tube. This technology was successfully tested, and bears some resemblance to the early days of manned spaceflight. From W. Hooper, *Rational Recreations*, London, 1782. Ann Ronan Picture Library.

ancient Near East could be worked. Were a comet to approach the Earth too closely, he argued, the gravitational tides might sweep the oceans (or just the Persian Gulf) up over a sufficient area of the land to account for the events described in the Bible. Halley also held that comets might from time to time actually impact the Earth, with still more horrible consequences. He seems to have been the first scientist to ask what might happen if a comet were to pass very near the Earth, a question now important to many areas of science, as we shall see.

Halley's interests were now more eclectic than ever. He devised the first weather map, creating a convention to indicate prevailing winds that you can find in televised weather reports today. He attempted to measure the size of the atom. He made valuable observations about magnetism, heat, air, plants, seashells, clocks, caviar, light, Roman history, aerodynamics, the habits of cuttlefish, and a method of keeping flounder alive for midwinter retailing. (These last two efforts were perhaps inspired by all those copies of *The History of Fish.*) By his own testimony, we know that Halley used opium. He gave a discourse on his personal experience of the drug at a meeting of the Royal Society, and hardly seems to have suffered much from the so-called amotivational syndrome sometimes associated with opium and other euphoriants. He also invented, developed, and tested one of the first practical diving bells. "By this means," he wrote, "I have kept three men 1¾ [hours] under the water in ten fathoms deep without any of the least inconvenience and in as perfect freedom to act as if they had been above." It worked so well that Halley was able to form a salvage company on the side that prospered, and eventually sold shares to the public.

That same year Halley published a paper on the means to determine the Earth's distance from the Sun—by measuring the timing of Venus as, on rare occasions, it transits across the Sun's disk. Much later, in 1716, he would publish another paper, imploring the astronomical community to organize cooperative international expeditions during the next transits. The first expedition of Captain James Cook in HMS *Endeavor* was intended specifically as a response to Halley's plea to measure a transit of Venus from Tahiti on June 3, 1769; for this reason alone (there are others) Halley plays a role in the history of the exploration of the Earth. From observations of the transits of 1761 and 1769, and with Halley's method of calculation, the astronomical unit was found to be nearly 93 million miles (150 million kilometers), a hairs-breadth from the figure we accept today. Halley gave us the scale of the solar system.

In 1691, Halley was being considered for the Savilian Professorship of Astronomy at Oxford. His confirmation required the approval of the Anglican Church, then as now headed by the Sovereign. But Halley was accused of something scandalous: "being guilty of asserting the eternity of the world." (By these standards, many scientists and all Hindus would be disqualified for the Savilian Professorship today.) His crime had been to consider the cause of the biblical deluge. Halley's conjectures on religious issues would today be considered irrelevant to his qualifications for teaching astronomy; moreover, the views ascribed to him by the authorities of the Church were a distortion of his true views. He never doubted that the Earth had been formed, or even

1893

Aristotle's opinion...that comets were nothing else than sublunary vapors or airy meteors...prevailed so far amongst the Greeks, that this sublimest part of astronomy lay altogether neglected; since none could think it worthwhile to observe, and to give an account of the wandering and uncertain paths of vapours floating in the Æther.

—Edmond Halley,
Transactions of the Royal Society of London, Volume 24, page 882 (1706)

. JOHANNIS HEVELII
COMETOGRAPHIA.

The cover of Hevelius' *Cometographia*, depicting three seventeenth-century savants debating the comparative merits of competing hypotheses on cometary motion, as indicated by the diagrams they display. While the scholars contend, a real comet appears in the sky above—apparently unnoticed by them, but tracked by their assistants on the observatory roof. Courtesy Ruth S. Freitag, Library of Congress.

created, and his efforts assumed the truth of the biblical time scales, or at least of the duration of the Flood.

Flamsteed—still the Astronomer Royal, or chief astronomical adviser to the Crown—seems to have played a disgraceful part in the affair. So miffed was he that Halley had disagreed with work he had done on the ocean tides that overnight he transformed himself from friend to archenemy. In a letter to Newton, Flamsteed insisted that the appointment be stopped, because otherwise the youth of Oxford would be corrupted by Halley's "lewdness." Even Newton, who was legendary for his prudishness, could not take this charge seriously. He urged Flamsteed to reconcile his differences with Halley, but the Astronomer Royal was unwilling. Now, the accusations went, Halley was not only lewd, but also a plagiarist, a thief of ideas. Throughout, Halley refused to be provoked. He never uttered an angry word in reply, and when the integrity of his work was attacked, he confined his rejoinders to the scientific merits of the argument.

The campaign by Flamsteed and the Church was successful, and Halley was denied the Savilian Professorship. During the interrogation on his alleged heretical beliefs, Halley made no attempt to curry favor. His inquisitor, a Chaplain Bentley, must have been outraged: Forty years later, he published a tract entitled *The Analyst, or a Discourse Addressed to an Infidel Mathematician*, widely understood to be a harangue against Halley.

However Bentley may have scolded the applicant, it is clear that Halley emerged from the experience in good spirits, and with his scientific convictions intact. He continued his investigations into the age of the Earth, this time using the salinity of seawater as a kind of clock that had begun ticking at the time the oceans were formed. He thought that regular measurement of ocean water would reveal increasing salinity as time went on. Rivers carry salt into the oceans at a rate that Halley crudely calculated. Extrapolating back to a time when seawater was fresh, Halley found that the world is much older than the Bible implies—not six thousand years, but at least a hundred million years old. Halley's method cannot be used to determine the exact age of the Earth, because seawater has long been saturated with salt. But it is a perfectly suitable way of arriving at a lower limit for that age. Moreover, it is a brilliant anticipation of a range of modern techniques for dating rocks (see Chapter 16), and gave heart to geologists and biologists of the next century, when they came upon evidence that the Earth and life are ancient beyond all human imagining. The age of the Earth—and the rest of the solar system—is now known, from a range of independent lines of evidence, to be a little over 4.5 billion years.

When he was thirty-nine, Halley began the work for which he is chiefly remembered. Newton had demonstrated that the comets, like the planets, moved in orbits that were conic sections (see the diagram on page 38), but *which* conic section was a matter of dispute. Newton himself was of the opinion that comets moved along open parabolic orbits; Cassini preferred circles; and the same Bishop of Salisbury whom Halley tried not to offend in his first published paper was leaning toward the ellipse. In addition, there were champions of the hyperbola. It was difficult to know who was right, because Earthbound observers with feeble telescopes could see comets only when they came close to the Sun. The little arc that their paths described during this briefest and swiftest part of their journey could be accounted for by almost any of the conic sections (page 38), although circles were more difficult to justify.

Halley approached this challenge—reconstructing a comet's whereabouts during the time it is invisible—with the discipline of a great detective. He steeped himself in every piece of recorded testimony on the subject, every surviving declaration made by a string of witnesses that stretched back to Pliny and Seneca. Through calculus and gravitational theory, Newton had provided the detection technology, which Halley mastered. And there was also an element of luck; it was Halley's good fortune to live in a century graced with an abundance of comets, and therefore a mass of recent, relatively accurate evidence on their behavior.

Where the Comet of 1682 was concerned, Halley regarded Flamsteed as the most reliable observer, and wrote to Newton, asking him to obtain the observations from Flamsteed, because "he will not deny it you, though I know he will me." Newton did as he was asked. Halley then compared the orbital characteristics, or elements, of the comets of 1531, 1607, and 1682, and found many striking similarities (see the box on page 56): in the tilt, or inclination of the cometary orbit to the zodiacal or ecliptic plane (Chapter 2); in the distance of the comet from the Sun at perihelion; in the region of the sky in which perihelion occurs; and in the place where the comet's orbit crosses the zodiacal plane (called the node). These similarities were already enough to suggest that the same comet was being seen in three different apparitions. When Halley compared the dates of the apparitions, he found something like a periodic return—just as Newtonian theory had promised if the comets were on elliptical orbits.* The case was almost broken.

But Halley worried that the differences in orbital ele-

---

*Cassini had suggested something similar, but, as we now know, for apparitions of quite different comets, which have not, at least so far, returned.

ORBITAL ELEMENTS OF THE COMETS KNOWN TO HALLEY

THE ASTRONOMICAL ELEMENTS OF THE MOTIONS, IN A PARABOLIC ORBIT, OF ALL THE COMETS HITHERTO OBTAINED

Passage of Perihelion, London Time				Longitude of Perihelion		Asc. Node		Inclination of Orbit		Distance from Sun at Perihelion
	(d.	h.	m.)	°	′	°	′	°	′	
1337, June	2	6	25	37	59	84	21	32	11	0.40666
1472, February	28	22	23	45	34	281	46	5	20	0.54273
1531, August	24	21	18	301	39	49	25	17	56	0.58700
1532, October	19	22	12	111	7	80	27	32	36	0.50910
1556, April	11	21	23	278	50	175	42	32	6	0.46390
1577, October	26	18	45	129	32	25	52	74	83	0.18342
1580, November	28	15	00	109	6	18	57	64	40	0.59628
1585, September	27	19	20	8	51	37	42	6	4	1.09358
1590, January	29	3	45	216	54	225	31	29	41	0.57661
1596, July	31	19	55	228	15	3.2	12	55	12	0.51293
1607, October	16	3	50	302	16	50	21	17	2	0.58680
1618, October	29	12	23	2	·14	76	1	37	34	0.37975
1652, November	2	15	40	28	19	88	10	79	28	0.84750
1661, January	16	23	41	115	59	82	30	32	36	0.44851
1664, November	24	11	52	130	41	81	14	21	18	1.02576
1665, April	14	5	16	71	54	228	2	76	5	0.10649
1672, February	20	8	37	47	0	297	80	83	22	0.69739
1677, April	25	00	38	137	37	236	49	79	3	0.28059
1680, December	8	00	6	262	40	272	2	60	56	0.00612
1682, September	4	7	39	302	53	51	16	17	56	0.58328
1683, July	3	2	50	85	30	173	23	83	11	0.56020
1684, May	29	10	16	238	52	268	15	65	49	0.96015
1688, September	6	14	33	77	0	350	35	31	22	0.32500
1698, October	8	16	57	270	51	267	44	11	46	0.69129

This table, taken from Halley's work, is the basis for his conclusion that the three apparitions here highlighted in blue refer to the same comet. See for yourself what Halley discovered when he compared the orbital "elements" or characteristics of the Comets of 1531, 1607, and 1682. Dates are given in days, hours, and minutes; angles in the sky in degrees and minutes of arc; and perihelion distance in Astronomical Units (the Earth is 1 A.U. from the Sun). *Asc.* stands for ascending; the ascending node is one of the two places in a cometary orbit where it crosses the zodiacal or ecliptic plane.

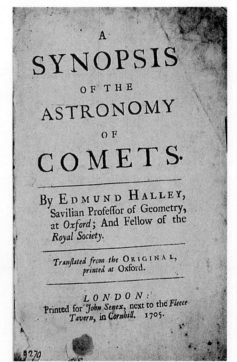

1894

ments from one apparition to another, while small, were much larger than what could be ascribed to errors of observation. The longitude of perihelion, for example, varied by more than a degree in the sky, yet the measurements were accurate to a few minutes of arc. Furthermore, the interval between the 1531 and the 1607 apparitions was more than a year longer than the interval between the 1607 and the 1682 apparitions. Thus, in addition to the metronomic regularity of an isolated comet in elliptical orbit around the Sun, there must, Halley believed, be some other influence or force that perturbed the comet one way in one apparition and another way in the next.

Newton had proposed that the variations in the periods of the comets were produced by the gravitational attraction of undiscovered comets. But Halley knew that Jupiter and Saturn were perturbed by one another, and thought it likely that a comet, of much lower mass than either of these giant planets, would be more perturbed by even moderately distant approaches to either planet than by close approaches to a comet. He made a crude estimate of the effects on the comet's motion of the gravity of Jupiter and Saturn, and found they fit well with the measured discrepancies. Halley concluded that the slight differences in the orbital elements of the comets of 1531, 1607, and 1682 had a ready explanation. These were visitations of the same comet, vulnerable, as all travelers are, to detours, delays, and road conditions.

Halley's investigation of comets constituted an enormous undertaking, requiring the painstaking calculation of the orbits of twenty-four comets that had achieved perihelion passage between 1337 and 1698. He was struck, as were Aristotle and Seneca, by the random character of their inclinations:

> Their orbits are disposed in no certain order....They are not confined like the planets to the Zodiac, but... move indifferently, every way both retrograde and direct.*

In their distance from the Sun at perihelion passage, the comets Halley studied ranged from about 1 A.U. to less than 0.01 A.U.—this was the Great Comet of 1680, that nearly grazed the Sun. And at aphelion, he found, the comets—including the Comet of 1682—ranged well beyond the orbit of Saturn, then the most distant planet known.†

Cover page of the 1705 work by Edmond Halley in which the orbits of comets are calculated, and the periodic return of the comet posterity has named after Halley is postulated. Courtesy The Royal Society of London.

---

*I.e., both clockwise and counterclockwise as seen from a vantage point high above the North Pole.

---

†In fact, because he fit the orbits to open parabolas, the aphelia were effectively at infinity.

Portrait of Edmond Halley in middle age by Richard Phillips. Date unknown. National Portrait Gallery, London.

1895

In 1705 Halley published the results of his "immense labor" in a paper entitled *A Synopsis of the Astronomy of Comets*. It was the first application of the laws of the universe, as revealed by Newton, by a scientist other than Newton to solve an astronomical mystery. This in itself was enough to guarantee Halley a place in the history of science. But he went much further, making a courageous leap that earns him a place in the larger history of civilization. For millennia, comets had been the almost exclusive property of the mystics, people who considered comets as portents, symbols, wraiths—but not *things*. Halley shattered the monopoly by beating them at their own game, a game that no scientist had ever before played: prophecy. He predicted that the comet seen in 1531, 1607, and 1682 would return more than fifty years in the future. And he did not hedge his bet. It would return, he stated flatly, at the end of 1758— from a particular part of the sky, with specific orbital elements. There is hardly a prophecy of the mystics that even strives for comparable precision. He was, for his time, remarkably free of nationalism, but this once, he indulged himself in a bit of chauvinism: "Wherefore, if according to what we have already said it should return again about the year 1758, candid posterity will not refuse to acknowledge that this was first discovered by an Englishman."

In the spring of 1696, Newton was appointed Warden of the Royal Mint. It required him to live in the Tower of Lon-

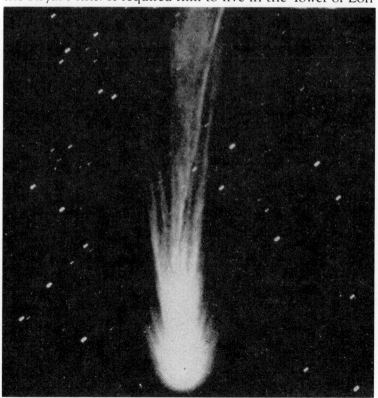

The spectacular appearance of Comet Morehouse (1908 III) as photographed by Max Wolf at the Heidelberg Observatory.

Halley's map of the Atlantic Ocean with lines indicating magnetic variation, ca. 1701. Courtesy The Royal Astronomical Society.

With the appearance of these magnetic charts and his already published work on the physics of wind, monsoons, and evaporation of seawater, Halley can truly be termed the founder of modern geophysics, a fact which was recognized by the Royal Society in 1957 during the International Geophysical Year, when it named its permanent scientific base in Antarctica Halley Bay.

—Colin Ronan,
Edmond Halley: Genius in Eclipse,
1969

don. His task was to prevent clipping or shaving of the coinage of the realm. Months later, Newton appointed Halley —who had just begun his study of comets—deputy comptroller of the mint at Chester. Under the Master and the Warden in the Tower were masters, wardens, comptrollers and the like in provincial facilities of the mint. Halley spent two miserable years at Chester supervising the mechanized production of milled coins out of the old handmade ones. When he and his local warden discovered two clerks skimming precious metals for their own gain, they spoke out— unaware that their own superior, the Master of the Mint at Chester, was getting a cut from the clerks. An acrimonious feud followed, and there were threats of a duel; but the mints were closed in 1698, and Halley was liberated to return to London.

He arrived just in time for a more agreeable assignment. The twenty-six-year-old Czar of Russia, who would later be known as Peter the Great, had come to England to learn how to Westernize his nation. He was ensconced at a grand country home near the Deptford shipyards, where he performed manual labor as part of an effort to gain firsthand experience of the British genius for shipbuilding. (One's powers of imagination protest when asked to contemplate some modern equivalent—an American president laboring in the spacecraft assembly facility in Tyuratam, say, or the General Secretary of the Soviet Communist Party installed for months in a hard-hat job at Cape Canaveral.) Peter had hoped to spend time with Newton, but Newton sent Halley in his stead. By all accounts, the English astronomer and the Russian Czar became fast friends, sharing passions for knowledge and brandy. For the duration of Peter's stay in England, Halley remained his chief science adviser and drinking partner. One disputed, though contemporary account depicts Halley raucously pushing the Czar of all the Russias through the streets of Deptford in a wheelbarrow in the dead of night. It was said that the two of them were inebriated, and that the escapade caused grave damage to the topiary hedges.

That same year Mary Halley gave birth to a son named Edmond, and her husband began yet another career, described in a contemporary account replete with italics:

In 1698 the King [William III] who had been inform'd of Mr. Halley's ingenious Theory of the Magnetic Needle, was desireous the variation shou'd, for the Benefit of Navigation, be carefully observed, in diverse parts of the Atlantic Ocean; for which purpose His Majesty, the 19th Aug. 1698, appointed Mr. Halley Commander of His Ship the Paramoor* Pink, with orders *to seek by*

1896

observation the discovery of the Rule of the variation of the Compass, and at the same time *to call at his Majestys Settlements in America, making som observations there, in order to the better laying down the Longitudes and Latitudes of those places, and to attempt a discovery of what Lands lay to the South of the Western Ocean.*

Seventy years before Cook set sail (in pursuit of a Halleyan objective), Halley was commanding the first marine scientific expedition to be commissioned by a British monarch. The *Paramour* sailed to Spain, the Canary Islands, Africa, Brazil, and the West Indies, before an incipient mutiny on the part of Halley's second-in-command necessitated an unscheduled return to England. In the course of the court-martial that ensued, it was revealed that the disgruntled lieutenant was a closet magnetic theorist, whose own writings on the subject had earlier been found wanting by the Royal Society. He nursed his indignation, and when the landlubber Halley was commissioned his captain, he became unhinged. Halley assumed the role of navigator and took great pride in the fact that he was able to bring his ship home without loss of life.

Halley commanded two additional voyages on the *Paramour*—one, amidst many perils, up the coast of South America to Trinidad, bound for Cape Cod. They found the seas off Nantucket too high, and were forced to sail on to Newfoundland. There the *Paramour* took friendly fire from an English fishing fleet that mistook her for a pirate ship. Again the captain safely returned his ship to England, but this time was denied the satisfaction of seeing every member of his crew at the homecoming. The cabin boy had been swept overboard and lost during a wild storm off the Canary Islands. For the rest of Halley's life, he was unable to speak of the boy's death without tears.

Halley fulfilled his royal mandate by publishing "A *New and Correct* CHART *Shewing the* VARIATIONS *of the* COMPASS *in the* WESTERN AND SOUTHERN OCEANS *as observed in ye* YEAR 1700 *by his Maties* [Majesty's] *Command."* The map contained a dotted-line convention that Halley devised to indicate points of equal variation in the Earth's magnetic field, the method that remains in use in magnetic maps today. Halley expanded his Atlantic chart to a world chart that remained in print, through many editions, for a century.

Halley's third and final voyage as Captain of the *Paramour*, in 1701, kept him closer to home. His purpose was to study

*Or *Paramour*. Spelling conventions were less rigid in Halley's time.

the tides of the English Channel, although it has been suggested that his hidden agenda included a reconnaissance of the French coast on the eve of the War of the Spanish Succession. The following year, Queen Anne dispatched Halley on diplomatic missions to European monarchs.

When Halley returned to England he was, perhaps to his surprise, offered a Savilian Professorship at Oxford—but of geometry, not astronomy. Thirteen years after his vitriolic campaign against Halley's appointment to another Savilian chair, Flamsteed wrote a carping letter to a mutual friend, denying both Halley's suitability and his prospects for the chair, and complaining, "He now talks, swears, and drinks brandy like a sea captain." But by now Halley was too well-respected for Flamsteed's venom to do much harm, and in 1704 he was appointed.

His inaugural lecture was a loving tribute to the geometrical achievements of his colleagues. Newton, of course, was singled out for the most lavish praise. Halley dedicated much of his tenure as Savilian professor to a rediscovery of the ancient founders of geometry, among whom was one Apollonius of Perga, a mathematician and astronomer who flourished during the second half of the third century B.C. In the great city of Alexandria, Apollonius did for conic sections what Euclid had done for geometry: he was the first to describe the parabola, the hyperbola, and the ellipse (see page 38). Halley, who used the properties of these curves to determine the orbits of the comets, wished to repay his debt to Apollonius by giving the work of the ancient mathematician new life. But no copies of Apollonius' work survived in the original Greek—largely due to the burning of the Library of Alexandria. The only copies were in Arabic. So at age forty-nine Halley taught himself Arabic. He worked initially in collaboration with David Gregory, the man who had been given the Savilian chair in astronomy for which Halley had been passed over. When Gregory died soon after the Apollonius project had begun, Halley carried on alone in a task that had already defeated a number of full-time Orientalists. But Halley succeeded where they had failed, and he astonished the premier Orientalist of the day with his accuracy and insight. Knowing the geometry probably helped.

During this same period, he reedited the most interesting papers from the *Philosophical Transactions* into a three-volume work for a popular audience. He thought nonscientists might be curious about the physical and biological world.

Meanwhile, the miserable Flamsteed remained at the Royal Observatory at Greenwich in his capacity as Astronomer Royal, where he was expected to share his observations with the astronomical community. This he steadfastly refused to do. For years he was permitted this clear dereliction of duty, but by 1704 it had become intolerable. Newton, by now president of the Royal Society, visited him at Greenwich in an attempt to discover the state of the observations. After thirty years as Astronomer Royal, Flamsteed had hardly published a thing. Newton was given the impression that Flamsteed's life's work, *The British History of the Heavens*, was nearing completion, and Newton returned to London to arrange for publication. But Flamsteed was lying; he was years from finishing.

The Astronomer Royal's procrastination and arrogance brought out the worst in Newton. There is plentiful evidence in the correspondence of both men attesting to an active mutual hatred. Nor were they each other's only enemy. There were legitimate grievances on both sides, but Newton had the upper hand and used it shamefully. For the next ten years he seemed to derive an ugly pleasure from torturing Flamsteed, who was by this time ill and desperate.

Halley was now presented with the perfect opportunity to avenge himself against the man who had, through unreasoning malevolence, campaigned to harm him. But vengeance was one of the few subjects that failed to engage Halley's interest. Indeed, at Newton's request, he worked on the manuscript of *The British History of the Heavens*, correcting errors, making many needed calculations, and helping to see the book through publication—but all this against Flamsteed's explicit wishes. In June 1711, Halley wrote to him,

> ...Pray govern your passion, and when you have seen and considered what I have done for you, you may perhaps think I deserve at your hands a much better treatment than you for a long time have been pleased to bestow on
>
> Your quondam friend, and not yet profligate enemy (as you call me),
>
> Edm. Halley

The book extended the map of the northern skies from 1,000 to 3,000 stars, including many too faint to be seen without a telescope, and was prized by astronomers for centuries. Nevertheless, Flamsteed was enraged by Halley's version of the *Historia Coelestis*, which appeared in 1712. By 1714 he had managed to burn nearly every copy in existence. The official version, with the word *Britannicae* appended to the title, was not published until 1725, in a posthumous edition.

Despite Flamsteed's contrary opinion, a review of the correspondence suggests that Halley had remained circumspect toward Flamsteed until the latter's death in 1719. Then fate intervened to thrust on Halley a kind of satisfaction that he had denied himself, appointment as Flamsteed's successor. Halley became Astronomer Royal. But when he arrived to take over his duties, he found the Royal Greenwich Observatory denuded of astronomical instruments; they had all been Flamsteed's personal property, his widow said. It was true. He had bought every last sextant and quadrant with his own money.

Halley was now sixty-three years old and as curious and passionate about science as ever. To read his paper, *An Account of the Extraordinary METEOR seen all over England on the 19th of March 1719*, is to encounter a kind of naked enthusiasm which has been effectively eliminated from the literature of science today. He begins by announcing "This wonderful luminous *Meteor*," which, he laments, "it was not my good Fortune to see." But others saw it and he gives their accounts. Sir Hans Sloan, vice-president of the Royal Society, was one of the lucky ones. Without warning, he sees something in the night sky much brighter than the Moon, first near the Pleiades, and then down below Orion's Belt. It was so bright that Sir Hans was forced to avert his eyes. He estimates that it moved across 20 degrees of the sky in about half a minute or less. All in all, pretty thorough testimony from a man unprepared for astronomical observation. But Halley is dissatisfied: "It were to be wished," he chides, "that Sir Hans had more especially regarded the Situation of the Track of this Meteor among the fixt stars, and let us know how much it past above the Pleiades and how much under the Belt of Orion...." Halley cannot help himself; he is desperate to know, and his desire is even greater than his celebrated kindness. He wants to know everything about the meteor: its altitude, velocity, what sound it made, how big it is, of what it is made. We feel privileged to know the answers to his questions (Chapter 13) and wish it were possible to share them with him.

At the age of sixty-five, in an act of consummate optimism, Halley undertook an ambitious study of the eighteen-year cycle of solar eclipses. Halley, the inventor of the actuarial table, could not have been unaware of the improbability that he would live long enough to complete the project. He confounded the odds, and finished when he was eighty-four.

His productivity and his longevity were extraordinary in another respect. Physicists are sometimes said to be like mayflies, with only a brief creative period; indeed, a strikingly large fraction of major discoveries are made before the age of thirty-five. This is more true for theoretical than for

He seems to be not one but all mankind's epitome.

—Herbert Dingle,
The Halley Lecture,
Oxford University,
1956

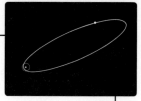

1897

experimental physics, and more true for physics than for astronomy. Perhaps past thirty, the mind loses some of its capacity to conceptualize on a grand scale. But in the last decades of Halley's life, he made major theoretical advances in our understanding of nature on its grandest scale: the universe. He discovered that the so-called fixed stars actually moved with respect to one another. The discovery may have been stimulated by his work on Flamsteed's book. It took another hundred years of instrumental development in astronomy before Halley's discovery of stellar proper motion could be confirmed. In another paper of his last years, he anticipated the discoveries of a much later time, arguing for a limitless universe without a center. Halley was, at last, guilty of believing in infinity.

Mary Halley died when Halley was eighty years old. Shortly after, he suffered a stroke and the loss of his son. Despite these blows, he continued making astronomical observations and attending scientific meetings until a few weeks before his death on January 14, 1742, at the age of eighty-six. His last words were a request for a glass of wine. He drank it while sitting in a chair. When the glass was drained, he died without a sound.

He had wished to lie next to Mary forever. His daughters had this tribute (translated here from its original Latin) engraved on their parents' tomb:

> Under this marble peacefully rests, with his beloved wife, Edmond Halley, LL.D., unquestionably the greatest astronomer of his age. But to conceive an adequate knowledge of the excellencies of this great man, the reader must have recourse to his writings; in which almost all the sciences are in the most beautiful and perspicacious manner illustrated and improved. As when living, he was so highly esteemed by his countrymen, gratitude requires that his memory should be respected by posterity. To the memory of the best of parents their affectionate daughters have erected this monument, in the year 1742.

A portrait of Edmond Halley at age eighty, just before the death of his wife, Mary. Painting by Michael Dahl. Courtesy The Royal Society of London.

Edmond Halley was not just a man who discovered a comet. In fact, discovering a comet was one of the few scientific activities in which he did not engage.

There is a widely held belief that the price of deeply understanding the complexities of nature is paid out in the currency of alienation. Something like an inverse square law exists, it is said, between scientific genius and the capacity for love. If the life of Isaac Newton is pointed to as the most dramatic illustration of this theorem, then the life of his friend Edmond Halley, who more than anyone else made manifest Newton's genius, may be offered as its most inspiring exception.

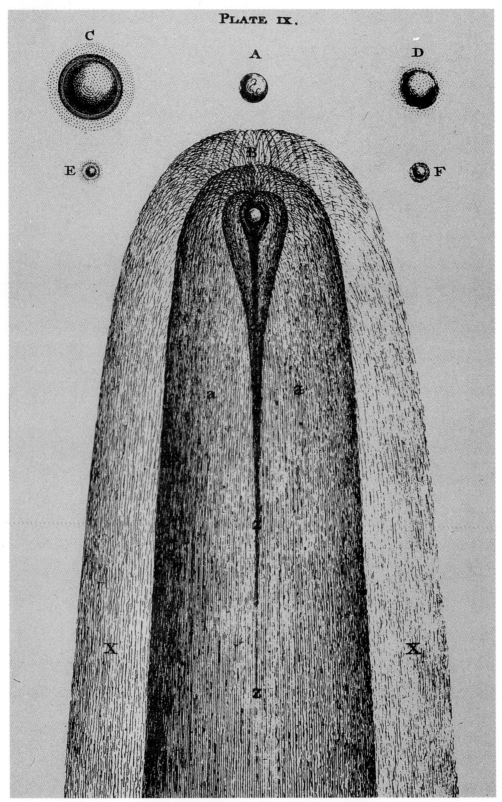

Drawing of the nucleus, coma, and tail of the Great Comet of 1680, and the nuclei of five other comets. From Thomas Wright of Durham, *An Original Theory or New Hypothesis of the Universe*, ed. by Michael A. Hoskin (London, 1750, and New York, 1971). Courtesy Michael A. Hoskin.

Chapter IV

THE TIME OF THE RETURN

Guardian and friend of the moon, O Earth, whom the comets forget not,
Yea, in the measureless distance wheel around and again they behold thee!

—Samuel Taylor Coleridge,
Hymn to the Earth,
1834

CHINA INVADED TIBET AND TURKESTAN; FRENCH troops seized the Ohio Valley; Britain declared war on France; Prussia defeated Austria, whereupon Austria defeated Prussia; a Russian army occupied Germany; and an Indian revolution against the British army of occupation was ruthlessly suppressed. In such respects, the decade of the 1750's was almost indistinguishable from many others. But it was also a time of enlightenment. Diderot's *Encyclopaedia* was published in France and Samuel Johnson's *Dictionary* in England; Hume, Rousseau, and Voltaire wrote seminal works; Bach died and Mozart was born. Lomonosov founded the University of Moscow. The Prussian Academy of Sciences in Berlin and the first mental asylum in London both opened their doors; *Tristram Shandy* was being written; Hokusai was born in Tokyo; and a little-known Virginia surveyor named George Washington married a widow named Martha Custis.

In science, this was the decade that would end with the return of a comet—if Edmond Halley's prediction could be believed. In the first half of the decade, two extraordinary scientific works were published, each bearing on the nature of comets, and each presenting a view of the universe surprisingly in advance of its time.

Thomas Wright of Durham was an astronomer by temperament, although entirely self-taught. Born in 1711, in the north of England, the son of a carpenter, he was prevented from continuing his early schooling "by a very great impediment of speech." He also seems to have been expelled from school on account of his behavior. He described himself as "very wild and much addicted [to] sport." Following the practice of the time, at age thirteen he was apprenticed—to a clockmaker, where he spent so much of his time poring over the astronomical literature that his father thought him mad. Unlike Edmond Halley's father, Wright's apparently attempted to influence his son's course of study by burning his books. Shortly after, the young man was dismissed from his apprenticeship, the culmination of a considerable scandal; fell in love with a clergyman's daughter, but discovered his plans for a secret marriage "prevented, and Miss lock'd up"; and arranged, in his anguish, a passage to the West Indies, but was prevented from embarking by his outraged father.

With this promising start, he taught himself surveying and navigation, became a tutor to the children of the aristocracy, turned down a professorship at the Russian Imperial Academy in St. Petersburg, and began writing books on astronomy. The most extraordinary of them, called *An Original Theory of the Universe*, was just that. Published in 1750, it is the first known statement of the true nature and geome-

Thomas Wright of Durham as depicted in an engraving in *Gentlemen's Magazine*, January 1793. The portrait is circumscribed by a serpent with its tail in its mouth, symbolizing eternity. Compare with the drawing at the top of page 71. Ann Ronan Picture Library.

try of the Milky Way—not a road of the gods, not divine milk splashed across the heavens, not an architectural support holding up the sky, but a flat disk of stars each like the Sun, all suspended in the ocean of space. From the time of Democritus, there had been a few people who had guessed that the Milky Way was made of individual stars too faint and distant to be seen individually; and this notion had been vindicated by Galileo with the first small telescope. So by Milton's time it was possible for a poet to describe the Milky Way as a Galaxy, "powder'd with stars." But the idea of the Milky Way as a flattened concentration of stars in which the Sun is embedded was first proposed by Wright. He even imagined that the stars revolved about the center of the Galaxy "as do the planets around the Sun."

While there are mystical elements in Wright's writings, and certainly not all that he proposed in his *Original Theory* has stood the test of time, his vision of the Milky Way is a landmark in the history of astronomy. The work is the more remarkable since its author had never acquired a formal education. This vision—a galaxy full of moving stars—turns out, as we shall see later, to be central to any understanding of the nature and origin of comets.

But comets were discussed in their own right in *An Original Theory*. Using his talents as draftsman and surveyor, Wright designed elegant diagrams of the solar system, with comets much in evidence. He delighted in showing the cometary orbits tabulated by Halley in their correct sizes and orientations [see page 71], providing many readers with a jolting first look at the comparatively small scales of the planetary orbits. Wright attempted to show the relative sizes of the worlds then known [see page 70]. The massive planets, Jupiter and Saturn, dominate the drawing. The nine moons, all that were then known, and the inner planets—Mercury, Venus, the Earth, and Mars, along with the Earth's moon—are shown as insignificant beside the two giant planets then known. Jupiter had already been observed to be cloud-covered, and Wright could not resist showing some land and ocean peeking through the overcast. He felt those worlds beckon.

With this fine beginning, he attempted to draw the comets to scale as well, from the observations available in his time. The result is shown in the frontispiece to this chapter. For scale, A represents the Earth; and C, D, E, and F, the nuclei of the comets of 1682, 1665, 1742, and 1744, respectively. In fact, what was being measured in Wright's day—and this is still largely true in our own—was not the nucleus of the comet, but the coma. The nucleus—Wright used the term—is the bright solid object at the center of the comet, the presumed source of the fine particles or gas that consti-

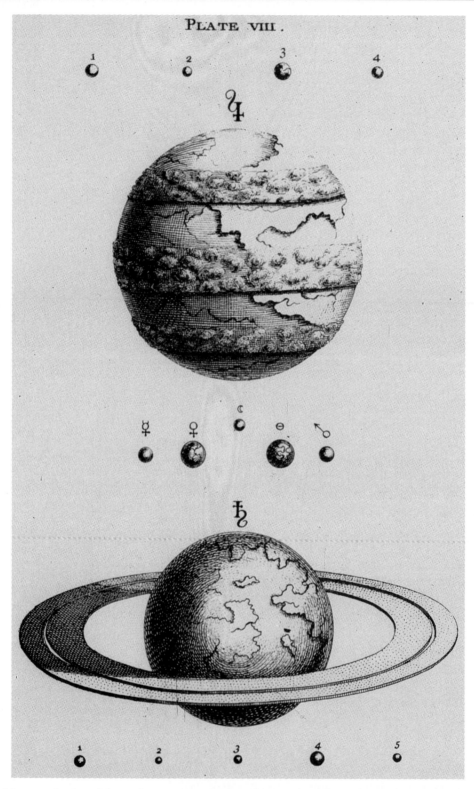

Portraits of Jupiter (*top*) and four of its moons, Saturn (*bottom*) and five of its moons, and the planets Mercury, Venus, Earth, and Mars—with the Earth's Moon thrown in for good measure in the middle. The Cassini Division in the rings of Saturn is shown, but the details on the surfaces of Jupiter and Saturn are wholly fanciful. From Thomas Wright of Durham, *An Original Theory or New Hypothesis of the Universe,* ed. by Michael A. Hoskin (London, 1750, and New York, 1971). Courtesy Michael A. Hoskin.

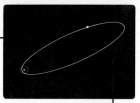

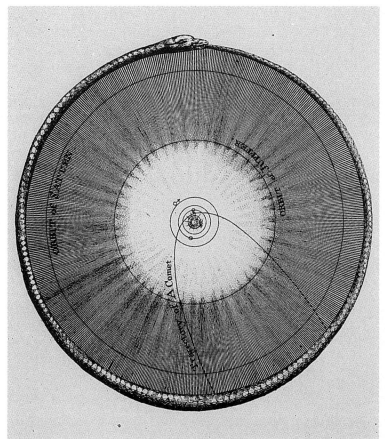

Left: The orbit of a comet cuts through a solar system circumscribed, just outside the orbit of Saturn, by a serpent. From Thomas Wright of Durham, *An Original Theory or New Hypothesis of the Universe,* ed. by Michael A. Hoskin (London, 1750, and New York, 1971). Courtesy Michael A. Hoskin.

Below: Thomas Wright's elegant depiction of the solar system as known in his day. At the center are the planets Mercury, Venus, Earth, and Mars, each represented by its astronomical symbol, and then, just outside the Sun's rays, the orbits of Jupiter and Saturn. Of the three comets shown, the Comet of 1680 is the one Newton had first computed the orbit of, and the Comet of 1682 is the comet Halley had predicted would return—less than a decade after the publication of this figure. From Thomas Wright of Durham, courtesy Michael A. Hoskin, ibid.

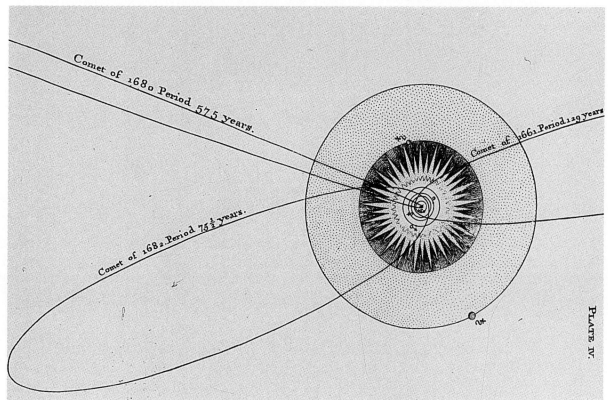

The scale of the solar system, according to Thomas Wright. Compare with pages 6 and 7. From Thomas Wright of Durham, *An Original Theory or New Hypothesis of the Universe*, ed. by Michael A. Hoskin (London, 1750, and New York, 1971). (In Figure 3 the cometary rosette is reminiscent of the Bohr atom.) Courtesy Michael A. Hoskin.

tute the comet's tail. But the coma, the cloud of matter around the nucleus, shields it from our view. The almost total absence of detail in Wright's drawings of cometary "nuclei" could have provided a hint that the nucleus is surrounded by coma. The nucleus might also be much smaller than the coma; it might even be too small to make out any details on, even if it were not enveloped in a shroud of matter. As the figure indicates, the comas of comets in the vicinity of the Earth can be as large as the Earth, or even larger.

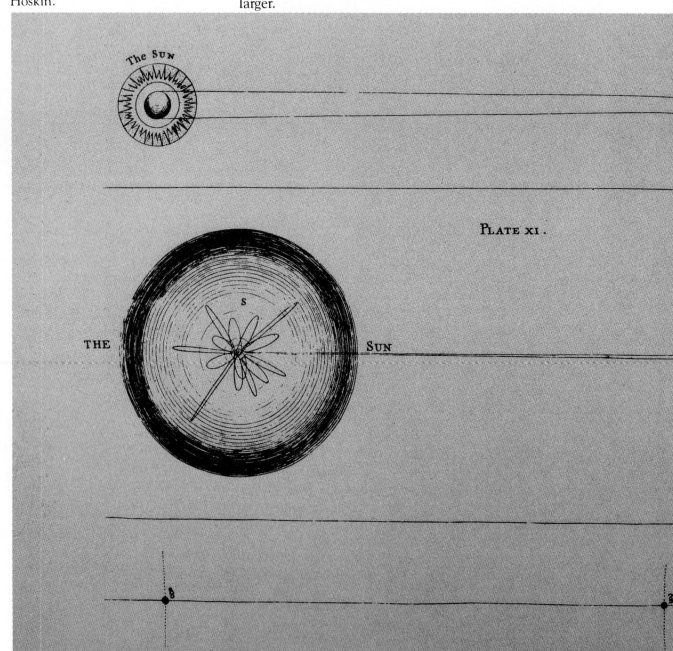

1900

Prominent in the frontispiece is Wright's depiction of the Great Comet of 1680, the very comet that Newton had, in the *Principia*, so brilliantly shown was orbiting the Sun along a conic section and obeying the law of universal gravitation; it had also helped to stimulate Halley's interest in comets. We do not know whether Wright merely drew what others had described in words or had access to drawings made at the telescope, but what is depicted here is close to the appearance of many comets. (See Chapters 7, 9, and 10.)

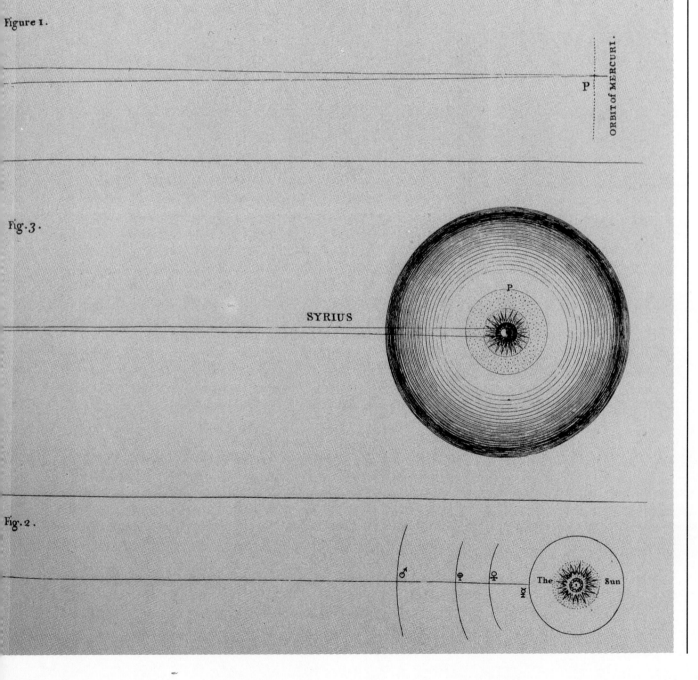

Figure 1.

Fig. 3.

Fig. 2.

Wright described the region labeled *aa* as "the comet's natural atmosphere," the converging central lines representing denser matter. XX is a representation of "the inflamed atmosphere and tail dilated near the Sun." The "nucleus" of the Comet of 1680, as we have already remarked, is merely the inner coma. Above it, are three concentric shells of matter that we know from observations of subsequent comets are commonly ejected, in succession, from the nucleus. The traceries of fine lines in this outer coma give a curious appearance, as if fountains of material were pouring out into space on the sunlit side of the cometary nucleus.

Then, to sum up the scale of the solar system, Wright produced for his readers' amazement and delight a three-part diagram (see pages 72 and 73). In his Figure 1, the size of the Sun is set to scale with the orbit of Mercury. In Figure 2, to scale with the orbit of Mercury are arcs representing, successively, the orbits of the remaining planets known in Wright's time—Mercury, Venus, the Earth, Mars, Jupiter, and Saturn. And in Figure 3, the entire planetary system is represented by a central dot, with a rosette of cometary orbits being the only hint of where the planets might lie.

Wright took the outer boundary of the solar system to be a little beyond the farthest reaches of the orbits of the most distant comets then known; and drew the "least possible distance" between the Sun and what was then considered to be the nearest star, Sirius. (We now know that both the outer boundaries of the solar system and the distance from the Sun to Sirius are vastly greater than was imagined in Wright's day.) He conjured up "a new-created Mind or thinking Being, in a profound State of Ignorance," who sees both Sirius and the Sun from a remote vantage point, and then arrives in the solar system and observes the disposition of comets and planets. What would this being imagine is in orbit around Sirius? Wright cheerfully gives the answer: "Why, Planets such as ours." And comets.

Portrait of Immanuel Kant by J.F. Bause, after a painting by V.H. Schnorr, 1789.

* * *

Wright's book influenced the perspective of astronomers and the future of astronomy in ways that can no longer be traced; but its most significant known influence occurred because of a review that appeared in the following year (1751) in a German magazine called *Free Opinion and Information for the Advancement of Science and of History in General.* One of its readers, living in the university town of Königsberg, was Immanuel Kant, a twenty-seven-year-old graduate student who had been gripped by the scope and elegance of Isaac Newton's work. He was later to become a towering figure in philosophy, but in the early 1750's he was mainly interested in science. In 1755, aroused by Wright's

vision of the universe, but still having read only a review of Wright's book, Kant published the *General Natural History and Theory of the Heavens*. In it, he makes clear his debt to Wright. Kant's book was published four years before the expected return of Halley's Comet.

In their intellectual lives, Kant and Wright were kin. But in their personal lives there were many differences. Beyond his native intelligence, Kant had few natural advantages. Unlike Wright, he was encouraged by his parents. In poor health his whole life, with a deformed chest, and barely five feet tall, Kant subjected himself to a strict regime of regular exercise, mainly walks. His father was a saddle-maker. Where Wright had at least *planned* to run away with the vicar's daughter, Kant seems not to have enjoyed a close relationship with any woman besides his mother. While Wright was prepared to seek his fortune in far-off America, Kant never ventured more than a hundred kilometers from Königsberg in his life. Where Wright precipitated scandals, Kant's behavior was proper, even a little austere. Where Wright was thrown out of school, Kant was a highly successful student, admired by his teachers. Wright and Kant were different men, molded by different cultures, but they were equally swept up in the great Newtonian vision of innumerable stars and innumerable worlds all moving in solemn obedience to a great universal law of gravitation that could reconstruct their motions in the distant past and predict their positions in the far future.

Kant's *Theory of the Heavens** blazed many trails. He accepted Wright's view of the Milky Way as a flattened volume of space bounded by two parallel planes and filled with stars. Kant was the first person to consider the origin and evolution of the Galaxy, a central topic of modern astrophysics. And he made a daring additional leap—he supposed that the Milky Way was one of innumerable other galaxies, each filled with stars and, it may be, planets and life, a cosmic perspective not fully demonstrated until the 1920's. Wright had reached for such a vision, but had not grasped it. Kant correctly proposed that the spiral nebulae, such as M31 in the constellation Andromeda (see page 76), were distant Milky Ways. That we live in a universe of galaxies, each composed of a multitude of suns, is perhaps the central revelation of modern astronomy.

There are many other delights to be gathered from the pages of the *Theory of the Heavens*, including the first state-

---

*The full title was *Universal Natural History and Theory of the Heavens, Or an Essay on the Constitution and the Mechanical Origin of the Whole Universe Treated According to Newton's Principles*—an ambitious mouthful, but Kant delivered.

The great galaxy M-31 "in" (i.e., beyond) the constellation Andromeda, which supplies the foreground stars in this image. Some two million light-years away, it is the nearest spiral galaxy like our own Milky Way. No individual stars in M-31 are here resolved, although it contains hundreds of billions of stars. If this were a photograph of our own Milky Way, somehow obtained from far outside, the position of the Sun would be so far from the center as to be off the edge of the picture. M-31 is shown accompanied by its two smaller satellite galaxies. Courtesy Hale Observatories, Carnegie Institution of Washington and California Institute of Technology.

ment that the solar system had been formed from a cloud of diffuse interstellar matter. The idea is known today as the Kant-Laplace hypothesis, and the cloud of matter is now called a solar nebula. Naturally enough for someone writing on astronomy in the decade of the predicted return of Halley's Comet, Kant discussed comets:

> The most distinctive mark of the comets is their eccentricity....Their atmospheres and tails, which on their great approach to the Sun are spread out by its heat,...[were] in ages of ignorance unusual objects of terror, [and were] regarded by the common people as foretelling imaginary fates....It is not possible to regard the comets as a peculiar species of heavenly bodies entirely distinct from the race of planets.

He pictured comets condensing from "primitive matter in regions of space far away from the center and which is feebly moved by [gravitational] attraction." He imagined that, unlike the planets, the comets formed with their orbits at all inclinations, a situation he described as "lawless freedom." "Hence," he said, "the comets will come to us without restriction from all quarters."

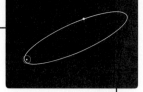

When he discusses the density and, in effect, chemistry of the comets, he gets off to a breathtakingly good start:

> The specific density of the matter out of which the comets arise is more remarkable than the magnitude of their masses. As they are formed in the furthest region of the universe, it is probable that the particles of which they are composed are of the lightest kind; and it cannot be doubted that this is the chief cause of the vapor heads and tails by which they are distinguished from other heavenly bodies.

But now a wrong turn:

> This dispersion of the matter of the comets into vapor cannot be attributed mainly to the action of the heat of the Sun; for some comets scarcely reach as near the Sun as the distance of the Earth's orbit; and many stop between the orbits of the Earth and Venus, and then turn back. If such a moderate degree of heat resolves and attenuates the matter on the surface of these bodies to this degree, must they not consist of the *very* lightest stuff, which can suffer more atten-

A photograph, from our position embedded inside the Milky Way, of the center of the Galaxy, beyond the nearby constellation Sagittarius—which provides the foreground stars. We see the great clouds of gas and dust, cutting diagonally through the picture and marking the plane of the Galaxy. Photo by David Talent. Courtesy National Optical Astronomy Observatories.

uation by heat than any other matter in the whole of nature?

Kant's problem is to find a material which turns from solid into vapor at the Earth's distance from the Sun. He had only to look out his window, perhaps at the very moment he was writing these lines, to see a cloud of vapor rising off the frozen River Pregel, and he would have had his answer: Ice. Ordinary water ice. Not an exotic substance, at least on Earth. No special celestial stuff. Just ice.

Of course it is unfair to Kant—whose vision on these matters was remarkable in a remarkable age—to criticize him in hindsight. We are the beneficiaries of the work of large numbers of capable scientists who have lived between his day and ours. But picture him, quill pen in hand, wondering about what this extraordinary "lightest stuff" might be, trying to grasp the quintessential, when it was all around him. You want to reach across the centuries and give him a word of encouragement. Of course there may be a simpler explanation. Perhaps his study had no window. Or perhaps he wrote the third chapter of the second part of the *Theory of the Heavens* in the summer.

Elsewhere in the book, he almost arrives at the right answer by a wholly unexpected route. He imagines that at some time in the past the Earth itself was graced by a ring like the rings of Saturn—which, he correctly believed, are composed of individually orbiting worldlets, perhaps made of ice. He exults at how lovely the sky would have been:

A ring 'round the Earth! What a beautiful sight for those who were created to inhabit the Earth as a paradise! What a convenience for those on whom nature was designed to smile on all sides!

It strikes him that such a ring might explain the odd phrase in the book of Genesis, "the water upon the firmament"—waters in some sense intrinsic to the heavens. Kant notes that this idea "has already caused not a little trouble to commentators"—that is, those struggling to reconcile physics and the Bible. Indeed, Thomas Aquinas had devoted time to this very question in the *Summa Theologica*. Kant had a suggestion for the exegetes:

Might this ring not be used to help them out of this difficulty? The ring undoubtedly consisted of watery vapors and besides the advantage which it might furnish to the first inhabitants of the Earth, it had further this property of being able to be broken up on occasion, if need were, to punish the world which had made itself unworthy of such beauty, with a Deluge.

1903

We thus are presented with the image of a circumterrestrial ring system, made, like its Saturnian cousin, out of individually orbiting satellites, and composed of water. In the *Theory of the Heavens* he actually proposes bodies made of water—whether solid, liquid, or gas—in space, near the Earth. But he never considers that comets might be made of ice, although he comes tantalizingly close.

Kant argued that comets condense beyond the orbit of Saturn, in a cloud of bodies with high eccentricities and all inclinations; that when they find themselves in the inner solar system approaching the orbit of the Earth, the Sun's heat warms them and vaporizes their surfaces; and that the tails are made of this vapor and driven back by some electrical influence of the Sun. Even apart from Kant's near deduction that comets are made of ice, this is a hard description to beat for 1755.

Kant's book was printed when he was thirty-one years old, and dedicated to Frederick the Great. But copies were never distributed to the Prussian Emperor or to anyone else; the publisher went bankrupt just as the book was coming off the presses. The dedication, groveling in its submissiveness to authority, was typical of its day. It is reproduced in the box on page 80. In fact, Kant was not an enthusiastic supporter of Frederick, and later in the century would express strong sympathies for the American and French antiroyalist revolutions. Kant complained more than once that the state spent too much on war and too little on education.

He was also very cautious on religious matters. It bothered him that a natural explanation of the evolution of the solar system, based only on Newtonian physics, might be offensive to the prevailing faiths. He correctly predicted that adherents of the established religions would argue as follows:

> If the structure of the world with all its order and beauty is only an effect of matter left to its own universal laws of motion, and if the blind mechanics of the natural forces can evolve so glorious a product out of chaos, and can attain to such perfection of themselves, then the proof of the Divine Author which is drawn from the spectacle of the beauty of the universe wholly loses its force. Nature is thus sufficient for itself; the Divine government is unnecessary....

That is, the truth of how the universe works might be dangerous to know if it worked to undo sectarian teachings. The argument is with us still. In light of this Kant says, "I did not enter on the prosecution of this undertaking until I saw myself in security regarding the duties of religion." If he could not have reconciled his scientific ideas with conven-

The Philosopher and the King

Dedication of Immanuel Kant's
Natural History and Theory of the Heavens
to Frederick the Great

———————

To
THE MOST SERENE,
THE MOST POWERFUL KING AND LORD,
FREDERICK,
KING OF PRUSSIA, MARGRAVE OF BRANDENBURG, HIGH
CHANCELLOR AND ELECTOR OF THE HOLY ROMAN
EMPIRE, SOVEREIGN AND ARCH-DUKE OF
SILESIA, ETC.

MY MOST GRACIOUS KING AND LORD, MOST SERENE,
MOST MIGHTY KING, MOST GRACIOUS KING AND LORD!

The feeling of personal unworthiness and the splendour of the throne cannot make me so timid and fainthearted, but that the favour which the most gracious of Monarchs extends with equal magnanimity to all his subjects, inspires in me the hope that the boldness which I take upon me will not be regarded with ungracious eyes. With the most submissive respect I lay herewith at the feet of your Royal Majesty a very slight proof of that zeal with which the Academies of Your Highness are stimulated by the encouragement and protection of their enlightened Sovereign to emulate other nations in the sciences. How happy would I be, if the present Essay should succeed in obtaining the supreme approbation of our Monarch for the efforts with which the humblest and most respectful of his subjects has unceasingly striven to make himself in some measure serviceable to the good of his country. With the deepest devotion till death,
I am,
YOUR ROYAL MAJESTY'S
Most humble Servant,
THE AUTHOR.

KÖNIGSBERG
14*th March*, 1755.

tional religious doctrine, he says, he would have suppressed the former.

Despite such cautious attitudes, in 1788, for the first time in his life, Kant became embroiled in a controversy that was simultaneously political and religious: Frederick's successor, Frederick William II, began a campaign to uproot the pernicious teachings of the Enlightenment which had brought science and rationalism to European culture. Kant received a cabinet order, in 1794, deploring his "misuse" of philosophy. Who would know more about the uses of philosophy, the Prussian emperor or Immanuel Kant? His teachings were found to be insufficiently deferential to the prevailing theological wisdom, and he was warned explicitly: "If you continue to oppose this order, you may certainly expect unpleasant consequences to yourself." Kant soon complied and attempted to justify his submissiveness: "Recantation and denial of one's inner convictions is base, but silence in a case like the present is a subject's duty. And if all that one says must be true, it does not follow that it is one's duty to tell publicly everything which is true." On this issue, he was no giant.

Kant's biographer, Friedrich Paulsen, writing in 1899, makes a character assessment that is of some interest in light of the tragic history of Germany in the first half of the twentieth century:

> Perhaps we may say that there is an inner relationship between Kant's ethics and the Prussian nature. The conception of life as service, a disposition to order everything according to rule, a certain disbelief in human nature, and a kind of lack of the natural fullness of life, are traits common to both. It is a highly estimable type of human character which meets us here, but not a lovable one. It has something cold and severe about it that might well degenerate into external performance of duty, and hard doctrinaire morality.

Kant's celebrated philosophical endeavors began in an attempt to work out the general implications of the Newtonian vision of the world. Much of Kant's philosophy represented a running battle with the widely influential philosophy of Gottfried Wilhelm Leibniz, a stilted, formalistic, and anthropocentric world view with delusions of completeness. (Leibniz was one of the many with whom Newton engaged in an extended feud.) In his *Critique of Pure Reason*, Kant announced that he had worked a Copernican revolution in philosophy. Copernicus had shown that the apparent motion of the Sun, the Moon, and the stars was in fact due to the motion of the observer. Some issues—including immortality, freedom, and God—Leibniz had asserted were knowable; but Kant argued that by their very natures they could not be fully experienced by humans and were in fact largely unknowns; Leibniz had presented the illusion and not the reality of knowing. This was heady stuff, and the authorities of course considered it subversive.

Kant's role in the history of philosophy is generally considered magisterial, but we wonder—we recognize that this is a heretical view—whether he might have been more effective in the long run had he continued the remarkable scientific work of his youth, and left the metaphysics to others. Inscribed on Kant's tomb are his words:

The starry heavens above and the moral law within fill the mind with ever new and increasing admiration and awe....

* * *

Edmond Halley had predicted that the Comet of 1682 would return at the end of 1758. It is easy to see why his

Nicole-Reine Etable de la Brière Lepaute. Portrait by Guillaume Voiriot. Courtesy Michel-Henri Lepaute.

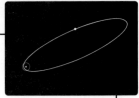

1904

prophecy caused little excitement at the time; 1758 was then more than half a century in the future. When Halley died in 1742, his obituaries made no mention that he had predicted the return of a comet. Instead, his voyages of exploration and the diving bell he invented were given great attention.

But by 1757 there were those who had become obsessed with the idea that the gravitational physics of Newton could actually be used to foretell the future. Among them was Alexis Clairaut, an eminent French mathematician who had published his first paper at the age of thirteen. He made a last-minute decision to try to improve on Halley's tables on the orbit of the Comet of 1682, and the predicted time of its return. It was imperative, of course, that the revised prediction appear before the comet did, "so that no one might doubt the agreement between the observation and the calculations." But the comet was fast approaching and the task was enormous, requiring meticulous calculation of the gravitational interactions of Jupiter, Saturn, the Earth, and the comet over a period of 150 years. Clairaut claimed that he engaged the astronomer Joseph Jerome de Lalande to help him. To hear Lalande tell it, it was the other way around. But Clairaut made no public mention of the third member of the team, without whom—as Lalande was later to admit—they would never have dared to attempt to beat the comet; it was she who deserved much of the credit.

We can only imagine the forbearance that Nicole-Reine Etable de la Brière Lepaute needed to get through her remarkable life. It was an age when upper-class women were valued for their appearance, their ability to oversee the running of the household, and their capacity for lively small talk. Madame Lepaute fulfilled these ideals, but she was also a first-rate mathematician. In this regard, she posed a problem for her colleagues that is made clear by Lalande's tribute to her in his *Astronomical Bibliography*. Writing soon after her death in 1788, he goes to great lengths both to exalt and to belittle her. Yes, she was vital to their work on the comet, but she was not pretty enough [see opposite]. Yes, her tables of parallactic angles and her accurate prediction for the whole of Europe of the annular eclipse of 1764 were important, but, no, she herself was significant mainly in terms of her male relatives. He *will* give her this much: "Her calculations never got in the way of her household affairs; the ledgers were next to the astronomical tables." For his part, Clairaut suppressed any reference to Mme. Lepaute's contribution, "in order to accommodate a woman jealous of Mme. Lepaute's merit, and who had pretentions but no knowledge whatsoever. She was able to have this injustice committed by a judicious but weak scientist whom she had

"Extra! Extra! Halley's Comet Returns on Schedule!" This is the substance but far from a literal translation of the front page of the *Hamburgisches Magazin* of late January 1759.

subjugated." Or such was Lalande's view of the matter. Scientific biographies were racier then.

But despite his cavils, Lalande's devotion to her is clear:

> Mme. Lepaute was the only woman in France who acquired true insights into astronomy....She was so dear to me that the day I walked in her funeral procession was the saddest I have spent since I learned of the death of my father....The times I spent near her and in the heart of her family are those I am most fond of, the memory of which, mixed with bitterness and pain, spreads some comfort over the last years of my life.... Her portrait, which I still have before my eyes, is my consolation.

Lepaute must have had her hands full in 1757. With Clairaut and Lalande, she worked day and night, often through meals, for six months in a desperate race with the comet. The enterprise was so taxing, Lalande later wrote, "that following this forced labor, I contracted an illness which would change my temperament for the rest of my life." Eventually they discovered that the comet would be detained by Saturn's gravity by 100 days. Jupiter meant a delay of at least 518 days. In the course of their calculations, they found that Halley had made a set of compensating errors which canceled each other out, and concluded that

Chart of the apparent path across the sky of the celebrated Comet of 1759. From *Histoire de l'Académie Royale des Sciences*, 1760 (Paris, Impr. Royale, 1766). Courtesy Ruth S. Freitag, Library of Congress.

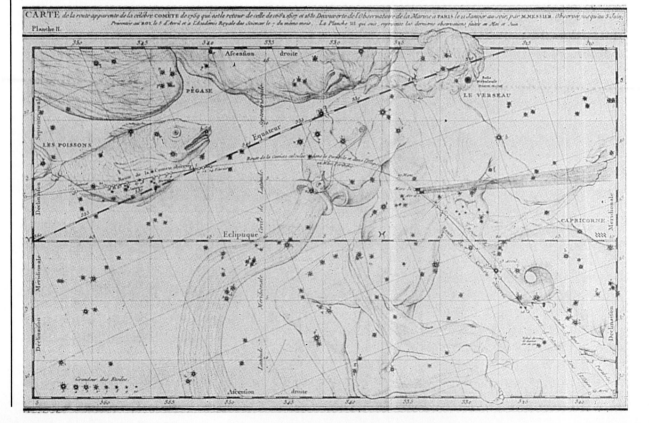

1904

Halley's estimate of the time of the return was essentially correct.

In November of 1758 they predicted that the comet would achieve perihelion passage in mid-April 1759, and might be visible some months before. On Christmas Night, 1758, a German farmer—one Johann Palitzsch—became the first to know that the long-dead Edmond Halley had successfully employed Newton's Laws to foretell the future. The comet was punctual, and it came from just the sector of the sky that Halley had foretold. Palitzsch, an avid amateur astronomer, one of many to make a contribution to cometary astronomy, rushed to tell the world. Halley's prodigal comet had returned. It reached perihelion on March 13 of 1759, within a month of the Clairaut-Lalande-Lepaute prediction. Science had succeeded where generations of mystics had failed. Newtonian prophecy had been fulfilled.

Many soon recognized what Halley and his French successors had accomplished. They had established a program, a goal, an ideal for the future of all of science: "the regularity which astronomy shows us in the movements of the comets," Laplace concluded, "doubtless exists also in all phenomena."

Halley's Comet in 1759 as painted by Samuel Scott. From the *Illustrated London News*, Vol. 235, October 31, 1959. Courtesy Ruth S. Freitag, Library of Congress.

A cometary nucleus fragments as it passes by the planet Jupiter, seen here nearly pole-on. Each of the large fragments will develop its own tail as this swarm of irregular pieces of ice approaches the inner solar system. Painting by Michael Carroll.

Chapter V

ROGUE COMETS

Thou too, O Comet, beautiful and fierce,
Who drew the heart of this frail Universe
Towards thine own; till, wrecked in that convulsion,
Alternating attraction and repulsion,
Thine went astray, and that was rent in twain;
Oh, float into our azure heaven again!

—Percy Shelley,
Epipsychidion,
1821

THE TRIUMPHANT RETURN OF HALLEY'S COMET IN 1758 powerfully supported, in the minds of people all over the world, the Newtonian view that we live in a clockwork universe. In the predictable motion of the planets, and in the periodic apparitions of Comet Halley (and, later, its brethren), many people saw the hand of God. Seeking new comets and making preliminary determinations of their orbits became a fashionable pastime. In the time of the American and French revolutions, optimistically proclaimed "The Age of Reason," the regular motions of the comets represented a continuing reminder of the gradual emergence of the human species from rank superstition, while at the same time the majesty and elegance of a Divine Purpose was considered evident in every cometary orbit.

As a larger sample of comets was examined, however, some odd idiosyncrasies and a disquieting departure from Newtonian regularity were uncovered. A class of short-period comets was found, circuiting the Sun once every few years, entirely among the planets in the inner solar system. For example, Comet Encke, discovered in 1786, comes once each orbit closer to the Sun than does the innermost planet, Mercury. In 1819, J. F. Encke was studying the repeated returns of the comet that now bears his name. Since the period is only 3.3 years, he had a number of orbits to contemplate. Encke found, to his considerable surprise, that in every perihelion passage the comet was arriving a couple of hours early—even after perturbations by Jupiter and the other planets were properly taken into account. Here was a substantial mystery, which Encke was destined never to solve. It put the new astronomy in an awkward position: comets had been touted as the proof of a precise and universal law of gravitation, but at least one comet chose not to play along. Even Newton could not make all the comets run on time. The phrase "defying the law of gravity" traces back to these times. The approach of most scientists of the day was that Newtonian gravitation was valid, although some additional force was here at play. But what?*

On February 27, 1826, a major in the Austrian army named Wilhelm von Biela found himself in South Africa looking up at the sky, whereupon he discovered what seemed to be a new comet. Ten days later, it was independently discovered in Marseilles by a French astronomer, Jean

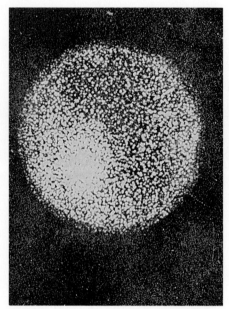

Drawing of the coma of Encke's Comet, observed on November 30, 1828. Courtesy R. A. Lyttleton, from his book *The Comets and Their Origin,* Cambridge University Press, 1953.

---

*Among the erroneous possibilities offered were friction by an enormous quantity of otherwise undetected interplanetary dust; the resistance of a postulated "luminiferous aether," through which light waves were once thought to propagate; and a small departure of gravity from the inverse square law.

Felix Adolphe Gambart. Both calculated the orbit and deduced a period of a little under seven years. Both recognized that comets observed in 1772, 1805 and 1826 were the same body. An unpleasant and petty controversy arose between them over the priority of the discovery. Should it be called Comet Biela, or Comet Gambart? Meanwhile, Gambart (and others) made the disquieting prediction that *his* comet would strike the Earth on its next return, around October 29, 1832. Although the comet appeared on schedule, it did not impact the Earth. We can imagine some anxious moments, and for more reasons than one, around the Marseilles observatory in the fall of 1832.

On its next predicted perihelion passage, in 1839, the comet's position in the sky at closest approach to Earth was very near the Sun, and in the glare no comet was seen. But the 1846 apparition was more favorable, and as astronomers peered through their telescopes in amazement, it was discovered that there was not one but two comets on almost identical trajectories, each with its own tail. During the next weeks the relative brightness of the two comets varied, first one and then the other shining more brilliantly. For a while there was even a common coma enveloping both. The finding was so bizarre that the first astronomer to note this twinning dismissed it as some internal reflection in his telescope. How a comet could reproduce itself was a mystery worth pondering, although, at least, it provided a Solomonic resolution to the battle over priority between Biela and Gambart. In the 1852 apparition two comets were again observed, now some two million kilometers apart, although still traveling in approximately the same orbit. They were never seen again. Comets that arrive early or late, comets that split or reproduce themselves, comets that disappear— all of this undermines the notion that comets are wholly and exclusively subservient to the Newtonian clockwork.

Although Comets Biela/Gambart never returned, they nevertheless provided a further astonishment for the perplexed astronomers of Earth. A few decades after the comet was lost forever, there began on Earth the November meteor shower called the Andromedids—thousands of brilliant "falling stars" illuminating an autumn night. When the orbits of the Andromedid meteors were traced back, it was discovered that they had precisely the trajectory of Comet Biela/Gambart. Somehow, both comets had disintegrated, essentially simultaneously, leaving in their place only a multitude of fine debris which entered the Earth's atmosphere when the orbits of the comet and our planet crossed. Since then, most of the prominent meteor showers have been connected with cometary orbits. Meteors—called, the world over, falling or shooting stars—streak across the sky, their

Drawing of the appearance, seen through the telescope, of the comet discovered by Biela and Gambart, during its apparition of 1846. The comet had split in two since its earlier apparition in 1832. From Camille Flammarion, *Astronomie Populaire,* Paris, 1880.

trails dissipating in a few moments. Comets do not streak; the brightest of them can be seen with the naked eye for months. Despite these differences, it was beginning to seem that meteors and comets were related. The idea took hold that comets were in fact self-gravitating swarms of fine particles: When they were all collected together, moving as a swarm, they were seen as comets; when they individually entered the Earth's atmosphere, they were seen as meteors. A model of the cometary nucleus as a kind of orbiting gravel bank gripped the minds of astronomers.

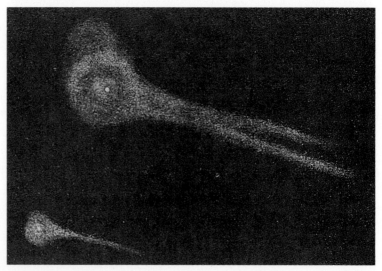

Another telescopic view of the splitting of Comet Biela. Drawing by Struve from Amédée Guillemin's *The Heavens,* Paris, 1868.

In 1744, de Cheseaux's Comet made a spectacular apparition. The tail was divided into six equal "rays," and for a while Europeans could see the tail of the comet above the horizon, while the head (and the Sun) lay below. It was commemorated in lithographs, scientific diagrams, and even on the coinage. A general expectation was being established that every few decades a bright comet approaches the Earth.

The first short-period comet on record is called Helfrenzrieder 1766 II. After making one foray past the Earth, it was lost forever. The second short-period comet to be discovered is Lexell's, which in the year 1770 came very close to the Earth. Lexell calculated the orbit and derived a period of only a little more than five years. The comet passed so close to the Earth that the transient gravitational grasp of our planet caused a decrease in its period by almost three days—although the period of the much more massive Earth did not change by so much as a second in a year. In 1776 it was not seen, but this was attributed to a greater distance from the Earth than during the previous apparition. Wait until 1781, the astronomers assured the public. But 1781 came and went without the comet. In this case, no meteor stream was subsequently found in the orbit of the comet. What had happened?

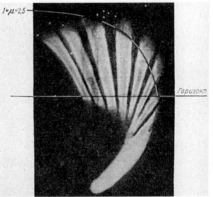

1906

A drawing of de Cheseaux's Comet of 1744, with the horizon indicated by the horizontal line. From S. V. Orlov, *On the Nature of Comets* (Soviet Academy of Sciences, Moscow, 1960).

The tail of de Cheseaux's Comet of 1744, seen above the horizon, while the head is below the horizon (cf. above). From Amédée Guillemin's *The Heavens*, Paris, 1868.

Obverse face of a 1744 German medal, probably struck at Breslau, depicting de Cheseaux's Comet with its six tails. The comet is properly shown as brighter than the brightest stars; indeed, in March 1744 it could be seen in full daylight. Courtesy American Numismatic Society.

Reverse of the German medal commemorating de Cheseaux's Comet. The words, from Romans 11:34, read "Who hath known the mind of the Lord?" Courtesy American Numismatic Society.

The question was answered by Lexell himself and by Laplace. In effect, they constructed a giant mathematical orrery,* and ran the motion of Lexell's Comet (and the planets) backward and forward in time from its solitary apparition in 1770. The calculations, especially in the era before computers, were not simple. They learned that the comet did not reappear in 1781 because two years earlier it had passed extremely close to Jupiter, possibly threading its way among the giant planet's four large moons. Earlier, Halley had found that, passing much further from Jupiter, the orbital properties of his comet would be changed a little by the tug of Jupiter's gravity. But here, Lexell's Comet had passed so close to Jupiter that its orbit must have changed dramatically. The comet had been thrown into some alternative trajectory, coming nowhere near the Earth—indeed, perhaps being ejected from the solar system altogether. In its close approaches both to the Earth and to Jupiter, Comet Lexell would have provided a magnificent view to a hypothetical observer astride the comet. Laplace could not help thinking that every now and then a comet might strike a planet, and this set him to wondering what the consequences would be on Earth if a comet plowed into it (cf. Chapter 15).

By Laplace's time, in the beginning of the nineteenth century, there were three categories of comets known: The short-period comets (like Lexell, Encke, and, later, Biela/ Gambart) had periods of a few years and lived entirely in the inner solar system. The long-period comets (like the Great Comet of 1680), with periods measured in centuries, had orbits that took them out beyond the farthest known planet.† But the bulk of the comets were "new"—with orbital periods so long that they could not be determined from the available data. Were these three separate kingdoms of comets, perhaps made of different building materials and having different origins? Or were they related, members of one kingdom evolving into another? All over the Western world there was revolution in the air, and the idea that absolute monarchies could be quickly transmogrified into something like real democracies made many people wonder what additional changes, hitherto deemed unthinkable, might be possible in other realms of nature. It was a mark of Laplace's

---

*A mechanical model of the solar system with the periods and/or distance of the moons and planets rendered to scale. Laplace and Lexell built no machinery, however; their orrery was purely computational.

---

†Halley's Comet, which has a period less than 200 years, is today classified as a short-period comet.

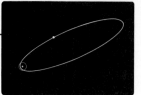

1907

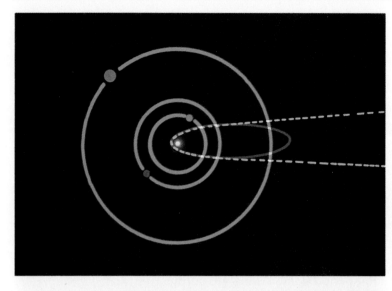

If we are able to track a comet only when it is close to the Sun, we may be unable to determine its orbit. Here the concentric circles show the orbits of the Earth, Mars, and Jupiter. In this schematic diagram, a comet on a parabolic or hyperbolic orbit arriving from interstellar space (dashed yellow line) is not readily distinguishable from a comet on an elliptical orbit (red closed curve).

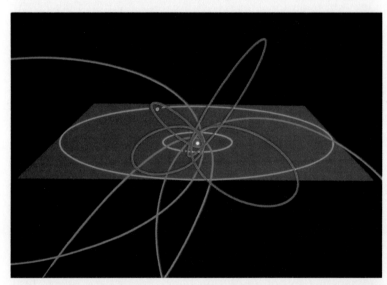

Inclinations of cometary orbits. The ecliptic or zodiacal plane, shown in dark blue, contains the nearly circular orbits of the planets (orbits of the Earth and Jupiter are shown). The short-period cometary orbits (displayed in red) tend to lie near the zodiacal plane. The long-period cometary orbits (blue, not circles) tend to be randomly distributed. Some short-period orbits are inclined at large angles to the ecliptic plane, and some long-period orbits happen by chance to lie in the ecliptic plane.

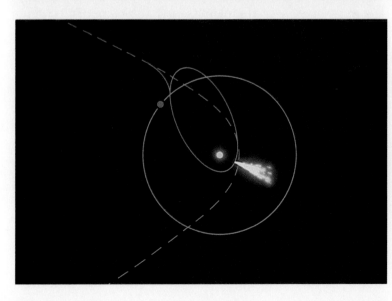

Conversion of a long-period comet into a short-period comet after passage by the planet Jupiter. Jupiter is the orange dot shown in circular orbit around the Sun. The comet approaches the Sun on the dashed trajectory from top left. If not for Jupiter, it would continue around the Sun and out into the distant reaches of the solar system, along the orbit extending bottom left. But Jupiter's attraction perturbs it into the elliptical orbit indicated by the solid line. The comet tail is fully developed as it achieves perihelion passage. Diagrams on this page by Jon Lomberg/BPS.

Marquis Pierre Simon de Laplace, the celebrated French mathematician, physicist, and astronomer, who played a major role in decrypting the nature and evolution of cometary orbits. Laplace's contributions to science were fundamental and varied; as a sideline, in 1780, with Lavoisier, he showed that respiration is a form of combustion.

genius how successfully he thought in evolutionary terms when most of his contemporaries still considered the main features of the universe as—necessarily, as a mark of God's handiwork—unchanging through the eons.

Laplace proposed that Jupiter's gravity was a kind of net, capturing new and long-period comets that wandered near and transforming them into short-period comets, henceforth to reside in the inner solar system. A modern rendition of the argument goes something like this: Sixty percent of the short-period comets have their aphelia, the far points from the Sun in their periodic orbits, near the orbit of Jupiter. An even higher percentage have one node of their orbits—the point where the plane of the cometary orbit intersects the plane in which the planets move around the Sun—close to Jupiter. In contrast, the orbits of the new and long-period comets show no connection with Jupiter. Also, some short-period comets—those restricted entirely to the realm of the terrestrial planets, for example—know nothing about Jupiter. But there are so many short-period comets with orbital characteristics tied to Jupiter that they are called the Jupiter family of comets.

Now what does Jupiter have to do with the family of comets associated with it? Some early astronomers thought this meant that Jupiter was the *source* of the comets, that they were somehow spat out, disgorged, from the inside of the largest planet. The most generous thing we can say about this view is that it has not stood the test of time. What other possibilities are there?

The comet is barreling in toward the Sun, approaching the orbit of Jupiter—as, it may be, it has done a dozen times before when Jupiter was on the other side of the Sun and nothing untoward transpired. But this time, by chance, as it crosses the orbit of Jupiter, Jupiter happens to be nearby. It is the most massive of the planets, and the comet is, by comparison, a little thing, a puff of gas surrounding a speck of matter. Jupiter's gravity attracts the comet—not enough to draw it into Jupiter (the comet, after all, is traveling at some twenty kilometers a second), but enough to deflect it toward Jupiter, and thus to change its orbit. The comet is still traveling in an ellipse about the Sun, but the gravitational encounter with Jupiter has dramatically changed its orbit.

A similar maneuver has been successfully negotiated by four spacecraft from Earth—*Pioneers 10* and *11*, and *Voyagers 1* and *2*—which in the 1970's were sent to Jupiter with so exquisitely crafted a trajectory that the gravitational acceleration of Jupiter swung each of them like a slingshot off toward a precisely preplanned point in the sky. *Voyager 2*, for example, swung by Jupiter in such a way that it was propelled toward a close encounter with Saturn two years later.

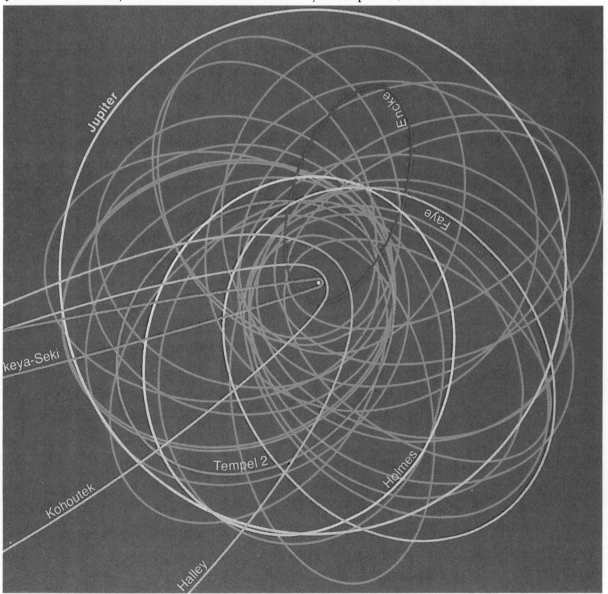

1908

(The close encounter of Saturn was also designed to swing it by Uranus in 1986, and the Uranus encounter to swing it by Neptune in 1989.) Now imagine the *Pioneer* or *Voyager* trajectories run backward in time. The spacecraft approaches Jupiter from the outer solar system; it races around Jupiter and is carried on its new orbit toward the Earth.

This kind of gravitational billiards explains the Jupiter family of comets. The inner solar system is sprayed with many comets, a few of which by accident come close to Jupiter. Some are immediately ejected by the encounter out of the solar system, some run into Jupiter or one of its moons, but many others have their orbits converted so that they become short-period comets of low inclination, with aphelion and node near Jupiter's orbit. Most short-period comets may have achieved their orbits by multiple

A few of the known cometary orbits, shown in relation to the orbit of Jupiter (large white circle). Among the comets shown are Encke's (in red) and Halley's. Note how frequently the aphelion of the orbit of a short-period comet lies near to the orbit of Jupiter, a clue to the origins of these comets. Courtesy National Aeronautics and Space Administration.

gravitational encounters with Jupiter, or even by multiple encounters with more distant planets and, eventually, with Jupiter itself.

A comet arriving from the depths of space in a direct or prograde orbit—one that goes around the Sun in the same direction as the planets (clockwise as seen from above the North Pole)—is more likely to suffer a large perturbation by an encounter with Jupiter. A comet coming in with a retrograde orbit—going around the Sun in the opposite direction—is more likely to experience a small perturbation. But only large perturbations suffice to carry the comet into the inner solar system. This is why the motion of short-period comets is direct. The comets supplied from the depths of space are as likely to be on retrograde orbits as direct ones, but only the comets in direct orbits tend to be captured. Jupiter's gravity makes distinctions.

It is estimated that there are about 2,000 long-period comets bigger than about a kilometer that cross Jupiter's orbit every year. The number of comparably-sized short-period comets in the Jupiter family is about 1,400. When we talk of the orbital evolution of the comets we see, we are talking about thousands of worlds.

Laplace showed that new and long-period comets could be converted into short-period comets by a gravitational machinery working before our very eyes. His work also seemed to demonstrate that interstellar comets, visitors from beyond the solar system, could be converted into short-period comets about the Sun.

A few comets have been seen in their passage through the inner solar system on very slightly hyperbolic trajectories—untied to the solar system, bound for the stars. It is natural to think of them as interstellar nomads, perhaps comets from some other star system, long wandering through interstellar space, and by lucky chance just passing through on our watch. Indeed, Laplace believed that both the short- and the long-period comets as well as many "new" comets were captured by the Sun from a population of free interstellar comets. If so, then some short-period comets might have evolved through an orbital cascade, a succession of changing trajectories determined by successive planetary encounters. The implication, which was hardly lost on Laplace, was that comets are ultimately denizens of the interstellar cold and dark, a fact fundamental to the modern understanding of comets.

Nevertheless, despite their unconstrained trajectories, these hyperbolic comets do not come from interstellar space, at least none observed so far. Using the same sort of mathematical orrery that Laplace pioneered, we can track the orbits of the apparently hyperbolic comets back in time.

Remarkably, every such hyperbolic comet is only slightly hyperbolic; if it were moving only a little more slowly it would be gravitationally bound to the Sun. When we track the orbits of such hyperbolic comets we find, in their recent past, an approach to one of the major planets close enough to perturb the cometary orbits. All of them seem to be comets that have been on long elliptical orbits about the Sun, and then slingshot out of the solar system by gravitational encounters with Jupiter or another of the giant planets. We see them on their exit trajectories. Not a single interstellar comet has ever been observed.

By the middle of the nineteenth century, the study of comets was an established part of professional astronomy. Their fundamental motions were well understood. Some new and long-period comets might eventually evolve into short-period comets, and in the tumult of gravitational encounters, comets might collide with the planets or the Sun or be ejected from the solar system. There was something chaotic, slightly unsettling about the motions of the comets. Some fell to pieces unexpectedly, and at least one delivered itself of small non-Newtonian feints and lurches, interrupting the stately circumsolar procession. But by and large comets were considered well understood. By now it was possible to illustrate public lectures with photographs of comets, and some of these stirred the multitudes. For a glimpse of what a professional astronomer, expert on comets and with a flair for lucid explication, was saying to the general public on the subject, the following is an excerpt from a talk delivered in 1882 by William Huggins at one of the public lectures (Friday-evening discourses), held then and now for the general public at The Royal Institution, London:

> With the aid of a telescope, in the heads of most comets a minute bright point may be found. The apparently insignificant speck is truly the heart and kernel of the whole thing—potentially it is the comet. It is this small part alone which conforms rigorously to the laws of gravitation…If we could see a great comet during its distant wanderings when it has put off the gala trappings of perihelion, it would be a very sober object, and consist of little more than nucleus alone…Under the Sun's influence, luminous jets issue from the matter of the nucleus on the side exposed to the Sun's heat. These are almost immediately arrested in their motion Sunwards, and form a luminous cap; the matter of this cap then appears to stream out into the tail, as if by a violent wind setting against it. Now, one hypothesis supposes these appearances to correspond to the real state of things in the comet, and that there exists a repulsive force of some kind acting between the Sun

The Great Comet of 1882, in what may be the earliest successful photograph ever obtained of a comet. Photograph by David Gill, in South Africa.

and the gaseous matter, after it has been emitted by the nucleus…Great electrical disturbances are set up by the Sun's action in connection with the vaporization of some of the matter of the nucleus, and…the tail is matter carried away, possibly in connection with electrical discharges, in consequence of the repulsive influence of the Sun…

A comet would, of course, suffer a large waste of material at each return to perihelion, as the nucleus would be unable to gather up again to itself the scattered matter of the tail: and this view is in accordance with the fact that no comet of short period has a tail of any considerable magnitude.

Virtually every one of these remarks of Huggins is in reasonable accord with the modern understanding of comets. Some were far ahead of the knowledge of his time. The subject was considered respectable and mature. Nevertheless, the central fact about the comets—their composition, and the nature of the spectacular variations in the appearance of comets—was at best only dimly glimpsed. In reading the literature of the time, we are struck by how rarely it was even acknowledged that these were important matters awaiting future discovery.

1909

Halley's Comet as a reckless driver streaking past the planets. Cartoon by Hermann Vogel, from *Fliegende Blätter*, May 1910.

Two manifestations of ice architecture: a comet over Antarctica. Painting by Kim Poor.

Chapter VI

ICE

The frightful ice that covers the whole face of the land.

—Hans Egede,
A Description of Greenland,
1745

ONE OF THE CENTRAL QUESTIONS ABOUT COMETS, and surely the key to many mysteries, is their composition. What is a comet made of? Are they all made of the same stuff? In the sixteenth and seventeenth centuries, it was still customary to think of comets as Aristotle had—as gases, vapors, "exhalations" from the Earth, and perhaps from the Sun and the planets as well. Newton, clear-sighted as usual, thought otherwise. He noticed that the Comet of 1680 came very close to the Sun; at perihelion, 0.006 of an Astronomical Unit, a little under a million kilometers. This should, he estimated, have heated the comet to the temperature of red-hot iron, from which he deduced that it could not be composed only of vapors and exhalations—because then its substance would have been rapidly dissipated during perihelion passage. Instead, he concluded that "the bodies of comets are solid, compact, fixed, and durable, like the bodies of the planets." Since the tail of this comet was "much more splendid" just after perihelion than before, Newton concluded that the heat of the Sun produces the tail: "The tail is nothing else but a very fine vapor, which the head or nucleus of the comet emits by its heat."

Fair enough. But what is the nucleus made of? Just what is this "very fine vapor" that constitutes the tail? This is the problem with which Kant and many others had tentatively grappled. When a comet comes as close to the Sun as the Great Comet of 1680 did, then virtually any common material should start to vaporize. But many comets begin to develop comas and tails when they are between the orbits of Mars and Jupiter. Heated only by sunlight, the temperatures of these comets, as they begin to pour vapor out into space, are something like a hundred degrees below zero on the Centigrade or Celsius scale ($-100°C$). Materials like iron that do not vaporize until they reach a high temperature are called involatile or refractory. Materials like ice that turn into gas after relatively modest heating are called volatile. The comets, therefore, must be composed of something quite volatile. But what?

The propensity of some comets to split suggests a nucleus not very strongly compacted. The forces that hold it together must be rather weak. As we have seen, every now and then comets depart from their scheduled arrival times in the inner solar system, or even exhibit a tiny darting motion, entirely at variance with the languid Newtonian sweep that the comets usually present to outsiders as they fall in towards the Sun. These erratic and unpredictable nongravitational motions of the comets recall Kepler's image of comets as fishes darting through the cosmic ocean. Encke described his comet as deviating "wildly"* from the motion predicted by Newtonian gravitation, and attributed

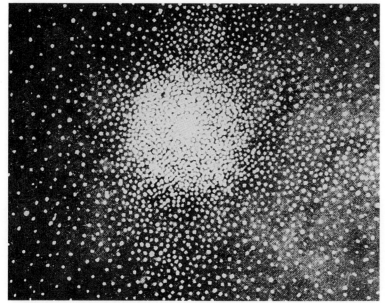

the anomalous and unpredictable movements to some resisting gas in interplanetary space, retarding its motion. But the accelerations are much too abrupt, and we now know there is not nearly enough material between the planets to have any detectable effect on the motion of comets. Some very different explanation is needed.

Comet Pons-Winnecke, as sketched by Baldet. Courtesy R. A. Lyttleton from his book *The Comets and Their Origin*, Cambridge University Press, 1953.

Until quite recently, the prevailing cometary fashions were dominated by the known association between comets and meteor streams—as when Comets Biela/Gambart disappeared, leaving the Andromedid meteor shower in their wake (Chapter 5). As late as 1945 the leading American college textbook in astronomy accepted without question the idea that comets are "loose swarms of separate particles moving on parallel orbits through interplanetary space." Some scientists believed this swarm of small meteors, imagined to make up the nucleus of the comet, to be gravitationally bound; others thought there was not enough mass in the nucleus to hold it together, and that instead an enormous number of small particles were traveling in, very closely, the same orbit through space. There was a certain tendency for advocates of this flying gravel-bank model to favor pointillist renditions of the comet head; some representative examples are shown in the drawings on this page.

The gravel- or sand-bank hypothesis deftly explained why an aging comet might one day be replaced by a cloud of fine particles. The spectra of meteors, as they burn up in the

*An exaggeration. The nongravitational forces change the period of Encke's Comet by no more than one day each orbit. The orbital period is 1200 days. So even to measure the effect requires an error less than 1/1200, or about 0.1 percent, not far short of perfect.

A pointillist rendition of the appearance of Halley's Comet in 1835. Courtesy R. A. Lyttleton from his book *The Comets and Their Origin*, Cambridge University Press, 1953.

Fred L. Whipple, leading proponent of the dirty ice model of the cometary nucleus. Photo courtesy Fred L. Whipple.

A summary describes briefly the observational success of the [icy conglomerate] comet model, both quantitatively and qualitatively. The surprising aspect of the model is its usefulness in spite of its vagueness…

—Fred L. Whipple,
"Present Status of the Icy
Conglomerate Model,"
Harvard/Smithsonian Center
for Astrophysics,
Preprint 1966 (1984)

Earth's atmosphere, show the presence of such materials as iron, magnesium, aluminum, and silicon, typical constituents of rocks on Earth. If meteors are made of rocky stuff, and comets in turn are made mainly of meteors, it followed that comets were rocks and stones. Then what about the coma and the tail? It was proposed that the sand particles were coated by some more volatile solid that evaporated as the swarm approached the Sun, or that gases were baked out of the stones as they were heated. But it was hard to see that much of this material—whatever it was—could be left after a single passage close to the Sun, if in the first place it was only a thin coating around a grain of sand, or the gas trapped near the surface of a rocky particle. And there were other difficulties—the jet fountains, for example, which have no ready explanation in a loosely bound swarm of gravel.

For Comet Encke, for example, the issue is in our time definitively resolved: when probed by radar from a large radio telescope on the surface of the Earth, it shows a single solid nucleus, not a swarm of particles. The size of the nucleus detected by radar—a kilometer or two in radius—is consistent with other estimates. The nuclei of several other comets have also been detected by radar, accordingly sounding the death knell for the orbiting gravel-bank hypothesis of the cometary nucleus.* But before 1950 even the idea of a compact cometary nucleus was disreputable, and the nature of the cometary volatiles only vaguely glimpsed.

Fred Whipple describes himself as an Iowa farm boy turned astronomer. He served as chairman of the Department of Astronomy at Harvard University and, for many years, as director of the Smithsonian Institution's Astrophysical Observatory in Cambridge, Massachusetts. He had for years been thinking about small objects in the solar system, including the physics of meteors entering the Earth's atmosphere, and the nature of comets (of which he has discovered half a dozen). By the late 1940's, Whipple was convinced that large quantities of matter poured out of comets near perihelion—much more than could be accounted for by ice coatings on sand, or the driving out of small amounts of vapor that might be trapped inside the individual grains. It was also clear that interplanetary space was too good a vacuum for comets to be able to refurbish their supply of volatiles when their orbits took them far from the Sun (as Huggins and many others had pointed out). The problem

*But there is, for another comet, evidence that the nucleus is accompanied by a swarm of debris; there may be some life left in the gravel-bank hypothesis, but as an addition to, not a replacement of, the cometary nucleus.

1910

was especially severe for comets like Encke, which become very warm as they pass close to the Sun, and which have made many perihelion passages.

As a convenient shorthand, Whipple called the refractory materials "dust" and the volatile materials "ice," and decided that the problem would be solved if there were a great deal more ice than the sand-bank model of the cometary nucleus admitted. He describes the idea as "obvious," although, remarkably, it had never been stated this baldly before. Such luminaries as Newton, Kant, and Laplace had all toyed with something similar, but it was Whipple who first stated the idea lucidly and coherently. He then showed that a number of other mysteries—including the splitting of comets, their dissipation into meteor showers, and the worrisome nongravitational forces acting on cometary motion —could all be explained if we revised our thinking and imagined the cometary nucleus as a ball of dirty ice, with mineral grains and perhaps other materials scattered throughout.

If comets are indeed made of dirty ice, then to understand comets we must understand something about ice. To begin, let's suppose that a comet is made of ordinary water ice. There are some 92 kinds of atoms occurring in nature, the most abundant of which are hydrogen, helium, oxygen, carbon, and nitrogen. These atoms combine with each other according to specific laws that are collectively called chemistry. Because there is more hydrogen than any other

Antecedents of Cometary Ice

Newton had obliquely implied that comets are made mostly of water (see Chapter 17, below), and Laplace mentioned casually that comets might be made of ice. They did not spell out their arguments, and these suggestions were mainly forgotten. But by the middle of the twentieth century the idea was in the air again. For example, in the *Annales d'Astrophysique*, in 1948, there is an article by the Belgian astronomer Pol Swings, who says (rough translation) "At large distances from the Sun, all the solids in a comet are at very low temperatures and all the 'gas' that they contain, except for hydrogen and helium, must be found in the solid state." A footnote, however, indicates that this insight earlier appeared in an article by the German astronomer K. Würm, writing in the *Mitteilungen der Hamburger Sternwarte Bergedorf*, in 1943. But in Würm's article, the comment is attributed to a remark made informally to Würm by the Czech/German chemist Paul Harteck. In 1942 and '43 Harteck was busy trying to build an atomic bomb for the Nazis. His comments on cometary ices were a distraction from other, more pressing duties.

Three molecules of water vapor, each containing an oxygen atom and two smaller hydrogen atoms. The molecules are not linked to their fellows, but are freely moving as part of a gas. They might be molecules of steam released from a teapot, or vaporized off a snowy field on a warm winter's day. Painting by Jon Lomberg.

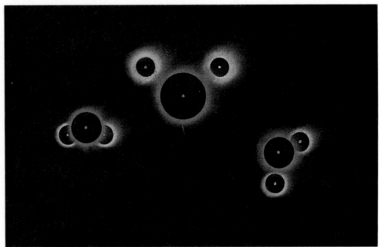

The molecular structure of ordinary ice. The circles represent the electron clouds of individual atoms, with the tiny atomic nucleus inside. Large orange circles represent oxygen atoms, smaller yellow circles represent hydrogen atoms. The molecular forces between these atoms assemble them into this hexagonal crystal lattice. We see two of the many successive, parallel planes of linked water molecules that comprise a microscopic fragment of ice. Painting by Jon Lomberg.

kind of atom in the universe, typical pieces of cold cosmic matter tend to be rich in hydrogen. Atoms such as oxygen, carbon, and nitrogen often are attached to as many hydrogen atoms as they can accommodate. Oxygen, for example, likes to combine with two hydrogen atoms, forming a molecule symbolized H_2O. The H, of course, stands for hydrogen, the O for oxygen, and the molecule in question, justly famous, is called water. A nitrogen atom likes to combine with three hydrogens, forming NH_3, also called ammonia; and carbon likes to combine with four hydrogen atoms,

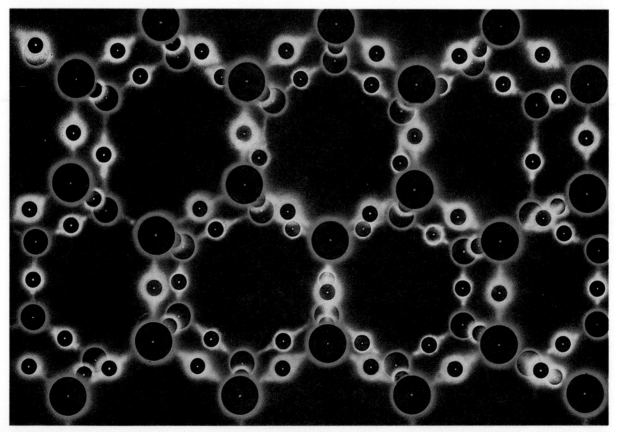

forming CH$_4$, a.k.a. methane. These abbreviations are a kind of shorthand picture of how the atoms are combined, what—if you could see it—the molecule actually looks like. There are many other simple combinations: CO, carbon monoxide; CO$_2$, carbon dioxide; HCN, hydrogen cyanide …and an enormous variety of more complicated* molecules, such as HCOOCH$_3$, CH$_3$CCCN and HC$_{10}$CH.

Atoms that are a little rarer in the universe also make chemical combinations. If the silicon atom is symbolized by Si, then ordinary quartz sand is made up of SiO$_2$, silicon dioxide. The molecular oxygen in the air is O$_2$ (as distinguished from an oxygen atom by itself, O). In this way the things of the natural world can be understood in terms of their constituent atoms. We humans are also aggregations of atoms, intricately and wonderfully assembled.

Now in a molecule like water, the atoms are not stuck onto each other at random positions or angles. An isolated water molecule always has the two small hydrogen atoms stuck onto the bigger oxygen atom at a precise angle, making the molecule look a little like a face with big ears, some-

*The first of these is called acetic acid. The names of the other two are almost unpronounceable, and much longer than their formulas. Diagrammatic representations of all known cometary and interstellar molecules are displayed in Chapter 8, pages 149–151.

Side view of the atoms in an ice crystal lattice. If you were to view *this* from the side, you would again see the hexagonal structure shown opposite. Painting by Jon Lomberg.

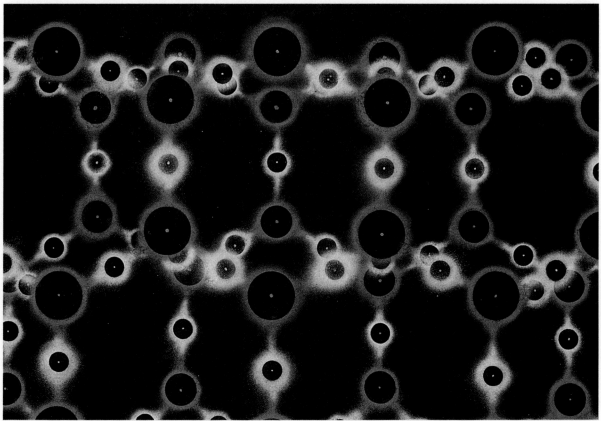

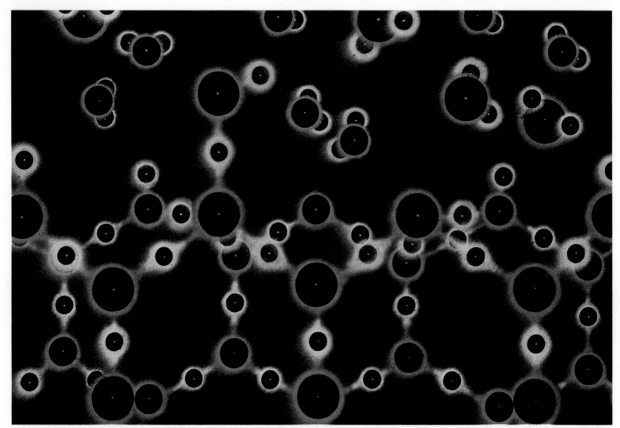

Evaporating ice. At bottom is a portion of the ice crystal structure, but at the warming top surface (middle of the diagram) the smallest fragments of ice—simple water molecules—are pouring off into the adjacent space. The process is called evaporation or sublimation. Painting by Jon Lomberg.

thing like Mickey Mouse (see page 106). The chemical forces that join atoms together do so according to precise and invariable rules that underlie the beauty and order of nature. These rules are set by the cloud of electrons that surrounds every atom, but for the moment we treat atoms as little impenetrable spheres. If you bring two of them near each other they will tend to combine according to certain definite laws, but if you try to push them too close together they will repel one another. A single isolated water molecule—exhaled from your mouth on a winter's day, for example—goes bobbing and tumbling off into the air, happily colliding with the other molecules in front of it, and bouncing off them. It does not readily combine with any of them; their constituent atoms are already preoccupied in passionate chemical embraces.

Temperature is nothing more than a measure of the motion of molecules. When the temperature is high, the molecules are in a riotous state, rushing about, tumbling, colliding, bouncing back—a frenzy of activity. As the temperature drops the molecules become more restrained, almost sedate. At sufficiently low temperatures, water molecules connect—they are moving so slowly that the short-range molecular forces can now engage, and adjacent water molecules join weakly with one another. When this happens—on a rainy day, for example—we say that the gas has

condensed into liquid water. At lower temperatures still (below the freezing point), the molecules condense not in a helter-skelter order, but in an elegant repetitive pattern called a crystal lattice. This is the hidden structure of ice (see pages 106 and 107). In such a lattice every water molecule is in a specific place, joined to its neighbors.

In two dimensions, the pattern is hexagonal, like the tiles on a bathroom floor. Every hexagon is made of six oxygen atoms and the accompanying little hydrogen atoms. If you had a superior sort of microscope, still to be invented, you might see such a structure. You could travel a million oxygen atoms to the left or the right, and find exactly the same structure. In three dimensions the crystal lattice is a kind of hexagonal cage. Beyond the oxygen hexagon in front of us is another oxygen hexagon connected by the little hydrogen atoms. Beyond that is another, and so on. You could rotate your head almost 120° and not see any discernible difference in the pattern. This hexagonal symmetry on the molecular level carries all the way up to the macroscopic level, and is responsible for the exquisite six-sided symmetry of snowflakes—a truth, oddly enough, first glimpsed by the astronomer Johannes Kepler.

If we were to look closely at such a lattice, we would notice that the constituent atoms were not stationary, but instead were vibrating and throbbing in place. If we decreased the temperature, the throbbing would moderate; if we increased the temperature, the throbbing would become more violent. There is a characteristic strength of the chemical bonds that hold the ice crystal lattice together. At a certain temperature the constituent atoms are throbbing so violently that some of the bonds become broken and a small piece of crystal lattice—an isolated water molecule—detaches itself from its fellows and goes tumbling off. Statistically, this occurs on occasion even at low temperatures. As the temperature increases, it occurs more often. At high enough temperatures the throbbing motion becomes so violent that the upper layers of ice become disengaged and large numbers of individual molecules go gushing off. If this happens on a cometary nucleus, the water molecules then find themselves in nearby interplanetary space. The process is called, variously, evaporation, vaporization, volatilization, or sublimation. Fundamentally, it is a change from water in the solid state, called ice, directly to water in the gaseous state, called vapor, without experiencing an intermediate liquid state.

If we had a piece of ice in a closed container and heated it up, at the characteristic temperature the water molecules would start spewing off in earnest. But they could not escape to space; they would bounce off each other as well as the

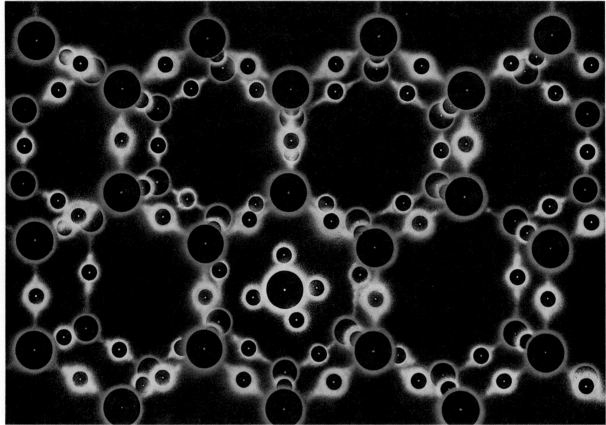

A molecule of methane sits trapped in one of the hexagonal cages of an ice crystal lattice. The carbon atom is shown with white outlines, attached to four smaller hydrogen atoms, comprising methane, CH_4. Ices with such trapped molecules inside them are called clathrates. Painting by Jon Lomberg.

walls of our chamber, and eventually they would all return to the surface of the ice, where they tend to stick. An equilibrium would be established between molecules of water coming off the ice and molecules of water bouncing back to it. Under these circumstances the rate at which the ice would be transformed into water vapor would be slow, and it is such circumstances with which we are familiar on Earth. An ice cube in a covered jar disappears more slowly than one in the open air. But on a comet there is no air, only a nearly perfect vacuum. So once the ice of the comet is sufficiently warmed, it begins to lose water molecules to space quickly, and forever.

Water ice isn't the only kind of ice there is. If you take ammonia gas or methane gas and sufficiently cool them, they will also form a cold, white crystalline solid. The temperature at which these gases form ices is called the freezing point, a different temperature for each material.* The freezing points of some common gases are shown in the adjacent table. The fact that water molecules crystallize at a compar-

---

* Depending on the local atmospheric pressure, gases can freeze solid without passing through an intermediate liquid stage; this is true for water vapor on Mars or an interstellar grain, but not on the surface of the Earth, where the atmospheric pressure is large.

Freezing Points of Ices

Water, H_2O	0°C.
Hydrogen Cyanide, HCN	−14°
Carbon Dioxide, CO_2	−57°
Ammonia, NH_3	−78°
Formaldehyde, HCHO	−92°
Methane, CH_4	−182°
Carbon Monoxide, CO	−199°
Nitrogen, N_2	−210°

Temperatures are given in degrees Centigrade, or Celsius.

atively high temperature is due to the strength of its chemical bonds; methane ice or ammonia ice fall to pieces at temperatures where water ice is quite stable and happy.

Now imagine some mixture of methane, ammonia, water, carbon dioxide, and other gas molecules, but with water in excess, as the general cosmic abundances suggest should be typical. You have to be about as close to a star as the Earth is to the Sun for water to be in the liquid state. Most of the universe is of course far from stars, and therefore well below the freezing point of water. On some tiny grain of dust out beyond Saturn or between the stars, the temperatures are very low, and a molecule of water hitting the grain will stick. As the grain passes over millions of years through the thin interstellar gas, it grows, and the crystal lattice bit by bit extends itself in all directions. As it does, other sorts of molecules may become trapped in the cage—a molecule of methane, perhaps, or ammonia, or something else (see opposite page). This kind of water ice structure, in which foreign molecules are trapped inside the cage, is called a clathrate.* The trapped molecule is not chemically combined with the ice, but merely imprisoned physically. You can see there is room for only about one big atom in the cage, and so methane clathrates tend to have around six water molecules for every methane molecule. In the vaporization of a water clathrate, the imprisoned molecules are released to space at the same time that the ice crystal lattice peels off.

If this grain is growing over millions of years it may be composed largely of water ice, because there is so much more water than anything else that will condense onto the

---

*Ice formed at very low temperatures tends to be amorphous, without the repetitive geometry of the crystal lattice, but there too clathrates form.

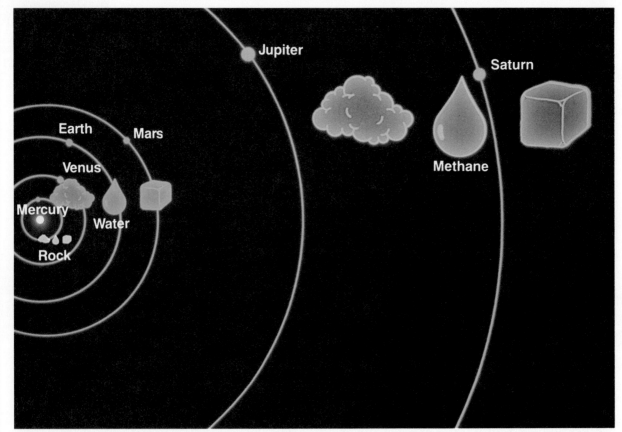

States of matter in the present solar system at various distances from the Sun. Concentric orbits are shown, counting outward from the Sun, of the planets Mercury, Venus, Earth, Mars, Jupiter, and Saturn. For three kinds of materials—silicates, water, and methane—the region in the solar system is shown in which they are gaseous (cloud), liquid (droplet), and solid (cube). Silicates are vaporized interior to the orbit of Mercury. Water is liquid in the vicinity of the Earth, and frozen at greater distances from the Sun than Mars. Methane can be liquid as far as the orbit of Saturn, but is solid at still greater distances from the Sun. Thus a mix of methane, water, and silicates making up a comet and entering the solar system from afar finds its methane first evaporating near the orbit of Saturn, water in earnest near the orbit of Mars, and silicates inside the orbit of Mercury. Diagram by Jon Lomberg/BPS.

grain. But other materials should be trapped as clathrates, and other ices should form on its surface—patches of CH_4 ice or NH_3 ice or CO_2 ice, as well as much more complex molecules. If the grains continue to grow and collide with one another, building structures of increasing size, eventually something like a small cometary nucleus will evolve. It will not be purely ice; there are many other kinds of material available, some of which we will mention later. But for now let us imagine the cometary nucleus as composed only of ices, and then picture it plummeting into the inner solar system. The temperature at its surface slowly warms. Because ice is not a good conductor of heat, the interior of the cometary nucleus remains for a long time at the temperature of the interstellar cold. But the outside becomes steadily warmer. Eventually it gets so hot that the chemical bonds holding the ice together begin to break, and the outer layers of ice go spewing off the comet into space.

Different ices experience this violent vaporization at different temperatures, and therefore at different distances from the Sun. As the comet crosses the orbit of Neptune, a patch of pure methane ice is feebly warmed by the approaching Sun; as chemical bonds are broken, a puff of methane gas is lost to space. A patch of ammonia ice would be lost as the comet crossed the orbit of Saturn. Carbon

dioxide ice would begin vaporizing in earnest somewhere between the orbits of Saturn and Jupiter. Ordinary water ice would not begin to vaporize significantly until the comet neared the asteroid belt, between the orbits of Jupiter and Mars.

But it is precisely in the vicinity of the asteroid belt that most cometary comas are observed to form. This in itself is evidence that water ice is a principal volatile constituent of the comets, and that water vapor and its degradation products are major components of the comas of comets. Much more direct evidence for cometary water now exists, as we shall see.

Occasionally, we do see comets outgas, forming at least temporary comas and even tails, when they are exterior to the asteroid belt. It is tempting to attribute such activity to the vaporization of other kinds of ices, so-called exotic ices such as CH_4 or CO_2. But the further from the Sun that these outbursts occur, the less likely it is that they will be noticed by the handful of astronomers on Earth. Imagine a comet with a surface composed of patches of different sorts of ices—water ice, methane ice, ammonia ice, carbon dioxide ice—and in an orbit that at first brings it only to the vicinity of Uranus, say, or Neptune. Each perihelion passage it will outgas methane ice (and, if any, nitrogen ice and carbon monoxide ice). But these puffs of gas, even if on a significant scale, will remain unnoticed and unrecorded on Earth. After many perihelion passages the methane—at least in the outer layers of the cometary nucleus—will have been entirely lost to space: the comet will have become devolatilized in methane. Each time it loses methane it becomes, relatively speaking, richer in water. If now such a comet is perturbed—by a close approach with the planet Neptune, say—into the inner solar system, it will lose ammonia, if it has any, when passing Saturn. But it will be most easily detectable when it comes interior to the orbit of Mars, heats up sufficiently that the strong lattice structure of ice becomes disrupted and an enormous cloud of the abundant volatile water comes pouring out. Thus, the prominence of water ice in comets is due to three factors: (1) the high cosmic abundance of water; (2) the possibility of loss of other volatiles in earlier incarnations of the comet's orbit; and (3) the fact that the water comes spewing off only in the inner solar system when the comet is close enough for Earth-bound observers to see.

Whipple and others showed that if a comet is made of ices, it would be able to supply copious quantities of molecules and small particles to form the coma and the tail. A cometary nucleus might lose a meter of material or more during each perihelion passage. If it started out with a radius

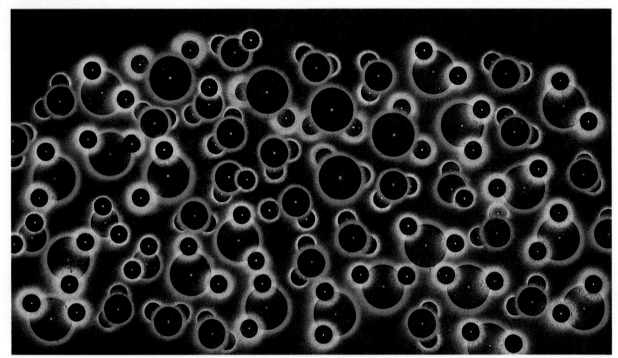

Molecular structure of liquid water. The water molecules are not arranged in a rigid crystal lattice, but are randomly oriented and free to move. Painting by Jon Lomberg.

of one kilometer, its substance would all be spent after a thousand perihelion passages. All that would be left would be materials like the involatile mineral grains—some of which, sooner or later, might be swept up by the Earth, producing a meteor shower. Think of two comets, both with perihelia near the Earth's orbit, but with different aphelia, and different periods of revolution around the Sun. Suppose one of them is a short-period comet and takes five years per orbit. Then from the moment it arrives in the inner solar system to the moment it is entirely vaporized and converted into meteors is five thousand years. The long-period comet, with a period of a hundred years, say, takes instead a hundred thousand years to lose its volatiles. These are typical lifetimes to be expected for comets, provided there is a fresh ice surface exposed directly to sunlight each perihelion passage. The layers of the cometary onion peel off near every perihelion until eventually there is nothing left. Therefore, the population of short-period comets must be resupplied from a more distant repository of comets, just as Laplace and others had calculated (see Chapter 5).

Whipple also realized that his dirty ice model could, in a most natural way, explain the strange nongravitational motion of some comets like Encke. Take an ordinary air-filled toy balloon, tie the nozzle closed, and let it fall. Its downward trajectory is slow and steady. But if you take the same balloon and hold it closed above your head, the nozzle squeezed between thumb and forefinger, when you let go it briefly darts across the room making sudden starts and turns and occasional rude noises. This is the rocket effect. When

Liquid Water

The kind of water ice we have described is the most common—those hexagonal crystal lattices characterize snowflakes and icebergs, glaciers and street slush on Earth. The structure is really quite remarkable. Ice has a lower density than liquid water because of the large void volumes or holes in the structure. This is why ice floats on liquid water. This is also why ice expands on freezing. In the liquid (see page 114), the water molecules are moving too vigorously to have settled down into the formal ice lattice. They are busy tumbling and colliding, and leaving little in the way of interior voids. But when the temperatures fall, the motion becomes less vigorous and the chemical bonds between adjacent water molecules assert themselves. The hexagonal lattice is formed and a material significantly less dense is formed. The density of liquid water is 1.0 grams per cubic centimeter, but the density of ordinary ice is 0.92 grams per cubic centimeter, a circumstance otherwise virtually unknown in nature. For other substances, the solid is almost always more dense than the liquid and promptly sinks. Floating icebergs and ice-covered rivers will be unheard of on planets where the dominant liquids are ammonia, say, or hydrocarbons such as methane.

To prevent ice from evaporating directly into vapor, to detain it in an intermediary liquid stage, requires an atmosphere. Collisions with the overlying gas molecules then inhibit evaporation. With no atmosphere, there are no collisions, no such inhibition, and therefore no liquids. Even Mars, with an atmospheric pressure thousands of times that at the surface of a comet, cannot today maintain open bodies of liquid water. There can be no liquid water on a cometary nucleus because, even when its coma is most prominent, a cometary nucleus is still, by Earth standards, surrounded by a high vacuum. You could imagine an interior pocket of the comet in which, if the temperatures were high enough, liquid water could readily form. But if the only source of heat is sunlight from without, the temperatures will get colder the deeper inside the comet you go. Modern speculation about life in comets, discussed later, is closely tied to the possibility of liquid water deep inside the cometary nucleus, perhaps even oceans of underground water.

the air rushes out of the nozzle, the balloon darts in the opposite direction. The reason is called Newton's Third Law of Motion: for every action there is an equal and opposite reaction. A rocket works on exactly the same principle. The exhaust blasts down on the launch pad and the rocket lifts up into the sky. The rocket exhaust does not push against the ground or against anything at all; rockets work equally well—indeed, better—in the vacuum of space. Another familiar example is the recoil of a rifle—the bullet goes forward, and the stock drives back into your shoulder.

Akin to the air in the balloon, the fuel in the rocket, and the bullet in the gun is the ice in the comet. Imagine an iceberg tumbling toward the Sun, its surface covered with patches of rocky material, interspersed with the more volatile ices. The surface of the comet is heated as it approaches the Sun. Some of the ice warms up and vaporizes. You can imagine a little gush of methane or ammonia gas into space —perhaps uncovering some deeper vein of the material, or perhaps only some rocky matrix material. But these jets of gas do not arise uniformly from all over the surface of the cometary nucleus at once. As a patch of methane ice vaporizes (action) at the orbital distance of Neptune, the orbit of the comet shudders slightly (reaction). Closer to the Sun, patches of ammonia or carbon dioxide can produce similar rocket effects. Detailed studies show that the tiny amount of nongravitational motion in the short-period comets can readily be explained by the rocket effect from subliming veins of water ice on the surface of the comet.

It is warmest in the afternoon on Earth rather than exactly at noon when the Sun is highest, because it takes a little time for the ground to warm up. The same is true on the comet, and it is therefore the afternoon side that heats up most and drives the bulk of the vaporized ice into space. Depending on which direction the comet rotates in, this will either accelerate or decelerate its orbital motion.

Sometimes a comet is structurally weakened—by collision, or by rapid rotation, so it is straining against the cohesive forces—and it splits into two or more pieces. The new frozen volatiles now exposed to sunlight produce still further jets of gas and dust, and some additional small tumbling and darting of the cometary fragments.

Something similar may even occur in comets that never approach the Sun closely. The most famous case is Comet Schwassmann-Wachmann 1. It lives between the orbits of Jupiter and Saturn and undergoes episodic outbursts, occasionally brightening by a thousand times in just a few days. When there are no such outbursts it appears to be a dark, reddish object, quite similar to asteroids rich in organic matter. One view of the outbursts of Schwassmann-Wachmann

A fountain of cometary matter jetting out from the nucleus of Comet Halley in 1910. Drawing made at the telescope of the Helwan Observatory, Egypt, May 25, 1910. Courtesy National Aeronautics and Space Administration.

1912

1 is that there are deeply buried repositories of exotic ices being gradually heated by the Sun. Sunlight eventually volatilizes a pocket of ice, and water vapor gushes off the surface. Tiny grains of ice and dust, carried along by the gushing water vapor, produce a temporary cloud around the object—a distant coma—that we see only as a temporary brightening. The circular orbit of this comet speaks to a long residence time in this part of the solar system. So why are there any exotic ices still left near the surface? Perhaps this is not a case of exotic ices, but of something else—collision of the comet with boulder-sized objects in its orbit, say, or collisions among the components of a multiple nucleus still undetected directly. The true explanation of the outbursts of Schwassmann-Wachmann 1 remains a mystery. It makes us wonder about other objects in the outer solar system—for example, Chiron, a small world that lives between Saturn and Uranus, or the icy moons of the giant planets. Might one of them someday abruptly wrap itself in clouds and brighten dramatically?

When the German mathematician F. W. Bessel witnessed the jets issuing from the nucleus of Halley's Comet in 1835, he speculated that a small nongravitational motion of the comet might result. The suggestion languished for 115 years, until Whipple revived it in a modern context. We can today see jets of matter spraying from the nucleus of a warm comet, irregularly turning on and off like the attitude control jets of an interplanetary spacecraft. Every time one of these great fountains goes shooting off into space, the cometary orbit makes a little turn. It is even possible, as with Comet Swift-Tuttle (page 118, top), to map the positions of the jets on the cometary nucleus.*

Thus, the dirty ice model in one swoop explains the timing and extent of the formation of cometary comas and tails, the nongravitational motion, and the explosive jetting from cometary nuclei that were noted at least since the time of Thomas Wright. As we shall see, there is now, as well, direct evidence for water ice as the principal constituent of the comets that we observe. Fred Whipple's "obvious" explanation, in the best tradition of science, produces a windfall of accurate prediction from a modest investment of hypothesis.

So it looks very much as though comets really *are* giant iceballs racing around the Sun. How much ice is that? Imag-

---

*Periodic comet Swift-Tuttle, 1862 III, is the source of the Perseid meteor stream. The jets mapped on the top of page 118 are propelling future Perseid meteors into nearby space where, in time, they will fill much of the orbit of the comet, as discussed in Chapter 13.

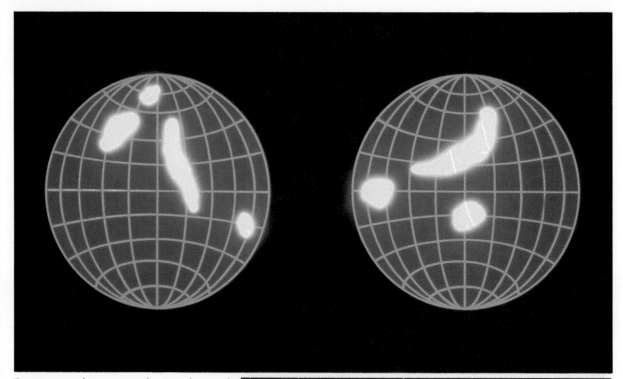

Positions of jets on the nucleus of Comet Swift-Tuttle. The comet is here imagined spherical, and the bright, white regions shown are the locales of jetting near perihelion in 1862 as determined by Zdenek Sekanina of the Jet Propulsion Laboratory. Diagram by Jon Lomberg/BPS.

Dirty ice comprising a cometary nucleus. Painting by Michael Carroll.

ine that you had snowplows stationed all over the Earth whose function it is to sweep up every flake of snow that falls in a year. And then imagine that all this snow is somehow packed into a rough sphere, carried out into space, and put into cold storage. You would then have something like a cometary nucleus ten kilometers across. You could make a hundred ordinary nuclei with this much snow. Put another way, a typical cometary nucleus contains about as much snow as falls each year in Eastern Europe, or the Northern United States. It does not seem so very much.

In a way, Whipple's snowball hypothesis is a disappointment. Newton and Halley converted comets from terrifying omens to a commonplace of nature, obedient to a divine mandate, unseen but manifest. But if comets are merely snowballs in orbit, is there any mystery left? Or are comets now demoted, becoming prosaic to the point of tedium? Remarkably, the new understanding of comets suggests that they are the key to the origin of the solar system, and to the surface characteristics of most of the worlds we know.

An Earthbound astronomer puzzles out the anatomy of a comet. Sheet music cover, 1910. Courtesy Ruth S. Freitag, Library of Congress.

Chapter VII

THE ANATOMY OF COMETS: A SUMMARY SO FAR

There was another sign seen in Heaven, and behold a great red dragon...
And his tail draweth a third part of the stars in Heaven.

—The Revelation of St. John the Divine
12:3-4

IN SEEKING TO UNDERSTAND WHAT COMETS ARE, where they come from, and what significance they conceivably might hold for us, we are the beneficiaries of thousands of years of patient observation and record-keeping by people all over the planet. An official scribe of Han Dynasty China or Seleucid Babylon, matter-of-factly noting the properties of a comet, helps us millennia later to test Newtonian gravitational theory, or the constancy of the arrival rate of new comets.

In all of human history, less than a thousand individual comets have been recorded, and only a few hundred have been seen in more than one passage by the Sun and Earth. Pliny wrote that naked-eye comets were visible for between a week and six months, which is still generally true. In most cases the comet is faint, barely visible as a bright smudge in the sky—a wisp of light, seemingly, a small fragment of the Milky Way broken off and on its own. The visibility of such comets greatly improves with a small telescope, or even a pair of binoculars.

Comets do not streak across the sky; they rise and set with the stars. The notion of streaking stems partly from a confusion with meteors, and partly from the immediate impression given by drawings and photographs: in our everyday experience, objects with such shapes are usually streaking. When we see a picture of a comet some of us are are immediately reminded of a woman with long, straight hair being blown back behind her, the reason, as we have said, for the very name comet, derived from the Greek word for hair. But the comet does not live down here, where there is air to blow things back. The comet lives in the nearly perfect vacuum of interplanetary space. And the tail does not always trail behind; instead, after perihelion passage, as the comet leaves the Sun, the tail precedes the nucleus. Something else is going on.

As a comet approaches the Earth, it typically increases both its brightness and the length of its tail. It may disappear into the Sun's glare as it negotiates perihelion passage, and then once more is visible—either brighter or dimmer, depending on the relative geometry of Sun, Earth, and comet. The tails of some naked-eye comets have stretched from horizon to zenith.

Somewhere between the orbits of Jupiter and Mars a bright comet may first appear to Earthbound naked-eye observers as a point of light—a star of fourth or fifth magnitude, surrounded by a perceptible haze. But such apparitions are rare. Comets vary greatly in brightness, of course; most are visible only through large telescopes, some can be seen with the naked eye, and occasionally—once every few years —there is one so prominent as to cause a general stir. About

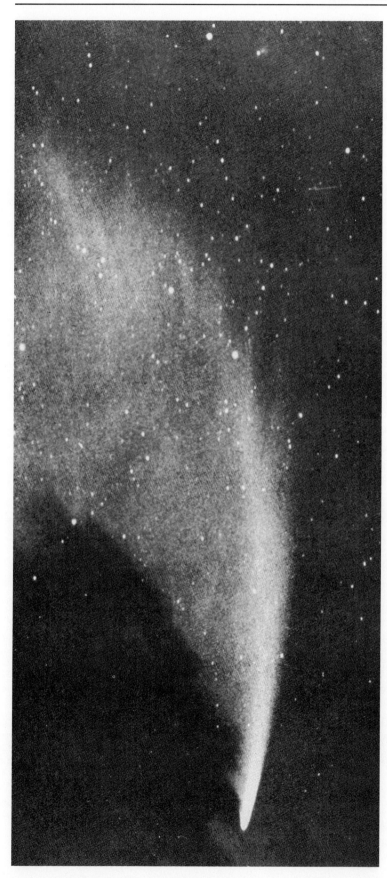

The Great Daylight Comet of 1910. From Henry Norris Russell, *The Origin of the Solar System* (Princeton University Press, 1935).

once a human lifetime, on the average, a comet appears that can be seen in the daytime sky, even very close to the Sun. The Great Comet of 1910 (1910 I)—which in people's memories is sometimes confused with Halley's Comet which arrived later that year—was such, and was called the Great Daylight Comet.

Most comets are found by astronomers, professional or amateur. Sometimes, during a total eclipse, a comet is discovered near the Sun, bathed in the light of the solar corona—a comet previously unknown, hitherto invisible, bleached out in the Sun's glare. After the eclipse, the comet becomes invisible again. But such instances are infrequent.* More often, comets are found in the vicinity of the Sun, as their motion carries them out of the glare, and comets are also discovered far from the Sun; a time exposure at night of a field of sky brings out some faint, nebulous object that is not on the standard charts. Amateurs—who can become ardent about the search for new comets—sometimes systematically scan the heavens in strips, using special telescopes able to see large patches of the sky in a single view. One recent comet was discovered by an amateur astronomer in his living room, gazing out the window with binoculars into the usually unpromising British sky. Some amateurs have found more than a dozen separate comets in a lifetime of dedicated searching.

With atmospheric pollution and city lights, the practice is now becoming less fashionable, but once there were many people who knew the map of the sky like the back of their hands, and who could tell at a glance—when out on a brisk postprandial constitutional—that there was a new point or smudge of light up there where no star had been seen before. Sometimes exploding stars, novas, are discovered in this way. And sometimes comets. Occasionally a bright comet is first detected with the naked eye by observers who are not astronomers at all. The classic case is the Great Daylight Comet of January 1910, which was discovered by three laborers on the South African railroads. There were then, as in Halley's time, few astronomers scanning the skies of the Southern Hemisphere.

Every night, on average, somewhere on Earth, there is at least one astronomer peering through the telescope at a comet. Almost always this is not part of the discovery process, but a segment of a much more intricate research program to understand the nature of comets. An astronomer might be photographing the comet, or passing its light into

---

*It is possible, however, to contrive such an eclipse artificially, at spacecraft altitudes, by using an opaque disk to block out the Sun.

1914

a spectrometer to uncover something about its composition and motions, or measuring the heat given off by the coma. Most often the discovery was not made by a human being at the eyepiece of a telescope, but from a photograph of the sky, taken through the telescope for quite another purpose.

Naming the Comets

Comets are often named after their discoverers, as Comet Ikeya-Seki, or Comet West. There is even a Purple Mountain Observatory Comet, discovered in a time when individual achievement was out of favor in the People's Republic of China.

Sometimes comets are named not after their discoverers, but after those who first recognized that the comets seen in two or more apparitions were really the return of the same comet. Halley's Comet is such a case, as is Encke's.

Comets are also designated by a Roman numeral, indicating the sequence of perihelion passages for the comets of a given year—as, 1858 VI, or 1988 I.

What do you look for when you think you've discovered a comet far from the Sun? You see a fuzzy patch of light. Might it be an artifact of the photographic emulsion? Take another picture. Could it be some other kind of faint patch of light—a nebula or a distant galaxy—not listed on your chart of the sky? Take another picture. If it moves with respect to the stars at reasonable speeds, then you have probably discovered a comet. If it isn't fuzzy, it might be an asteroid.

Discoveries of comets by professional astronomers are mainly made when the comet is far from the Sun, either on approach or—more rarely—after perihelion passage. Almost invariably (except when a known comet is being "recovered") such discoveries are made as an incidental by-product of some very different study. Discoveries by amateur astronomers, in contrast, are most often made when the comet is close to the Sun—within a few hours after sunset or before dawn. Because comets may brighten erratically, a comet that had been indetectable in a large astronomical telescope when it was far from the Sun may sometimes be discovered with a very modest instrument, or even with the naked eye, when it is close to the Sun—as was the case for the Great Daylight Comet of 1910.

A good fraction of comets with orbits reliably established as periodic are "recovered"—seen on some future approach to the Earth. But how can you be sure that what you're seeing is the same comet observed years ago? The comet generally has no distinctive identifying characteristics, no

The components of a comet. This page, top, a typical cometary nucleus, a few kilometers across, is shown for scale near the diffuse outer edge of a typical cometary coma. Opposite page, a coma in the inner solar system is shown next to an image of the Earth. The coma is composed of diffuse gas and fine particles, with the nucleus an insignificant and in this case invisible dot at the center. This page, bottom, a well-developed cometary tail extends from the orbit of Earth to the orbit of Mars. Diagrams by Jon Lomberg/BPS.

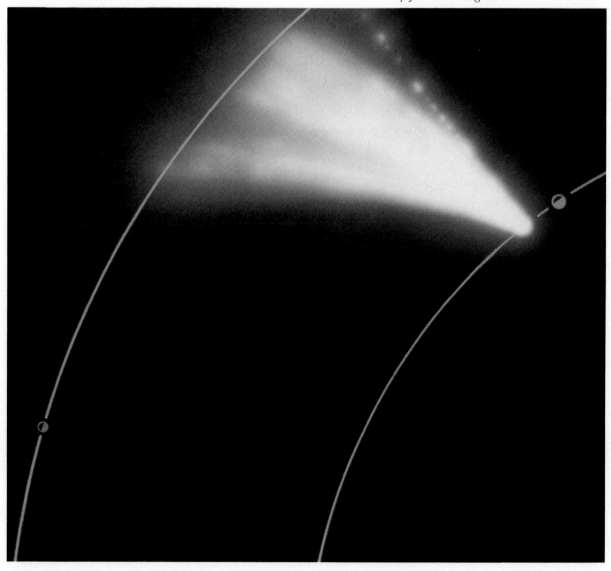

Nucleus of the Comet of 1618 drawn at the telescope on three different dates by Cysatus (top); and the Comet of 1652 (December 27) as drawn by Hevelius. What these observers, working with earliest astronomical telescopes, actually saw was the coma, not the nucleus of the comets in question; the details within the coma are probably spurious, artifacts of the imperfections of seventeenth-century lenses. From Amédée Guillemin, *The World of Comets*, Paris, 1877.

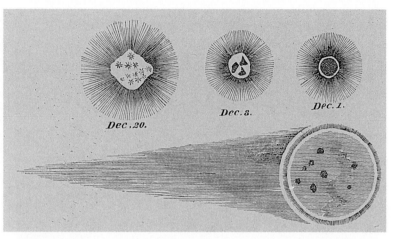

regimental tie, no tartan. Nevertheless, there are ways. Following Edmond Halley's pioneering work, the orbital characteristics of the earlier comet and the present one are compared: the period, eccentricity, distance of perihelion from the Sun, and inclination of the orbit, for example. For the 1986 return of Halley's Comet astronomers had calculated precisely where the comet should be at every point in its orbit. It was recovered with a large telescope on October 16, 1982, more than three years before perihelion, when it was beyond the orbit of Saturn. The comet was less than 1 percent the apparent size of the Moon from its predicted position.

Discoveries of comets are sometimes made that remain unconfirmed—usually because the orbital periods are undetermined or too long. Occasionally, short-period comets with well-determined orbits are not recovered in their next apparition. Many such reports are probably not due to an error made by the astronomer, but to a comet, ordinarily too faint to see, which is undergoing an explosive outburst of gas and dust. When the activity is over, the comet returns to obscurity. These unrecovered comets serve to remind us that there is a vast population of undiscovered comets.

From the cumulative weight of astronomical observation, an overall picture of the anatomy of comets has emerged, and even a little of their physiology. We have mentioned various components of this picture already; let us now take stock.

Comets are by far the largest and most variable objects visible in the solar system.* A tiny nucleus produces a substantial coma and an enormous tail, often much larger than

---

*Charged particles in Jupiter's magnetotail—formed by the interaction of the solar wind with Jupiter's magnetic field—reach to the orbit of Saturn; this planetary magnetotail is larger than any known comet tail. But it was discovered only when a spacecraft flew through it.

1916

the Sun. But for two comets at the same distance from the Sun, one may have an immense tail, and the other none at all. Following the spirit and precedent of Thomas Wright (Chapter 4),we show the relative scales of a typical cometary nucleus, coma, and tail on pages 126 and 127. When it can be seen at all, the nucleus of a comet appears through the telescope as a starlike point of light. Generally it is a few kilometers across, and yet this tiny ball of ice can generate a visible tail that is longer than the distance between adjacent planetary orbits in the inner solar system. A one-kilometer object with a tail a hundred million kilometers long is like a solitary mote of dust dancing in the sunlight in Washington, D.C., with a tail that reaches Baltimore.

In the earlier scientific literature there are reported observations of cometary nuclei (see page 128), generally reported to be hundreds or thousands of kilometers in diameter. Almost certainly, these measurements—made near the closest approach of the comet to the Earth—were in fact of the brightest part of the coma; the nucleus itself, much smaller, must have been lying hidden inside. A few observers have claimed that they could see a background star winking off and on as the nucleus passed in front of it.

Sometimes, by chance, a comet travels through the line of sight from the Earth to the Sun. If the nucleus were huge —1,000 kilometers across, say—it might appear as a small moving black dot silhouetted against the solar disk. But in

Through violent jetting a comet fragments into many small pieces. The vaporization of the ice destroys the comet. Painting by William K. Hartmann.

The size of a cometary nucleus (something like Comet Halley's, for example) compared to the depth of the ocean. Diagram by Jon Lomberg/BPS.

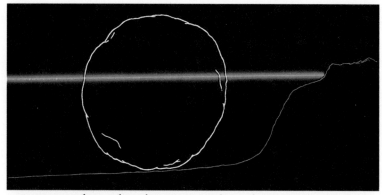

every case where the observations have been attempted—as, for example, when the Great Comet of 1882 or Halley's Comet in 1910 transited the Sun—the nucleus was too small for any trace of it to be seen.

Occasionally, a comet comes close to the Earth without a thick coma, and a small, bright point within the coma can be seen. This may be the cometary nucleus, but it may be only a dense inner coma of dust.

When the comets come close to the Earth they veil themselves from view by inquisitive astronomers. In general, they only undress in the dark—beyond the orbit of Jupiter. If you can catch the comet when it is far from the Sun, before it has generated a coma, then you might catch a glimpse of the naked nucleus. But because comets are so small, when they are at great distances from the Sun we can only see them as points of light no matter how bright they are, and no surface detail is discernible. There is an intrinsic limit to the resolution of ground-based telescopes, and to our ability to distinguish fine detail. So how can we know even the size of a naked nucleus? Astronomers have found another way.

The amount of sunlight reflected back from the nucleus to the Earth is measured: then, if only we knew how dark the nucleus is, we could calculate how big it has to be to return the amount of sunlight we measure. Bigger bodies reflect more light. The more reflective the surface, the smaller the nucleus must be to return the measured amount of light; conversely, the darker it is, the larger it must be. For dark, dirty cometary nuclei, for which there is other evidence, the calculated sizes are several kilometers across, or less.

The most direct way to determine the size of a cometary nucleus is to send a burst of radio waves out from the Earth to intercept the comet and be reflected back. Even a dense planetary atmosphere, like that on Venus, is transparent to such radar probing; these radio waves will pass through the coma with no difficulty at all. The few comets that have come close enough to Earth since the advent of advanced radar astronomy show nuclei between two hundred meters

and a few kilometers across, in good agreement with the best estimates made by other techniques.

A cometary nucleus even five kilometers across, almost the size now estimated for Halley's Comet, doesn't sound like much; but if it were gently placed at a typical locale on the Earth's ocean bottoms (see opposite), it would stick up smartly above the ocean surface, making—at least until it melts—an unusual variety of tropical island.

The number of comets that have been so measured represents less than 1 percent of the known comets. How likely is it that much smaller or much larger comets exist? A comet a few kilometers across is the size of a city. Are there then comets as small as a house, or ones as large as Luxembourg, say, or Brunei? Very small comets in the inner solar system might be impossible to detect; they also would have an extremely brief lifetime as their ices vaporized and the comet dissipated. As comets split and break up, there are surely fragments released that are the size of a house or smaller, but they do not long survive. Oddly, astronomers do not find more and more small comets by looking at fainter and fainter objects. Some think this is merely because smaller comets are harder to find, but others suspect there is a real lack of comets smaller than a football field (a hundred meters across). No one knows why.*

There is indirect evidence that occasional long-period comets are considerably larger, perhaps a hundred kilometers across or even more. The most celebrated example is the Great Comet of 1729. It was visible to the naked eye,

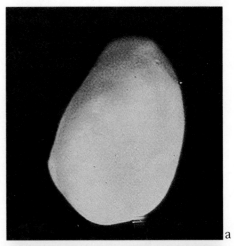

*Although there is an intriguing possibility that this paucity of small comets, if real, may be an echo of events in the building of the solar system, 4.6 billion years ago (Chapter 12).

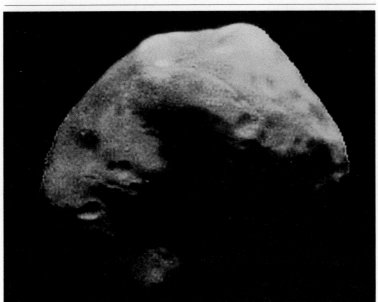

A family portrait of some of the irregular small moons in the solar system. Because their gravities are so weak they are not compressed into a spherical shape. (a) Deimos, the outermost moon of Mars. (b) Three views from different vantage points of the Saturnian moon Hyperion. (c) Some of the small moons of Saturn. (d) Phobos, the innermost moon of Mars. The shapes of cometary nuclei are expected to resemble what is depicted here. Photographs obtained by the Viking and Voyager spacecraft during their historic reconnaissance missions. Courtesy National Aeronautics and Space Administration.

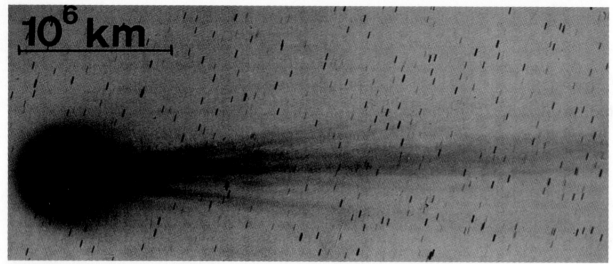

A photographic negative of Comet Ikeya (1963 I). Since the comet was moving slightly against the background of more distant stars, the stars appear as short, dark lines in this time exposure. The transparency of the tail is indicated by the stars that can be seen through it. The horizontal scale is one million kilometers long. Photograph taken by E. H. Geyer, Boyden Observatory, South Africa, February 24, 1963, courtesy K. Jockers.

although its perihelion was in the exterior of the asteroid belt, almost at the orbit of Jupiter. If it had instead come all the way to the vicinity of the Earth, it would have been bright enough to read a newspaper by at midnight.

For it to have been so bright so far away, it must have been both very large and outgassing great quantities of exotic ices. In the vicinity of Jupiter, it is too cold for significant vaporization of ordinary water ice. As we have seen, an iceberg of frozen nitrogen or carbon monoxide or methane begins vigorous evaporation near the orbit of Pluto, or beyond. Comets made of such materials would have spent much of their substance before ever coming close enough to Earth to be seen. Comets made of ammonia or carbon dioxide would vigorously evaporate between the orbits of Jupiter and Saturn. But unless such comets are particularly massive, their outgassing and jetting should remain indetectable from Earth.

Comet Kohoutek was exceptionally bright when it was still far from the Sun, from which widely publicized conclusions were drawn about how brilliant it would be as it passed the Earth in December 1973. But the apparition was far from spectacular; Comet Kohoutek was visible with the naked eye from the surface of the Earth, although it was much more clearly seen by the Skylab astronauts in Earth orbit. Apparently Kohoutek was so bright so far from the Sun because of the vaporization of exotic ices, a process entirely completed when the comet came close to the Earth.

From the periodic appearance of jets, it has been possible to calculate the rotation period of the cometary nucleus, even though it is sequestered inside the coma. The spin rates of dozens of comets have now been measured by this or other methods. A typical comet turns out to rotate once every 15 hours, not too different from the length of the day

A succession of concentric shrouds thrown off from the daylight hemisphere of Comet Donati, 1858 VI, drawn October 4, 1858, by G. P. Bond at the telescope at the Harvard College Observatory.

Comet West, 1976 VI. Left, the coma and tail as photographed in ordinary visible light by a ground-based telescope. Right, in ultraviolet light emitted by atomic hydrogen, photographed during the brief moments above the atmosphere of a high altitude rocket. Both pictures are to the same scale. Revealed in this comparison is the immense cloud of hydrogen gas, ordinarily invisible, that accompanies comets in the inner solar system. Courtesy Paul D. Feldman.

on Earth. The direction of the spin axis in the sky seems to be entirely random: cometary spin axes do not, for example, mainly point toward the North Star. For some comets, observations ranging back in time over a century or more can be used to determine rotation rates. Comet Encke, for example, has not changed its rotation period much in 140 years—although its period of revolution is erratic due to the rocket effect in ice (Chapter 6).

Because the typical comet is so small, its gravitational pull is tiny. If you were standing on the surface of a cometary nucleus, you would weigh about as much as a lima bean does on Earth. You could readily leap tens of kilometers into the sky, and easily throw a snowball to escape velocity, as we imagined in Chapter 1. The Earth and the other planets tend to be almost perfect spheres because, as Newton showed, gravity is a central force—it pulls everything equally toward the center of the world, itself held together by the force of gravity. The mountains sticking up above the spherical surface of the Earth represent less of a deviation from a perfect sphere than does the layer of paint or enamel on the surface of a typical globe that represents the Earth. If you were able to pile a sizable mountain on top of Mount Everest, it would not just sit there, poking in solitary magnificence into the stratosphere. The additional weight you had added would crush the base of Everest, and the new composite mountain would collapse until it was no larger than Everest is today. The Earth's gravity severely limits how much deviation from a perfect sphere our planet is permitted.

On a comet, on the other hand, the gravity is so small that odd, lumpy, potato-like figures would not be squeezed into a sphere. Such shapes are already known for the small moons of Mars, Jupiter, and Saturn. (See the illustrations on page 131.) On a typical comet you could build a tower reaching a million kilometers into space, and it would not be crushed by the comet's gravity—although it would certainly be flung off into space by the comet's rotation. Both the small moons and the cometary nuclei stay lumpy because their gravities are low. But they *get* lumpy for different reasons—the moons because of the history of collision with other bodies, and the comets because they accreted irregularly, or because they have vaporized surface ices irregularly. ices irregularly.

What is the pressure at the center of a cometary nucleus? Pick one that has a radius of a kilometer. Everyone has the sense that the pressure from the weight of the overlying rocks one kilometer beneath the Earth's surface is considerable. We know about coal mine shorings collapsing and miners being crushed. But if the gravity were thirty thousand

times less, then the overlying rocks would weigh thirty thousand times less. Put another way, the pressure at the center of such a comet is the same as it is 1/30,000 the comparable depth on the Earth. But 1/30,000 of a kilometer is 3 centimeters, about an inch. So the pressure at the center of a comet is roughly the same as that under a down comforter, or even a thin blanket. Even fragile structures can survive at the nucleus of a comet.

Suppose some comets had interior structure—a rocky inner core, say, or even an underground lake. How would we know? When a comet passes very close to the Sun and is split apart, its interior is suddenly exposed to space. Is there then a different kind of ice observed evaporating? Do we see previously undiscovered molecules in the spectrum of comets after fragmentation? A different ratio of gas to small dust particles? The answer to all these questions seems to be no. The interiors of these comets, at least, seem to be made of the same stuff as their exteriors—although involatile cores are hard to exclude. It is true that only a few comets have been examined as they split, and it is possible to imagine a quite different population of comets with interiors very different from their exteriors that have not in the past century come close to the Sun and therefore never been subjected to this test. There might be a reasonably abundant

Explosive jetting on the surface of a cometary nucleus. The evaporation of ice during successive perihelion passages leaves a lag deposit of rocky material behind. Painting by William K. Hartmann.

1919

population of larger comets that are far from uniform throughout. We simply haven't seen any yet.

From space, the atmosphere of the Earth is only a thin blue band hugging the horizon, compressed by gravity. On a comet, by contrast, the pathetically low gravity permits the atmosphere to stream over distances much larger than the size of the nucleus itself, producing comas tens of thousands of kilometers across. Indeed, much of a cometary atmosphere is not bound to the nucleus at all; the velocities in the spectacular jet fountains approach a kilometer a second —far in excess of escape velocity. At the surface of a cometary nucleus, the atmosphere is as thin as in the black sky 75 kilometers above the Earth's surface.

The gases around a cometary nucleus do not form a permanent atmosphere as on Jupiter, or the Earth. Instead, these gases are in transition between their production from subliming ices and their escape to interplanetary space. The gravity is simply too low to hold even the heaviest gases to the comet. Circumstances on a comet are unlike the situation on Earth, where the vaporization of the winter snows or the outgassing from a volcano or a fumarole generally makes negligible changes in the overall composition or pressure of the massive terrestrial atmosphere. On the comet, by contrast, the responsiveness of the thin cometary atmosphere to changes in the gases arriving from the nucleus should produce dramatic changes with time in the coma and the tails. This is exactly what is observed.

The coma is often an asymmetrical hood that develops from the sunward hemisphere of the cometary nucleus. It sometimes has a sharp boundary, and successive shells are regularly ejected by successive outgassing events. William Huggins described the coma as "a luminous fog surrounding the nucleus." In the inner solar system, there is also an extended halo of atomic hydrogen, and OH, from the breakdown of water vapor by sunlight, that envelops each comet and that glows strongly in the ultraviolet; generally, the halo is larger than the Sun. Until the advent of orbiting telescopes in the early 1970's, no one had seen the hydrogen corona. Even in visible light, the head of a comet—the nucleus and coma, apart from the tail—can be larger than the Sun; the Great Comet of 1811 is an example. As the comet approaches the Earth's orbit the size of the coma shrinks, although the comet becomes more active.

The new and long-period comets tend to be brighter and bigger than their short-period counterparts, because they are fresh from the outer solar system, and loaded with volatile ices that have not yet experienced the Sun's heat. After they vaporize much of the ice on successive perihelion passages, they become smaller and less active.

1920

The gas streaming off the nucleus is a kind of wind blowing into the interplanetary vacuum, and entraining cometary dust. Some comets are very dusty, others comparatively clean. Even in an aging cometary nucleus, there are still localized jets of gas and sources of fine dust. While unpredictable, there is a tendency for both jetting and nongravitational forces to increase the closer the comet is to the Sun, consistent with the vaporization of ices and the rocket effect.

Dusty comets have been observed to pour tons of fine particles into interplanetary space every second, and for most comets several times more water is lost than solids. Besides tiny individual particles of dust swept off into space in the jets, there may also be fragile clumps. A cloud of particles at least centimeters across surrounding a cometary nucleus has been found by radar. The relative amount of dust and ice near the surface of a cometary nucleus probably varies from comet to comet.

Comet Schwassmann-Wachmann 1, orbiting the Sun always beyond Jupiter, has produced over a hundred known outbursts, an average of about two a year. In successive perihelion passages, some comets seem to lose a meter or so of ices each perihelion passage for a number of orbits, and then close down shop. It may be that the refractory, rocky stuff that has not gushed away to space with the ices is now dominant, preventing sunlight from penetrating to the cold interior, and also preventing buried ices, even if warmed, from making their way to the surface. After successive evaporations, the surface of some comets may resemble the lag deposit on a glacier, the involatile rocky material covering a deep layer of still frozen ice. But for most comets, layer after layer of ice is lost in successive perihelion passages. Since the material in the coma and tail is never recaptured by the comet, it gradually dwindles, successive layers are peeled off and lost to space, and its interior parts are exposed to view. One way or another, every comet we see is dying.

> We think ourselves unhappy when a comet appears, but the misfortune is the comet's.
> —Bernard de Fontenelle,
> *The Plurality of Worlds*,
> Paris, 1686

The Earth and Moon pass through a comet tail, with the Milky Way in the background. Painting by Rick Sternbach.

Chapter VIII

POISON GAS AND ORGANIC MATTER

Hast thou ne'er seen the comet's flaming flight?
The illustrious stranger passing, terror sheds
On gazing nations from his fiery train
Of length enormous; takes his ample round
Through depths of ether; coasts unnumber'd worlds
Of more than solar glory; doubles wide
Heaven's mighty cape; and then revisits earth,
From the long travel of a thousand years...

—Edward Young,
Night Thoughts,
1741

WILLIAM HUGGINS SCARED THE WORLD. IT WAS entirely unintentional. No one could have foreseen it. He was minding his own business, which happened to be astronomy. But because of Huggins, in 1910 there were sustained national panics in Japan and Russia lasting for weeks; a hundred thousand people in their nightclothes filled the rooftops of Constantinople; Chicago apartment dwellers anxiously stuffed rags under their doors; and Pope Pius X condemned the hoarding of cylinders of oxygen in Rome. A dispatch from Lexington, Kentucky—typical of reports from all over the world—announced that "excited people are tonight holding all-night services, praying and singing to prepare themselves...and meet their doom." Through fear of the imminent catastrophe, a smattering of people in many lands took their own lives. But all this is getting ahead of the story.

Huggins was one of the first astronomical spectroscopists —scientists who decompose light into its constituent colors, or frequencies, and are able to deduce the motion and composition of a distant object. Spectroscopy traces back to another towering accomplishment of Isaac Newton. By

The light given off by a comet enters a telescope and passes through a prism (rear), where its constituent frequencies are untangled, and a rainbow spectrum of visible light is produced. The five colors shown schematically here correspond to five prominent emission bands of C_2—the molecule composed of two carbon atoms, symbolized at bottom. In addition, cometary spectra show emission bands of other molecules, and the characteristic absorption features of the sunlight they reflect. Diagram by Jon Lomberg/ BPS.

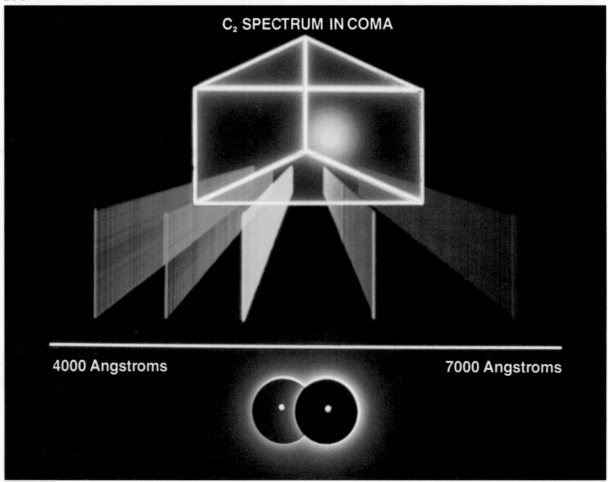

C₂ SPECTRUM IN COMA

4000 Angstroms   7000 Angstroms

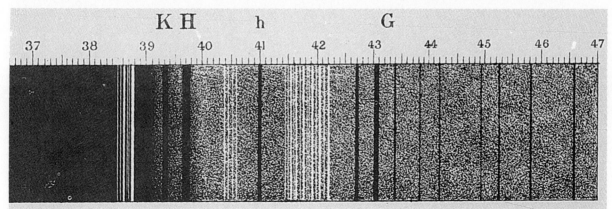

1921

passing a ray of sunlight, admitted into a darkened room, through a prism of glass Newton showed that ordinary white light is really a mixture of light of many colors. Different colors of light were bent by different angles when passing through the prism, so they could be spread out, or dispersed, against a surface on the other side of the prism. At first the surface was something like a white piece of cardboard. Much later, it became a photographic emulsion. The machine built around the prism was called a spectrometer, and the rainbow pattern generated was called a spectrum (plural, spectra).

When a spectrometer of higher dispersion was used, it turned out that sunlight was composed, in addition to its rainbow continuum, of a set of irregularly spaced dark lines, representing missing frequencies. It was soon determined that these lines were produced when light from the hotter, deeper layers of the Sun is absorbed by a cooler higher atmosphere of gas surrounding the Sun. Each chemical element absorbs a different set of frequencies and produces a different set of dark lines. In the laboratory, when a spectrum of a mix of chemical elements (and simple molecules) was obtained, the dark lines of all the constituent atoms would be seen. Soon it became possible to identify some of the spectral lines in sunlight, by comparison with laboratory experiments on the absorption spectra of common materials on Earth.

Each chemical element leaves its own unique signature in the spectrum. You could accumulate a catalogue of such signatures and then recognize in the spectrum of the Sun evidence of its constituent atoms. Scientists found themselves, to their own astonishment, measuring the constitution of the Sun and stars. Spectroscopy revolutionized the science of the time, and astronomy most of all.

Huggins was the beneficiary of a spectroscopic technology just reaching maturity. He put his spectrometer at the focus of large telescopes and looked at everything he could.

A cometary spectrum by William Huggins showing dark solar absorption lines (designated K, H, h, G, etc.) superimposed on three clusters of bright lines representing emission from the comet. From *Proceedings of The Royal Institution*, Volume 10.

Nineteenth-century spectra of emission bands in (2) olive oil, (3) ethylene, and (4), (5) in a pair of comets. Comet Winnecke II seems to have more in common with olive oil than it does with Comet Brorsen I. All the comparison actually shows is that olive oil, ethylene, and cometary nuclei, when heated or irradiated, give off such molecules as C_2. Diagram from Camille Flammarion, *Astronomie Populaire*, Paris, 1880.

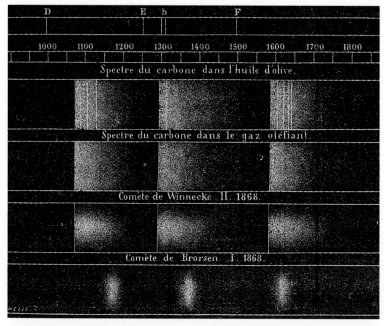

He was the first to show that the stars were made of the same chemical elements as the Earth and the Sun. He proved Edmond Halley's conjecture that certain interstellar nebulae were vast clouds of glowing gas. When Comet Winnecke passed near the Earth in 1868, it was a matter of course for Huggins to examine the spectrum of its coma. He confirmed a finding made for another comet by Donati in 1864 that there were three *bright* bands in the blue part of the cometary spectrum. Huggins also found that, in the spectra of the comets there were two components—a continuum of color interspersed with dark absorption lines that he correctly guessed was merely reflected sunlight, and the three bright bands that had been seen by Donati. If a dark line meant that some atom or molecule was absorbing light, then a bright line, or a set of them, meant that something was emitting light. But what?

The only way to answer this question was to look at the emission spectra of a wide variety of materials in the laboratory. In 1868 Huggins made a surprising discovery. He sparked ethylene (C_2H_4)—something like the natural gas used in household ovens—examined it in the spectrometer, and found exactly the same three bright bands as had been detected in emission from the comets. He concluded:

> There could be no longer any doubt of the oneness of chemical nature of the cometary stuff with the gas we were using, in fact, that carbon, in some form or in some state of combination, existed in the cometary matter.

The three bright bands are now known to be produced by

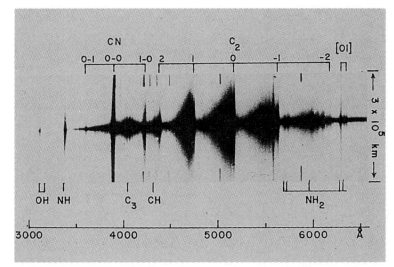

A modern cometary spectrum. Comet Kobayashi-Berger-Milon shows prominent spectral features due to C_2 and CN; the spectral signatures of other molecular fragments are also displayed. At bottom is a scale of wavelengths that runs through the visible spectrum. Observations from the Wise Observatory, Mitzpe Ramon, Israel. Courtesy S. Wykoff and P.A. Wehinger.

the molecular fragment C_2, two carbon atoms bonded together. It is produced when you pass a spark through C_2H_4 and break the molecule apart.

When the Great Comet of 1881 appeared, Huggins again obtained a spectrum, in which we can today recognize not only the presence of C_2, but also C_3, CH, and CN—a rich harvest from the first cometary spectrum to be photographed. (Huggins himself essentially identified C_2, CN, and argued for hydrocarbons of which the molecular fragment CH is the simplest exemplar.) We now know that it is hard to find a comet *without* the spectral features of C_2, C_3, and CN in their comas. Huggins was struck by the fact that the material in the comets was similar to organic matter of unquestioned biological origin on Earth. Many scientists cautiously concluded that the carbon compounds found by Huggins in the comas of comets were, as one of his contemporaries wrote, "the result of the decomposition of organic bodies." Only rarely voiced explicitly was the key question: Were such "organic bodies" biological?

These discoveries of Huggins and his successors interested virtually no one—until 1910, when it seemed the Earth was about to brush against the tail of Halley's Comet. The molecular fragment CN, a carbon atom attached to a nitrogen atom, had by then been detected in the comas and tails of many comets, and then confirmed in Halley's Comet as well. It was called cyanogen. But when chemically combined in a salt it had another name—cyanide. Since only a grain of potassium cyanide touched to the tongue is sufficient to kill an adult human, the idea of the Earth flying through a cloud of cyanide generated a certain apprehension. People imagined themselves choking, gasping, and dying, millions asphyxiated by the poison gas.

The global pandemonium about poison gas in the tail of Halley's Comet was, sadly, fueled by a few astronomers who

LIVRE V

LES COMÈTES

The sedate title page of the introduction to the cometary section of Camille Flammarion's *Astronomie Populaire* (Paris, 1880).

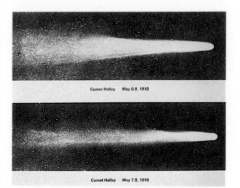

Two views on successive nights in 1910 of Comet Halley, when the Earth may have grazed the tail of the comet. Courtesy International Halley Watch.

should have known better. Camille Flammarion, a widely known popularizer of astronomy, raised the possibility that "the cyanogen gas would impregnate the atmosphere [of the Earth] and possibly snuff out all life on the planet." With vaguely similar pronouncements by Gambart and Laplace preceding him, and a traditional dread of comets tracing back into prehistoric times, it is not surprising that statements like Flammarion's helped to produce a worldwide comet frenzy.

In fact, it was not even clear that the Earth would pass within the tail of Halley's Comet. In any case, the tails of comets are extraordinarily thin, a wisp of smoke in a vacuum. The cyanogen is in turn a minor constituent in the tails of comets. Even if the Earth *had* passed through the tail in 1910 and the molecules in the tail had been thoroughly mixed down to the surface of the Earth, there would have been only one molecule of cyanogen in every trillion molecules of air—a good deal less than the pollution caused even far from cities by industrial and automobile exhaust (and much less than what would happen in the burning of cities in a nuclear war). Also, the Earth had, half a century earlier, passed well into the tail of the Comet of 1861 with no apparent ill-effects.

All this, or something like it, was made clear by the global astronomical community, but—like the 1979 worldwide anxiety that the Skylab satellite would fall on the heads of innocent passersby—the assurances had little effect. Why should the world population be so susceptible? This was certainly a time of burgeoning industrial pollution and attendant respiratory disease. But it also seems possible that the fear was not so much about poison gas up there in the comet as down here in the national arsenals. In H.G. Wells' *War of the Worlds* (1898), Martians invading the Earth used chemical weapons. This was one of many novels of the period, depicting a future war, in which devastating poison gas attacks were vividly portrayed. However, far from being merely the inventions of science fiction authors, these works reflected a dismaying reality. At least until a few years before the 1910 apparition of Halley's Comet, new approaches to chemical warfare were being actively pursued in European military establishments. At the Hague conference of 1899, proposals for banning certain chemical weapons were offered, but they were successfully opposed by John Hay, the American Secretary of State. A resolution forbidding asphyxiating gases in artillery shells was adopted, but the United States would not sign. When the First World War came, four years after the comet, 120,000 tons of poison gas were used, mainly by Germany, France, and to a lesser extent, the United States, producing one and a quar-

Not all public interest in the 1910 apparition of Halley's Comet was connected with poison gas. Courtesy Ruth S. Freitag, Library of Congress.

ter million casualties. More than a quarter of all American casualties in World War I were due to poison gas. Only a small fraction of the poisons employed were cyanides.

But in 1910 the comet passed, no one was asphyxiated, and the Earth seemed none the worse for the experience. The most sensitive tests revealed not a trace of increased cyanogen in the air—a source of relief to many, although no comfort to the discoverer of cometary cyanogen. Less than a week before the closest approach of Halley's Comet to the Earth, Sir William Huggins, the man who had frightened millions by minding his own business, died at eighty-six.

Since then, compelling evidence has emerged for the presence of frozen water and silicate minerals in addition to organics in the comets. A list of the variety of molecules found to date is given in the figure on page 149. But most of these simple molecules and molecular fragments are not

"Waiting for the End of the World,"
by R. Jerome Hill, in *Harper's Weekly*,
May 14, 1910.

present as such on the cometary nucleus; rather, they are bits and pieces of molecules, broken off the comet and released into space by radiation from the Sun. They were originally parts of larger, so-called parent molecules, whose identities are conjectured, but which remain largely unknown.

In ordinary visible light, the spectra of comas tend to be dominated by the blue emission of C_2. This molecule is unfamiliar because it breaks up or combines with other molecules upon collision. If you had some in the air before you, it would not last long. But in a comet, the densities are so low that a long time will pass before C_2 collides with another molecule. Imagine a C_2 molecule sitting in space, exposed to radiation from the Sun. A photon of blue light hits it and raises it to what is called an excited state. It cannot get rid of this additional energy by collision, since there are no collisions. For a short while, it simply sits there in the coma, throbbing, until it does spit out a blue photon, but in a random direction. The C_2 molecule has no memory of where the Sun is. As a result, blue frequency light from the Sun strikes the coma, and some of it that otherwise would

Sir William Huggins (1824–1910). Courtesy Mary Lea Shane Archives of the Lick Observatory.

be lost into space is reradiated back to Earth. The process is called fluorescence.

Accordingly, since the dust and ice in the comet reflect or scatter sunlight, there are two sources of light from the comet—scattered sunlight and fluorescence, both of which Newton anticipated. Other molecules may be much more weakly fluorescent, or may fluoresce in other wavelengths of light that are not so readily detectable from the surface of the Earth. So the prominence of certain molecules such as C_2 in cometary spectra does not necessarily mean they are very abundant in comets. But their abundance is high enough to indicate that their parent molecules—carbon, or organic molecules—must be reasonably common in the cometary nucleus from which the coma derives. Much more common is the molecular fragment OH, produced by tearing water (HOH or H_2O) apart.

Comets were not observed in the ultraviolet part of the spectrum until rocket and, later, orbiting observatories were aloft, above the Earth's atmosphere, in the beginning of the 1970's. These first observations showed that comets have an envelope of hydrogen gas that extends for millions of kilometers from the cometary nucleus away from the Sun. Accompanying the hydrogen is OH, the two together evidently produced by the dissociation of water making up the cometary nucleus. The identification of ionized water, H_2O^+ (a water molecule with a negatively charged electron missing) in comets forges another link in the chain of evidence for water as a major constituent of the cometary nucleus. These results both qualitatively and quantitatively provide powerful additional support for Whipple's dirty ice model of the cometary nucleus.

Subsequent ultraviolet observations by orbiting observatories, and by rockets specially fired above the Earth's atmosphere for the purpose, have revealed a number of new molecules and molecular fragments—including S and CS, the first discovery of sulfur and its chemical products in comets. The most likely parent molecule of CS is CS_2, and it in turn probably derives from more complex sulfur-containing organic molecules. In most cases the cometary spectra invite us to trace backward from simple molecular species that we can readily identify to progressively more complex and more uncertain parent molecules, among which may be complex organic molecules of the sort that are found in the interstellar grains and gas, discussed below.

Infrared spectra of both the coma and the tail of comets (for example, Comet Kohoutek) show a spectroscopic emission feature due to silicates, the principal constituents of rocks. Beginning with observations of the Great September Comet (1882 II), the spectroscopic signatures of metals have been found in comets that come very close to the Sun. Such atoms as chromium, nickel, and copper, which are never observed in other comets, can be detected. The reason is clear: As Isaac Newton had first pointed out, sungrazers come so close to the Sun that even iron would become red hot. The individual mineral grains evaporate and boil, and a gas composed of metal atoms is poured off into the surrounding coma for astronomers on Earth to view. In the cold and dark from which the comets come, before their substance is boiled off by sunlight, the metals are probably chemically bound to the silicates.

Even for comets that never venture very near the Sun, the constituent dust grains and organic molecules find themselves in a kind of blast furnace during perihelion passage—compared to the staid and safe environment far from the Sun. High energy solar photons and charged particles

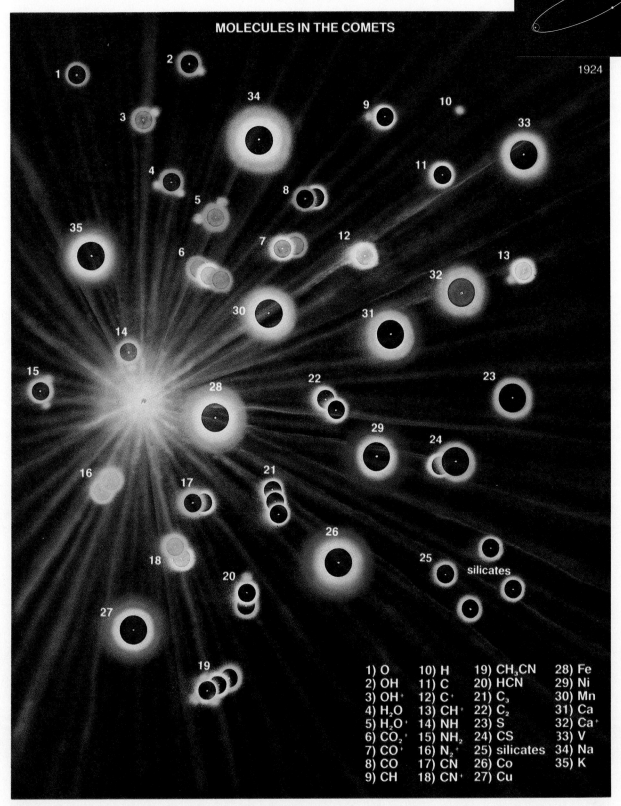

MOLECULES IN THE COMETS

1924

1) O	10) H	19) CH_3CN	28) Fe
2) OH	11) C	20) HCN	29) Ni
3) OH^+	12) C^+	21) C_3	30) Mn
4) H_2O	13) CH^+	22) C_2	31) Ca
5) H_2O^+	14) NH	23) S	32) Ca^+
6) CO_2^+	15) NH_2	24) CS	33) V
7) CO^+	16) N_2^+	25) silicates	34) Na
8) CO	17) CN	26) Co	35) K
9) CH	18) CN^+	27) Cu	

The composition of molecules and molecular fragments identified in comets (this page) and—especially striking in their variety and complexity—in the interstellar gas (next page). There is a considerable similarity between cometary and interstellar molecules, but the molecules thus far identified in the comets tend to be much simpler. Paintings by Jon Lomberg.

MOLECULES IN THE INTERSTELLAR MEDIUM

1925

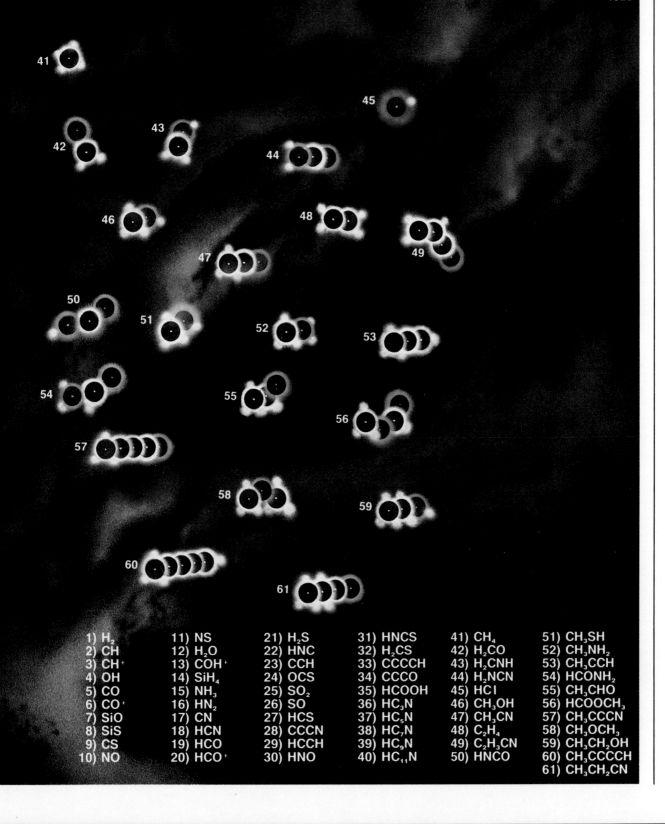

1) H₂
2) CH
3) CH⁺
4) OH
5) CO
6) CO⁺
7) SiO
8) SiS
9) CS
10) NO

11) NS
12) H₂O
13) COH⁺
14) SiH₄
15) NH₃
16) HN₂
17) CN
18) HCN
19) HCO
20) HCO⁺

21) H₂S
22) HNC
23) CCH
24) OCS
25) SO₂
26) SO
27) HCS
28) CCCN
29) HCCH
30) HNO

31) HNCS
32) H₂CS
33) CCCCH
34) CCCO
35) HCOOH
36) HC₃N
37) HC₅N
38) HC₇N
39) HC₉N
40) HC₁₁N

41) CH₄
42) H₂CO
43) H₂CNH
44) H₂NCN
45) HCl
46) CH₃OH
47) CH₃CN
48) C₂H₄
49) C₂H₃CN
50) HNCO

51) CH₃SH
52) CH₃NH₂
53) CH₃CCH
54) HCONH₂
55) CH₃CHO
56) HCOOCH₃
57) CH₃CCCN
58) CH₃OCH₃
59) CH₃CH₂OH
60) CH₃CCCCH
61) CH₃CH₂CN

are constantly bombarding cometary molecules, breaking them into pieces, tearing fragments off, ionizing, dissociating. Many of these molecules are so fragile that they last hours or less in the ultraviolet radiation of the Sun before they are torn apart. This is another reason that the molecules observed in comas and tails are not necessarily the most abundant molecules: what we see tends to be the molecules most resistant, most impervious to the radiation that bombards them.

Since methane ice (CH_4) is expected to be abundant in the outer solar system, it is natural to consider it the parent molecule for such fragments as CH_2. But, unless a new vein of methane ice is by chance exposed just as the cometary nucleus enters the inner solar system, CH_4 is unlikely at such high temperatures; and for this reason it is attractive to consider the methane trapped as a clathrate in the ice crystal lattice (Chapter 6). However, no amount of methane clathrate can explain such molecules as C_2 and C_3, which must derive from more complicated organics. Thus even the simplest cometary organic molecules such as CH may derive from more complicated organics and not from methane.

In addition to measurements in the visible, ultraviolet and infrared parts of the spectrum, there is also spectroscopy at radio frequencies. It does not use prisms and the like, but is perfectly able to distinguish absorption or emission lines due to particular atoms or molecules. So far as is known, the first attempt at radioastronomical observations of a comet was performed during the closest approach of Halley's Comet to the Earth on May 18, 1910, by Lee De Forest, the inventor of the triode electron tube, a mainstay in the development of modern radio. With an aerial and his new receiver, he observed from the roof of a building in Seattle, Washington, and surmised—probably erroneously—that the comet was the source of the enhanced crackling static he thought he had detected. In our time, radio observatories have found abundant OH in comets, fragments expected from the breakup of water molecules. In addition, they have provided some evidence for cometary HCN (hydrogen cyanide) and CH_3CN (acetonitrile), cyanides that may be parent molecules of the cyanogen that frightened millions in 1910. The molecules and molecular fragments that have been found in cometary spectra are shown on page 149.

In the 1970's radio spectroscopy opened a surprising new field when a variety of exotic molecules were found in interstellar space. You might, for example, be looking at some distant source of radio emission—the center of the Galaxy, say—and discover a range of new spectral lines produced by

[The Laputans'] apprehensions arise from several changes they dread in the celestial bodies. For instance…that, the Earth very narrowly escaped a brush from the tail of the last comet, which would have infallibly reduced it to ashes; and that the next, which they have calculated for one and thirty years hence, will probably destroy us.

—Jonathan Swift,
Gulliver's Travels,
1726

the diffuse gas between the radio source and your radio spectrometer. The gas between the stars is very thin, but if you can add up all the molecules in a line of sight thousands of light-years long, you might find even fairly rare molecules. This bonanza of currently known interstellar molecules, most of them organic, is shown on pages 150 and 151.

Astronomers tend to be nervous about the word organic—concerned that it might be misunderstood as a token of life on another world. But "organic" only refers to molecules based on carbon. And organic chemicals would be produced and destroyed even if there were no life anywhere in the universe. Circumlocutions such as "carbonaceous" are in common use, but we will here employ the correct chemical term, organic. While organic chemistry does not in any sense imply biology, nevertheless, if there are complex organic molecules produced somewhere in the universe, it may well have some bearing on the question of life elsewhere, or even life down here on Earth. Indeed, the origin of life must have involved pre-existing organic molecules that by definition could not have been made by living things.

Since comets mainly live far from the planets, in the interstellar realm, it is not unreasonable that they should be made of interstellar stuff. If we assume that comets are composed simply of interstellar silicates and ices—frozen H_2O, CH_4, NH_3, CO_2, and so on—and allow molecular fragments to come off under solar radiation, calculations show insufficient C_3 and CN produced relative to other molecules. Silicates and ices alone cannot explain all the organic fragments discovered by William Huggins. But if it is assumed that the comets ultimately derive from interstellar matter of the sort shown in on pages 150–151, then the molecules detected in the comas and tails of comets can be understood.

Cometary silicates are probably intimately mixed with—perhaps coated by—complex organic compounds. The relative abundance of atoms in comets, so far as is known, is quite like that in the interstellar grains and gas (see the charts on pages 154 and 155). There is less carbon detected in the spectra of comets than there is—again in terms of relative abundances—in the spectra of interstellar grains. One possible explanation is a large proportion of complex organic molecules (or just plain carbon) in the cometary nucleus which are involatile or which do not produce accessible spectral features. If this interpretation is correct, comets might be as much as 10 percent organics. Only a few percent of dark organics would darken and redden the cometary snows sufficiently to be in agreement with observa-

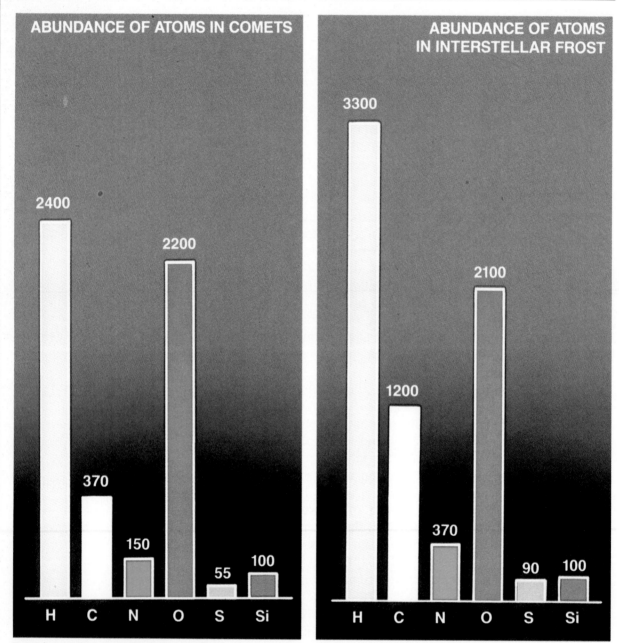

ABUNDANCE OF ATOMS IN COMETS

2400

370

150

2200

55

100

H C N O S Si

ABUNDANCE OF ATOMS IN INTERSTELLAR FROST

3300

1200

370

2100

90

100

H C N O S Si

Bar charts showing the relative abundances of key chemical elements in cometary ices (above left), interstellar ices (above right), and in the universe generally (opposite page). The atoms are hydrogen (pale yellow); carbon (white); nitrogen (blue); oxygen (orange); sulfur (bright yellow); and silicon (brown). Hydrogen is greatly in excess in the universe compared to interstellar frost or cometary nuclei. The comparison shows that cometary ices are very similar in atomic composition to interstellar frost. Diagram by Jon Lomberg/BPS.

tion. But the best evidence for organic matter and carbon in comets is obtained from recovered small pieces of cometary matter, a subject we treat in Chapter 13.

The picture that emerges from the work of Huggins and his successors at the spectroscope is this: A comet is a snowball filled with small mineral grains and complex organic matter. The organic molecules are distributed throughout the comet, although they may be concentrated in the sur-

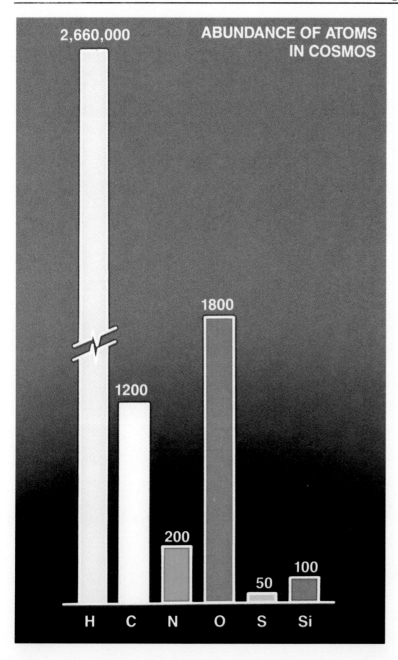

1926

face layers. The amount of organics is at least a few percent —perhaps as much as 10 percent—enough to darken dramatically and somewhat redden the snows. This chemistry is similar to what we know of the molecules between the stars, and is hard to understand unless the comet has been formed out of interstellar matter. Thus, more than any other body we have access to, comets may be emissaries from the void between the stars.

A comet and its train wander among the other lights of the night sky. From Grandville's *Un Autre Monde*, 1844. Courtesy Ruth S. Freitag, Library of Congress.

Chapter IX

TAILS

Hung by the heavens with black, yield day to night!
Comets, importing change of time and states,
Brandish your crystal tresses in the sky...

—William Shakespeare,
Henry VI, Part I, I,i

Stars seen through the more distant reaches of the multiply-tailed de Cheseaux's Comet, 1744. From Amédée Guillemin, *The Heavens*, Paris, 1868.

THIS IS A CHAPTER ABOUT NOTHING, OR SOMEthing closer to nothing than anything in our everyday life. A cubic centimeter is the volume of a sugar cube. If you select a cubic centimeter of air in front of your nose and look extremely closely, you would find thirty billion billion molecules, all very small, vigorously running into one another. If, on the other hand, you were to make an inventory of the atoms or molecules in a cubic centimeter of a comet's tail, you would find a thousand or less—hardly any at all. A bright comet tail is much nearer to a perfect vacuum than the best laboratory vacuum our technology can produce on Earth. But even with very few molecules in a cubic centimeter of comet tail, there are a great many cubic centimeters. It is sometimes said that, like a genie in the Arabian Nights, a cometary tail stretching from world to world can be bottled up in a brass lamp. Newton wrote, "a gaseous comet [i.e., the tail alone], with a radius a thousand millions of miles in length, if submitted to the same degree of condensation as the earth, could be easily kept in a good-sized thimble." But even with only one molecule per cubic centimeter, the enormous tail Newton imagined would require a thimble three kilometers across to contain it.

"Good-sized," indeed.

The tail is almost perfectly transparent: when it passes in front of a bright star you can always see the star through it. But still, it may seem astonishing that the air in the room should be transparent while the near vacuum in the comet tail is visible to the naked eye. The air is a *something*; we can feel it readily enough in a stiff breeze. The comet tail is very close to nothing. How can the something be invisible when the nothing is clearly seen? The key to answering this question is to remember that we see the comet against a black sky, while the daylight sky is almost uniformly illuminated. Newton did a very difficult experiment, remembering the brightness of a comet and comparing it with the brightness seen in a laboratory experiment:

> Nor is the brightness of the tails of most comets ordinarily greater than that of our air, an inch or two in thickness, reflecting in a darkened room the light of the sunbeams let in by a hole of the window shutter.

A great many widely separated particles reflect light as well as the same number of particles compressed by the overlying atmosphere. A little bit of air illuminated by sunlight and viewed against a black background is as bright as a comet tail in the night sky. Still, what is it that we see when we look at the tail of a comet?

If you are fortunate enough to be standing under a clear sky with a comet readily visible above you, note the orienta-

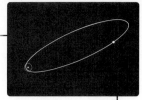

Stars seen through the tail of the Comet of 1843. From Amédée Guillemin, *Les Comètes*, Paris, 1875.

tion of the comet's tail with respect to the Sun. If it is just before sunrise, note how the tail streams away from the glimmer of dawn on the eastern horizon. If it is just after sunset, note how the tail points away from the Sun, now located just beneath the western horizon. And if you are fortunate enough to witness a great daylight comet, it will then be entirely obvious that the comet tail points away from the Sun. There are no significant exceptions to this rule. Although from comet to comet there is some variation in the exact angle the comet's tail makes with a straight line from comet to Sun, this regularity is as invariable as the observation that the horns of the Moon's crescent always point away from the Sun.

...and the cork goes ceilingward, the flow of comet wine spurts forth.
—Aleksandr Pushkin, *Eugene Onegin*

...comet wine / Viná Kométi: Fr: *Fin de la Comète*, Champagne of the Comet Year, an allusion to the [Great] Comet of 1811, which was also a wonderful vintage year.
—Vladimir Nabokov, Commentary on *Eugene Onegin*, New York, Bollingen Foundation, 1964

Illustration: Moet & Chandon advertisement from 1910. Courtesy Ruth S. Freitag, Library of Congress.

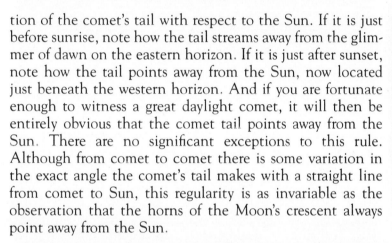

The position of the tail of the Great Comet of 1843 as it sped around the Sun. P marks the position of perihelion. The tail (also shown in the figure above) always points away from the Sun. From Camille Flammarion, *Astronomie Populaire*, Paris, 1880.

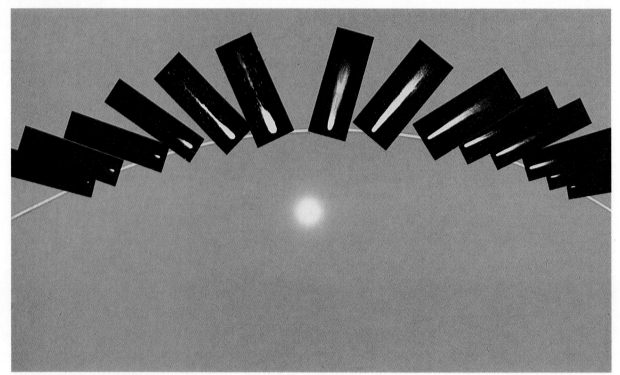

Photographs of Halley's Comet near its 1910 perihelion passage, located in the actual orbital positions and proper orientations. Diagram by Jon Lomberg/BPS.

Comet Tago-Sato-Kosaka, photographed January 1, 1970, at the University of Michigan Observatory by F. D. Miller. Courtesy National Aeronautics and Space Administration.

Comet tails point away from the Sun whether the comet is approaching the Sun or receding from it, a conclusion apparently first drawn by Chinese astronomers during the apparition of Halley's Comet in the year 837. After perihelion passage the comet flies tail first out of the solar system. The tails of comets are much more like the effluvia from an industrial smokestack blown back on a windy day than the long hair of a bicyclist flowing behind her as she coasts down a hill on a windless day: it is not the motion of the comet through some resisting gas that determines the orientation of the tails, but rather the comet's motion combined with something like a wind blowing out from the Sun.

There are two kinds of cometary tails: long, straight, faintly blue tails, pointing almost perfectly straight back from the Sun; and usually shorter, curved, faintly yellow tails. Before their nature was understood, they were called Type I and Type II tails, respectively (see opposite page, top). These designations are still in use. The Type II tails are yellow because they reflect sunlight back to us; but the Type I tails give off a blue light of their own (see page 164), although they are commonly not very prominent in visible light. A comet may have either type, or neither or both types of tails at a given time. The Type I tails often display an intricate dancing pattern of straight ray streamers, each ray narrower than the diameter of the Moon, but perhaps ten million kilometers long.

Type I tails are variable—not only from comet to comet, but from hour to hour, day to day, week to week, within the

The thin straight Type I tails and the curved Type II tails of Donati's Comet over Paris, October 5, 1858. From Amédée Guillemin's *Les Comètes*, Paris, 1875.

The Type I and Type II tails of Comet Bennett, photographed March 16, 1970, at the University of Michigan Observatory by F. D. Miller. Courtesy National Aeronautics and Space Administration.

same comet. Like lizards, comets can grow new tails. On the surrounding pages are examples, in drawings and photographs, of this propensity for regeneration. Comets are quick-change artists. Some of these pictures are printed as negatives, because black on white tends to bring out fine detail. In 1908, for example, Comet Morehouse—sizable fragments of which were reported to have left the nucleus for the tail, where they emitted their own subsidiary tails—astonished the astronomers gathered at a conference in Oxford:

The formation of the tail seems to be intermittent rather than continuous. There seem to be at intervals

The tails of comets appear to be composed of the most volatile molecules which the heat of the Sun raises from their surface and by the impulsion of his rays banishes to an indefinite distance...The different volatility, size, and density of the molecules must needs produce considerable differences in the curves which they describe: hence arise the great varieties of form, length, and breadth observed in the tails of comets. If we suppose these effects combine with others which may result from a movement of rotation in the comet . . . we may partly conceive the reason of the singular phenomena represented by the nebulosities in tails of comets.

—Pierre Simon, Marquis de Laplace, *The System of the World*, Paris, 1799

convulsions or explosions in the nucleus, producing big bunches or lumps of tail which travel away and leave the comet with a small tail for a time...the signs of a coming convulsion can be recognized.

A young British astronomer named Arthur Eddington used the newly perfected lantern slide in a talk on Comet Morehouse:

Here is the same comet a day later. Everything is altered completely; you cannot point to a feature in the tail on this photograph and say that it corresponds to a certain feature in the previous photograph, and that one has changed into the other; you cannot say that this is the tail of the previous day modified. As far as can be judged the tail is an entirely new one...

Type I tails sometimes exhibit "knots"—small condensations of matter brighter than their surroundings. Knots are sometimes observed to accelerate down the tail away from the Sun. In talking about Comet Morehouse, Eddington was clearly puzzled by the acceleration turning on and off capriciously. By taking a quick succession of photographs you can measure how fast the knots move. Their speed ranges up to 250 kilometers per second (540,000 miles an

The Comet of July 1819 and Signora Bietta

Oh what a fix! What a damned fix
I got into, if I'd only known, my husband and me!
I'm two weeks late with my...
So that...*you* know, Mrs. Bietta!

And you know what? All this trouble on account of the Comet
With that tail that amazed everyone so!
The same thing also happened to Amelli,
And to Gina, and to Bina, and to Babetta.

And Nunziada's a full month late,
You know what a little fool she is:
Just imagine, that sneaky Comet!

And do you remember? We were there
laughing at that enormous tail
and playing under it! Can it be?

—Anonymous nineteenth-century poem,
translated from the Italian by Gina Psaki

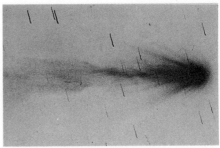

hour), or more, and their acceleration (which you would feel if you were moving with the knot) can be as high as 1 g, the acceleration that the Earth's gravity imparts to falling objects. There can also be knots with very little speed or acceleration. The motion of knots in comet tails is as unpredictable as the weather.

Visible light spectroscopy of the Type II tail shows light from the Sun reflected back to the observer without the tail adding or subtracting spectral features of its own. This is characteristic of dust; and in the infrared spectrum of some comet tails there is the signature of silicates, the main constituent of ordinary rocks on Earth. For this reason, the Type II tail is called a dust tail, even though we suspect there is an intimate association of sticky, dark organic matter with the fine silicate dust particles.

Type II tails are clearly made of innumerable fine particles. If only Newtonian gravitation were working, a collection of fine particles could not travel through space as if it were a solid body. Instead, each particle would be on a separate orbit about the Sun, a microplanet moving in an almost perfect vacuum. Because the initial speeds of the dust grains leaving the comet depend on the properties and orientation of the puff of gas that carried them off, some will be traveling a little faster and some a little slower than the comet. In

Comet Morehouse (1908 III) negative, with star trails, photographed at Royal Greenwich Observatory, September 30, 1908. Courtesy National Aeronautics and Space Administration.

Schematic representation of the pressure of sunlight blowing back the gases and fine particles in the cometary coma to produce the tail. Diagram by Jon Lomberg/BPS.

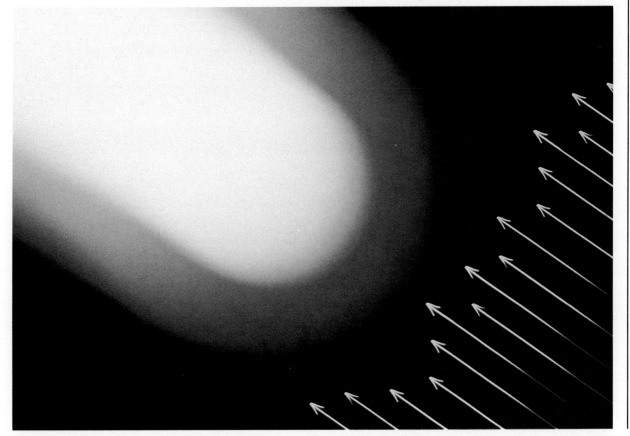

The straight blue ion tails and the curved yellower dust tails of Comet West (1976 VI). Photographed on March 8, 1976. Photo by Dennis di Cicco/Sky & Telescope Magazine. © 1976 by Dennis di Cicco.

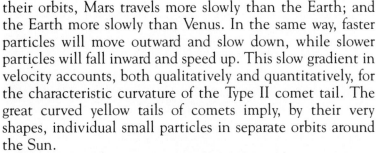

1930

their orbits, Mars travels more slowly than the Earth; and the Earth more slowly than Venus. In the same way, faster particles will move outward and slow down, while slower particles will fall inward and speed up. This slow gradient in velocity accounts, both qualitatively and quantitatively, for the characteristic curvature of the Type II comet tail. The great curved yellow tails of comets imply, by their very shapes, individual small particles in separate orbits around the Sun.

Tiny particles—clusters of still smaller motes of silicates and organics—are jetted off the cometary nucleus, and redirected back away from the Sun. Newtonian gravity will then produce the graceful curved tail. But what is this mysterious influence driving the particles back? The first person to guess the right answer was Johannes Kepler, who held that the tail of a comet is pushed away by the pressure of sunlight. The net result, he argued, was that the comet would ultimately dissipate into interplanetary gas.

Radiation pressure is not a factor in everyday life. Even very small people are not thrown to the ground by sunlight as on a cloudless day they step out of doors. The force of radiation pressure is the equivalent of the weight of a layer one atom thick at the surface of the Earth. Radiation pressure amounts very nearly to nothing. But if you are made of almost nothing in the first place, radiation pressure can push you around. The total force of the sunlight on us depends on how much area we present, but the resistance we offer to sunlight in its effort to push us around depends on our mass. The smaller a particle is, the more area it presents for its mass. Eventually, in free space, a sufficiently small particle will feel the force of radiation pressure driving it outward from the Sun more strongly than the Sun's gravity pulling it inward toward the Sun.

Both the radiation pressure force and the gravitational force vary inversely as the square of the distance from the Sun; so once a particle is small enough for radiation pressure to dominate, it is continuously accelerated out of the solar system. If radiation pressure wins over gravity at the orbit of Mercury, it is still winning over gravity by the time the particle is halfway to the nearest star. But dust particles must be very small—less than a ten-thousandth of a centimeter across, too small to see even with an ordinary microscope—for radiation pressure to eject them from the inner solar system. Therefore, of the dust particles that are blown off the nucleus of a comet by the jets, only the smallest particles are driven back into space by sunlight. The bigger particles—having achieved escape velocity from the comet, but not from the Sun—establish individual orbits about the Sun.

When the particles are much smaller than this, they are

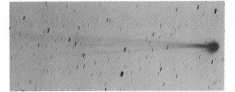

Comet Finsler (1937 V), photographed at Lowell Observatory, August 9, 1937. Courtesy National Aeronautics and Space Administration.

Photograph of Comet Kohoutek taken January 20, 1974, at the Joint Observatory for Comet Research, New Mexico. In this negative, the background stars are shown as points, not trails, because Comet Kohoutek was sufficiently bright not to require a long time exposure. Courtesy the Joint Observatory for Cometary Research, Laboratory for Astronomy and Solar Physics, NASA/Goddard Space Flight Center, and New Mexico Institute of Mining and Technology.

Turbulence in the tail of Comet Brooks (1893 IV). Courtesy Lick Observatory, University of California.

The aurora borealis flickering in the Alaskan night sky produces a pattern of lights resembling drapery. Courtesy National Aeronautics and Space Administration.

even smaller than the wavelength of most of the sunlight, and the particles slip through the crests and troughs of the lightwaves and again are not driven outward. Thus there is only a small range of particles with sizes in the vicinity of the wavelength of yellow light that are accelerated by sunlight out of the solar system. Interplanetary space should be depleted in particles of this size.

If the Type II tail is made of dust, what is the composition of the longer, still more prominent Type I tail? Aristotle had thought that comet tails bore some similarities to the aurora borealis, a view echoed by Kant:

> The Earth has in it something which may be compared with the expansion of the vapors of comets and their tails, namely, the Northern Lights, or Aurora Borealis.... The same force of the rays of the Sun, which makes the Aurora Borealis, would produce a vapor head with a tail, if the finest and the most volatile particles were to be found as abundantly on the Earth as on the comets.

The tail of Halley's Comet, with exaggerated brightness and extent, hanging over the city of San Francisco, April 1910. Photo © Jerred Metz, *Halley's Comet, 1910: Fire in the Sky.*

While it was not understood in the times of Aristotle or Kant, today we do know something about the aurora—the multicolored time-variable pattern of lights in the sky, sometimes resembling rustling drapery—seen chiefly in the arctic and antarctic regions of the Earth. The aurora appears in the polar regions because it is there that charged particles—especially protons from the Sun—are guided by the Earth's magnetic field. The protons pour into the atmosphere above the poles, tear the molecules of air apart a little bit, and induce them to glow. The aurora is variable because the supply of protons is variable. While the purported simi-

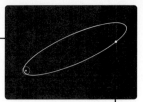

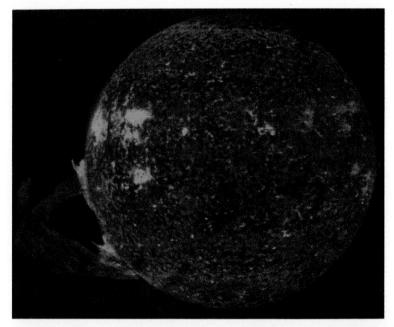

1931

An enormous prominence erupts off the surface of the Sun on December 19, 1973. This image was obtained, in the light of ionized helium atoms, from Earth orbit by *Skylab*. Helium is, after hydrogen, the second most abundant atom in the Sun. Courtesy National Aeronautics and Space Administration.

Another prominence of protons, electrons, and helium ions rushes off the Sun, producing a disturbance in the solar corona, seen in blue. Still more energetic eruptions propel blobs of charged particles off the Sun altogether, constituting a solar flare event. Courtesy National Aeronautics and Space Administration.

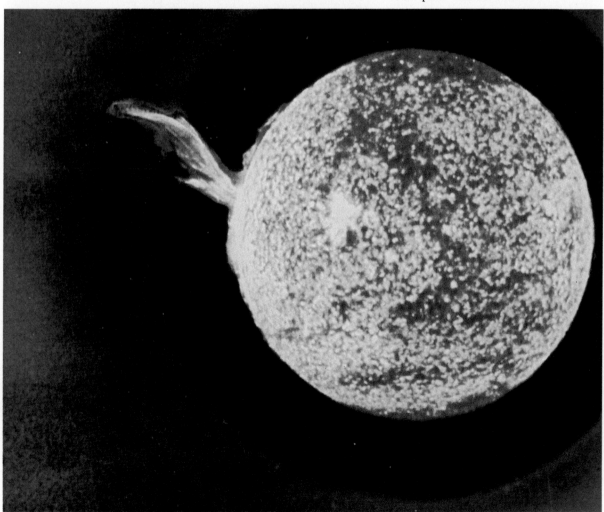

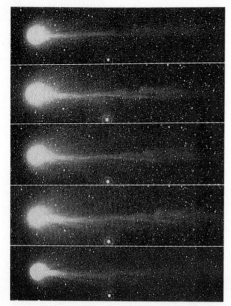

Five photographs taken approximately an hour apart of Comet Whipple-Fedtke-Tevzadze (1943 I) on March 8/9, 1943. Photographed in wartime Nazi Germany by C. Hoffmeister, Sonneberg Observatory. From N. B. Richter, *The Nature of Comets*, Methuen, 1963.

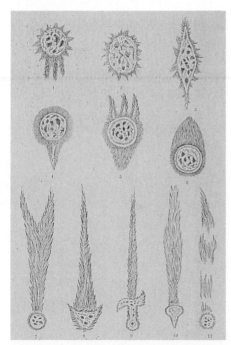

Forms of comets, drawn according to Pliny's written description. From Amédée Guillemin, *The World of Comets*, Paris, 1877.

larity in appearance between the aurora and the Type I tails of comets may be less than obvious, it is certainly true that electricity and magnetism are central to the understanding of both.

The spectra of comas mainly show, as we have said, such molecular fragments as C_2, C_3, and CN. But when the telescope points away from the coma and toward the tail, a completely different spectrum is seen, in which the lines of CO^+ are dominant. CO^+ is a molecule of carbon monoxide with an electron removed. Such electrically charged molecules are called *ions*. CO^+ preferentially absorbs blue light from the Sun, which, through fluorescence, is re-radiated again in all directions as the same blue light. This is why Type I tails are called ion tails, and why they glow bluely. If you could somehow turn off these particular frequencies of blue sunlight, the dust tail would look almost the same, but the ion tail would be invisible. There is only a relatively small amount of CO^+ in the tails of comets, but if you took the CO^+ away, the blue ion tail would be replaced by a much dimmer red ion tail, due to H_2O^+.

Molecules are ordinarily electrically neutral, the number of negatively charged electrons on the outside of the atom just balancing the number of positively charged protons in the inside. Consider a molecule of water sitting in interplanetary space near the Earth. It is struck by a photon of ultraviolet light from the Sun (or a proton), and one of its electrons is carried off into space. The molecule is therefore positively charged: it has more protons than electrons. The charged molecule is called an ion, and the process that carried off its electron is called ionization. If one electron had been stripped off, we would symbolize such a water molecule as H_2O^+, the plus sign indicating an excess positive electrical charge, or equivalently a deficit of one electron. If two electrons had been carried off, we could write this as H_2O^{++}. The same is true for CH^+, N_2^+, or CO^+.

All right, so a molecule of CO, let's say, scurries off the cometary nucleus into space, and is mugged by a light beam, losing an electron in the scuffle. Why should it now move straight back from the Sun? Why should the knots in the ion tails accelerate erratically? What emanation from the Sun does an ion see that a dust grain doesn't?

Radiation pressure is far too feeble to explain the acceleration of knots in the ion tail. Also, the acceleration of the knots varies with time, while the light put out by the Sun is extremely steady. There must be some other means besides the light it radiates to space, by which the Sun can affect the tails of comets. If we study the knots over a period of months, we sometimes find a periodicity in the accelerations closely equal to the period of rotation of the Sun as

seen from the comet. The conclusion is straightforward: the influence emanates from a particular region on the Sun, not from the Sun as a whole. The influence, whatever it is, reaches out to the comet only when the active solar region is oriented cometward. As the Sun rotates (once every 27 days, as seen from the Earth), the active region on the Sun eventually faces away from the comet, and the ion tail acceleration subsides, only to revive again as the region rotates around again.

It has long been known that the Sun episodically emits charged particles: A flare on the surface of the Sun is seen through the telescope and three days later a "magnetic storm" strikes the Earth, interfering with long-distance radio communications. Fortuitously, on March 29, 1943, the ion tail of Comet Whipple-Fedtke-Tevzadze exhibited a marked knot acceleration, and on the same day a major magnetic storm reached the Earth. From such observations, the astrophysicist Ludwig Biermann, working in Germany during and after World War II, calculated the properties of these bursts of charged particles from the Sun. His conclusion was that there must be a steady solar wind that blows outward from the Sun at all times, and that drives the ion tails of comets leeward no matter what section of the sky the comet happens to be in. But in addition, the episodic accelerations in the ion tails must be due to something else, now known to be enhancements to the solar wind pouring out of holes in the Sun's corona and sharing the Sun's rotation.

The solar wind was first directly detected by the Soviet *Luna 3* spacecraft in 1959, and later by the American spacecraft *Explorer 10* and, especially, the interplanetary spacecraft *Mariner 2* on its historic first voyage from Earth to Venus in 1962. The measurements showed that the solar wind consisted mainly of protons and electrons. There were only a few of them per cubic centimeter in the vicinity of the Earth, but they were rushing radially outward from the Sun at speeds of several hundred kilometers per second. Faster, higher density gusts were also found. In addition, occasional flares and other violent events on the Sun's surface propel globs of high energy protons and electrons out into space. This wind from the Sun carries with it a magnetic field that sweeps up ions in its path. The moment a molecule is ionized by sunlight, it is captured by the magnetic fields carried with the solar wind. However, those molecules that are not ionized, that remain electrically neutral, are unaffected by magnetic fields. Thus the ion tails of comets are weather vanes in the solar wind, always blowing straight back; but occasionally an irregular magnetized cloud from the Sun catches up with the tail and stirs up the ions.

1932

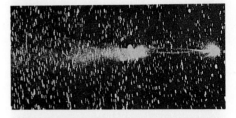

Tail disconnection event in Comet Morehouse (1908 III). These three pictures were taken on three consecutive days, September 30, October 1, and October 2, 1908. A major piece of the ion tail has been detached and lost downwind of the comet. Dashed lines are background stars trailed in this time exposure. Courtesy Yerkes Observatory, University of Chicago.

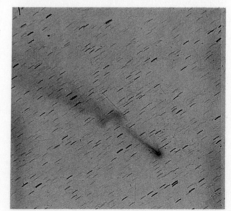

Comet Morehouse (1908 III), a view two weeks later of the tail disconnection event shown above. Indiana University photograph. Courtesy National Aeronautics and Space Administration.

The ion tail has more variability and often more structure than the dust tail. In a period of hours the thin, straight rays can change, merge, and dissipate. Evanescent features take complex forms, including sharp, right-angle discontinuities and corkscrew shapes millions of kilometers across. Some of the tail patterns (see page 165, bottom, for example) look almost as if the comet is veering erratically, like smoke bombs on the Fourth of July. But the comet is moving on an almost perfect elliptical orbit; the solar wind is responsible for the patterns. Sometimes the entire tail disconnects itself from the nucleus and, fading, drifts slowly back until it disappears entirely (see page 169). Typically, the nucleus forms a new tail after a disconnection event, and the old and the new tail may interact or even become briefly intertwined. The complexity and variability of the tails of comets has an almost biological feel to it. The prevailing opinion among astronomers is that the interaction of the electrically charged molecules in the ion tail with the highly variable solar wind accounts for most of the strange behavior of comet tails. The comet, of course, contributes some idiosyncrasies of its own—as, for example, the timing and magnitude of gas and dust outbursts. But, in detail, we are still far from understanding the changing forms in the ion tails of comets.

Astronomers spend long hours arranging photographs of the same cometary tail in their proper time sequence, and—with equipment ranging from a magnifying glass to a computer contrast enhancement system—examining every burp and wiggle in hopes of uncovering the underlying physical mechanisms. Since the flow of charged and neutral gases through a moving stream of protons carrying its own magnetic field is not something we often run into in our everyday lives, the subject is necessarily arcane. It involves elaborate three-dimensional magnetohydrodynamic calculations, and laboratory simulations in which electrically charged plasma interacts with some solid object (for example, a ball of wax) intended to duplicate the cometary nucleus.

The motion of comets in the Sun's gravitational field involves essentially the same physics as a rock thrown up into the air and falling back to the ground. There is a ready parallel in everyday life. The vaporization of ice off a cometary nucleus and into the interplanetary vacuum is not very different from the evaporation of snow into the Earth's atmosphere on a sunny day. Again, we easily understand the phenomenon because it is part of everyday life. But—beyond a distant relationship to the aurora borealis—the plasma physics of cometary tails does not have a ready terrestrial analogue, so naturally it presents to us facets that are

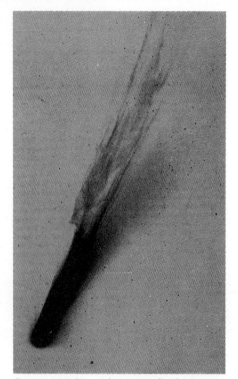

Comet Mrkos photographed August 3, 1957, with the 48-inch Schmidt telescope at Mount Palomar Observatory. Note the structureless faint dust tail and the extremely complex morphology of the ion tail. Palomar Observatory photograph.

Comet Humason (1962 VIII). Note the ray structure. Mount Palomar Observatory photograph. Courtesy National Aeronautics and Space Adminstration.

Comet Humason (1962 VIII). Drawing based on photographs by Elizabeth Roemer. Courtesy NASA.

1933

mysterious, that tend to resist our attempts to understand. However, as a result, comet tails are natural laboratories in which we can test our understanding of plasma physics.

Our knowledge of the ion tails of comets is new and in certain respects still tentative, but it fully satisfies the expectations of astronomers of an earlier epoch. The American astronomer E.E. Barnard wrote in 1909:

> In any attempt to explain certain of these features of the comet, we are forced to tread on very dangerous ground, for there is nothing in the adopted theories of comets that will explain them, and we must assume unknown quantities that perhaps violate all our ideas of the conditions existing in space in the vicinity of the Sun and planets. But, as there seems to be no other explanation, the very fact that we are driven to extremes in the search for a possible cause may lead to a knowledge of interplanetary conditions that would never become known without the aid of the wide sweeping tail of a comet.

Today we understand that the ion tails are solar wind socks, probes of interplanetary weather conditions which might otherwise escape our notice altogether.

After a comet begins to outgas and a coma forms, the solar wind and the magnetic field it carries collide with the cometary atmosphere. Molecules in the outer layers of the coma are ionized by ultraviolet light from the Sun. The solar wind sweeps these ions up, flows around the sunward segment of the coma, and carries the ions far downwind. A shock wave is created, similar to that in front of a high performance aircraft when it reaches supersonic speeds. This bow shock is established as much as a million kilometers windward of the comet. The ions whipped leeward by the solar wind may stretch out a hundred million kilometers away from the Sun, pumped to glow an eerie blue by sunlight. Occasionally, the tail is overtaken by a gust and more rarely by a small cyclone in the solar wind, and havoc is wrought in a tail that formerly stretched perfectly radially out from the Sun.

Because of their rapid and erratic variability, and technical and organizational limitations, there has never been an adequate time lapse motion picture of a comet tail. But now we have reached the point when this is possible, both from the ground and from spacecraft. Even color stereoscopic movies are now possible. Moreover, there are today a number of monitoring stations in interplanetary space that are routinely surveying the solar wind and its variations. We are now nearly ready to employ our knowledge of interplanetary weather in a comprehensive understanding of the long and graceful tails of comets.

Comet Morehouse (1908 III) photographed November 25, 1908, at the Royal Greenwich Observatory. Courtesy National Aeronautics and Space Administration.

A daylight comet portrayed on sheet music from 1909. Courtesy Ruth S. Freitag, Library of Congress.

Chapter X

A COMETARY BESTIARY

If a rare [comet] and one of unusual shape appears, everyone wants to know what it is and, ignoring the other celestial phenomena, asks about the newcomer, uncertain whether he ought to admire or fear it. For there is no lack of people who create terror and predict dire meanings.

—Seneca,
Natural Questions
Book 7, "Comets"

The astronomers…give more attention to the laws of their movements than to the strangeness of their form.

—Immanuel Kant, on comets,
*General Natural History and
Theory of the Heavens,* 1755

LIKE A BREACHING WHALE BEFORE IT PLUNGES INTO the ocean depths, a comet briefly luxuriates in the sunlight and then is gone. The comets dance in the night sky, capriciously altering brightness, size and form. Sometimes we make out extraordinary activity near the nucleus. Halos are produced, shed, and dissipated. Multiple fountains can be seen spraying gas and dust into space. Because a comet typically rotates once every few hours, the curvature of the streamers is marked. Everything is blown back from the Sun by radiation pressure and the solar wind.

Some comets propel a clockwork succession of shroud-like comas into space. Comet Donati is a famous example. The most ready explanation is that there is a patch of surface ice which explosively evaporates in sunlight; since the comet is rotating, a typical patch sees both night and day, and new comas are generated on the sunlit side once each rotation.

On these pages we have accumulated a kind of cometary bestiary, like the animal bestiaries assembled by medieval authors to amaze and delight, and even to instruct. Most of the animals displayed were real; many were exotic; a few, such as the unicorn, were the result of errors in transmission, garbled accounts—in this case, of the African rhinoceros. There were also some conscious hoaxes. We look back at the bestiaries with fondness, and perhaps a little amusement, recognizing that they are the forerunners of modern textbooks in zoology.

In a similar spirit, we offer here a range of images, photographed or drawn at the telescope, of the forms of comets; particular attention is paid to the jets within the comas. Sometimes the comet presents itself so that we are looking down the axis of rotation; other times, we are viewing obliquely. In all cases, it is easy to tell where the Sun lies. Some of the illustrations are reproduced black on white, to heighten contrast.

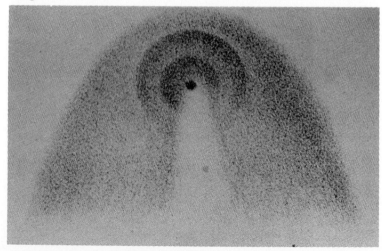

Successive envelopes of gas ejected from the nucleus of Comet Donati (1858 VI), drawn at the telescope on October 5 of that year by Schmidt. Courtesy National Aeronautics and Space Administration.

1935

Five (or possibly six) jets off the nucleus of Comet Tebbutt (1861 II), drawn by Schmidt. Courtesy National Aeronautics and Space Administration.

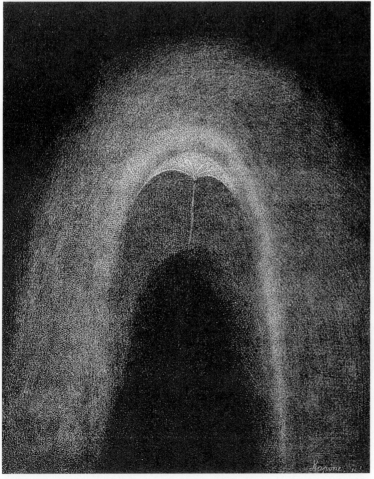

Multiple fountains playing off the sunward side of the nucleus of the Comet of 1861. Drawn by Warren de la Rue, July 2, 1861. From Amédée Guillemin, *The Heavens*, Paris, 1868.

Schematic representation of an oblique view (*left*) and a view looking down on the axis of rotation (*right*) of the same jetting cometary nucleus. Diagram by Jon Lomberg/BPS.

At the telescope, the astronomer is peering through an ocean of turbulent air, and the image produced is always distorted. One of the reasons that telescopes are located on isolated mountain tops, above much of the atmosphere, is to improve the seeing. At the focus of a large telescope, the astronomer's eye has a great advantage over the photographic plate: the astronomer can recall how, just a moment ago, the atmosphere steadied and fine detail in the comet could be glimpsed. By remembering such moments and drawing what was seen, the astronomer can often produce detail inaccessible to a camera attached to the same telescope. But the disadvantage of this method is that people are imperfect recording devices, and at the limit of resolution the eyes can play tricks. Nevertheless, by comparing the drawings of independent observers, by matching drawings with photographs, and by new image enhancement techniques, it is possible to verify that, by and large, the early cometary observers knew what they were doing.

Every one of these cometary forms represents a snapshot of the life and death of a comet, a glimpse of sunlight doing violence to a large lump of ice. Each of these displays is produced, at least in part, by streamers of gas and dust, rotation, radiation pressure, and the local weather conditions in the solar wind. The streamers and the coma are typically thousands of kilometers across, or more. Deducing the underlying events from the appearances left is what detectives do. It is also what field geologists do. Let your eye roam over the variety of cometary forms exhibited here, and see if you can construct a plausible explanation of what is happening in each case.

On page 175, top, we are looking obliquely at a nucleus while five or six streamers are being blown back by the Sun. On page 175, bottom, and elsewhere, we see that the foun-

Two views of the inner coma of Halley's Comet at different times of day on May 5, 1910. Drawn in Johannesburg by R.T.A. Innes (*left*) and W.M. Worsell (*right*). The right-hand image seems to imply rotation counterclockwise to our line of sight. Courtesy National Aeronautics and Space Adminstration.

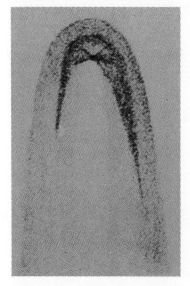

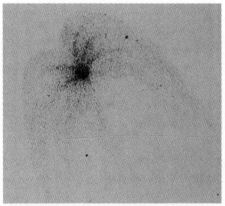

tains are restricted to the heated sunward hemisphere of the cometary nucleus. The direction of rotation can often be deduced from the curvature of the streamers (see page 176, bottom, for example).

In the drawing at the bottom of page 175, we are looking obliquely at the multiple fountains of dust erupting from the day side of the Great Comet of 1861. But imagine that instead the comet had approached the Earth from behind, its pole by accident pointed in our direction. Then we would be looking straight down on the axis of rotation, and the pinwheel of fountains would be rotating as we watched (see page 176, top).

Fortunately, the Earth's gravity is insufficient to drag passing comets to us. But every now and then, by chance, a comet will make a close pass by the Earth. It is only a matter of waiting. If you can wait long enough—a hundred million years, say—you are likely to see even very odd comets. For example, between the orbit of Saturn and Uranus is Chiron, a highly unusual object discovered in 1977 by Charles Kowal of the Hale Observatories in California. Named after the centaur who taught Jason and Achilles, Chiron is three or four hundred kilometers across, bigger than any known comet, although it is no larger than the bigger asteroids. Could it be the most visible member of a previously unknown population of massive comets that live mainly beyond Pluto? Chiron is very dark, and red. Other objects in the outer solar system are darkened and reddened by complex organic matter, and it is likely that this is also the case here. It certainly cannot have much uncontaminated fresh ice on its surface. Perhaps Chiron is a comet whose surface ices of methane and other exotic volatiles have evaporated leaving a dark organic matrix behind.

Every few thousand years, Chiron makes a sufficient number of distant passes by Saturn that its orbit is considered unstable. It may be slowly working its way in towards the Sun, and may one day become a short-period comet. Imagine, then, after repeated encounters with Saturn and Jupiter, that Chiron one day enters the inner solar system. If Comet Schwassmann-Wachmann 1 can still produce outbursts from its lonely post between the orbits of Jupiter and Saturn, if the Great Comet of 1729 could be visible to the naked eye when it was almost at the distance of Jupiter, then what would an object like Chiron look like if it were newly arrived from the outermost solar system and making a close pass by the Earth? Very likely, there is water ice present that would explosively evaporate as Chiron approached the Sun. A very dark comet, perhaps perceptibly red, hundreds of kilometers across, with multiple dust fountains and an immense tail, would be quite a spectacle as it passed by the

1938

The thought suggests itself that the curves [in tail streamers] may be spiral curves produced by a rotation of the nucleus of the comet whilst it is discharging the streamers.
—Arthur Stanley Eddington

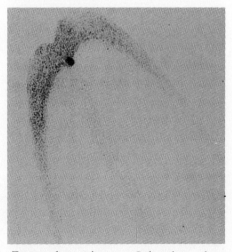

Halley's Comet in 1835 as drawn by Schwabe. Courtesy National Aeronautics and Space Administration.

Four days later, Schwabe drew the same comet but this time in a bat-like configuration. Courtesy National Aeronautics and Space Administration.

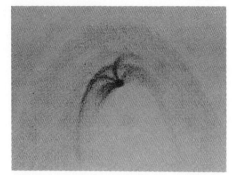

Halley's Comet again in 1910 by Ricco. Courtesy National Aeronautics and Space Administration.

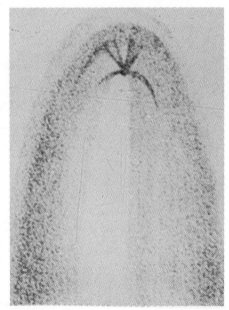

During the same apparition of Halley's Comet in 1910, Innes made this drawing. Courtesy National Aeronautics and Space Administration.

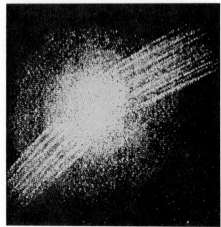

The Comet of 1823. Courtesy R. A. Lyttleton, op. cit.

Earth. There does not seem to be any historical record of such an apparition. It is doubtless a rare event.

So let us ask a more modest question: Is there any historical record of the close approach to Earth of any sort of rotating, jetting cometary nucleus? In the case of Comet Morehouse 1908 III, the inner coma with its jets was about four thousand kilometers across, a more or less typical value, roughly the size of the Moon. If such a comet came as close to the Earth as the Moon is, the coma would of course appear to us just as large as the Moon, about half a degree across. On May 11, 1983, Comet IRAS-Araki-Alcock (named after a robot and two humans) passed within five million kilometers of our planet, a dozen times the Moon's distance from the Earth. From the observed frequency with which new comets arrive in the inner solar system, it's possible to calculate how long we have to wait before a comet comes as close as the Moon. The answer is a few thousand years at most. If you're prepared to wait four or five thousand years, a cometary nucleus should pass you by at considerably closer range. Imagine the sky dominated by a dull, red irregular object, spitting out white canopies, its shimmering, curved fountains flowing into space, and all the material eventually swept back into a vast tail that extends from horizon to horizon. It would be a memorable event.

Except for an approach to the Earth from an unlikely sector of the sky, the comet would be seen by cultures all over the world. Surely, there would be a mythological framework —sometimes called a world view—into which this apparition would be fitted. People would naturally think the display held some portent or significance for them. Some cometary form should therefore have entered the art of many cultures, perhaps even in a central way. As time passed, the memory of the true events might fade, and the stories become vaguer, but the cometary form would still be a dominant motif in the art and records of the previous generations. If we saw such an apparition, and believed it was a message for us, we would not be disposed to ignore it. After thousands of years pass, the cometary symbol, whatever it was, might be wholly disconnected from its bizarre and awesome origins. In a prescientific, preliterate society, accounts of an unprecedented occurrence involving uncommon physics must necessarily, after thousands of years, take on lives of their own.

Because of extravagant claims made on meager evidence in a succession of popular books and articles over the centuries, the notion that historical catastrophes have been caused by collisions of comets with the Earth has become disreputable. But, at least from Halley's time, it has been apparent that close approaches and even impacts of comets

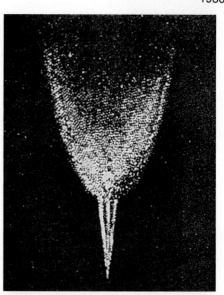

1936

with the Earth will happen if only we wait long enough. As in all scientific questions, this one comes down to the quality of the evidence. We are aware of the need to proceed with caution.

The temptation to speculate on this matter is easily understood. If comets have come close in the last few generations, then over historical time, a really big comet must have ventured very near, with an appearance far more spectacular than anything in living memory. So you scurry around in the art and literature of myth, or in the geological record, until you find something that seems to you attributable to a comet. Then you write a book. Thus, Ignatius Donnelly—congressman, lieutenant governor of Minnesota, passionate advocate of human rights and the conservation of nature—proposed in 1883 in a wildly popular work called *Ragnarok* that the thick deposits of clay all over the world fell from the skies when the Earth passed through a comet built along the lines of the orbiting sand-bank model (Chapter 6). But clays are readily produced by known geological processes, and most show no sign in their atomic or molecular makeup of extraterrestrial origin. Also, the sand-bank model of comets is now known to be invalid, although it was the expert consensus in Donnelly's time. Or, as another example, a Russian-American psychiatrist named Immanuel Velikovsky proposed that the plagues of Egypt, manna from heaven, and other snippets from ancient myth were generated by a comet that came too close to Earth. But a comet is hardly the only conceivable or the most probable explanation of such stories, and other elements of Velikovsky's hypothesis are immersed deeply in error.

Yet there are reasons to think that accurate depictions of extraordinary as well as ordinary cometary events were, at least on occasion, made by the ancients. For example, the historian Ephorus, in the fourth century B.C., reported a comet that in 371 B.C. split into two. Subsequent writers continued to quote Ephorus, although with considerable skepticism. By the time of Seneca and Pliny, there were no other examples of fragmenting comets, and yet a record that a comet had once split in two was safely transmitted to our time. Today we know—as with Biela's Comet and the sun-grazers—that fragmentation does occur. Ephorus is vindicated, and we learn that unusual cometary happenings can accurately be passed down through the millennia.

In our century, there have been extensive Newtonian calculations of the orbit of Halley's Comet, with predictions—or, rather, postdictions—of the date and position of the comet in the sky for every apparition back to 240 B.C. For every one of these calculated apparitions, convincing evidence exists in ancient chronicles that a bright comet was

The Comet of 1851. Courtesy R.A. Lyttleton, op. cit.

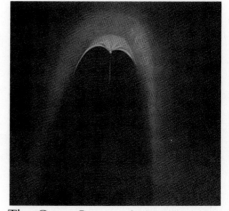

Comet Arend-Roland (1957 III) on April 25, 1957. Photograph by H. Neckel. From N.B. Richter, *The Nature of Comets*, Methuen, 1963.

The Great Comet of 1861 drawn by Warren de la Rue, on July 2, 1861. From Amédée Guillemin's *World of Comets*, Paris, 1877.

Comet Tebbutt (1861 II) as drawn by Secchi. Courtesy National Aeronautics and Space Administration.

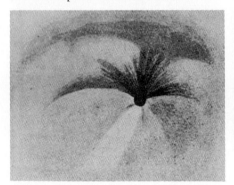

Comet Tebbutt (1861 II) as drawn by Secchi one day later. Courtesy National Aeronautics and Space Administration.

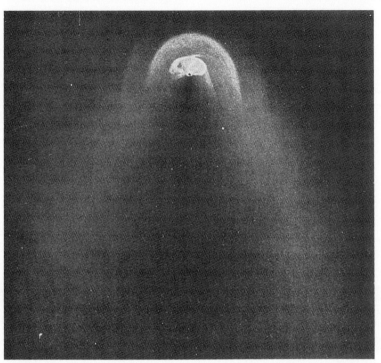

Complex, almost biological forms near the jetting nucleus of Comet Donati. Drawn at the telescope, Harvard College Observatory, October 5, 1858, by Bond. Courtesy National Aeronautics and Space Administration.

The Great Comet of 1861 drawn by Warren de la Rue, one day later. From Amédée Guillemin's *World of Comets*, Paris, 1877.

indeed in the predicted part of the sky on the date in question. In addition to improving our already high confidence in Newtonian gravitational theory, this concordance between theory and observation renews our respect for the precision and attentiveness of ancient Chinese, Korean, Japanese, Babylonian and European chroniclers. While this respect does not extend so far as to giving full credence to descriptions of comets encumbered with dragons, swords or faces (see page 18), it does suggest that in the historical depictions of the forms of comets there may be preserved for us something of the natural history of these idiosyncratic visitors from the depths of space.

Pliny noted the appearance of a comet "too brilliant to be looked at directly; it was white with silver hair and resembled a god in human form." What are we to make of this? For a comet to be so bright, it would have to be sheathed in a coma and pass close to Earth. Not impossible. The configuration of the coma can be complicated, and can suggest a human form (for example, Bond's drawing of Comet Donati, page 180, top right, which resembles a fetus). But a silver-haired comet god is certainly not widely dispersed in the myth and art of Earth.

Pliny described another kind of comet in these words: "Like a horse's mane, it has a very rapid motion, like a circle

1937

revolving on itself." The theme of rotation is occasionally connected with comets in the ancient records. Epigenes proposed that comets are born in whirlwinds, a view that Seneca dismisses for good reason, but a rotating, jetting cometary nucleus might look very much like a whirlwind. Among Seneca's objections are that whirlwinds have brief durations, and that their rotary motions would be dissipated by the speedy passage of the heavens around the Earth. Seneca did not know that the Earth turns. He says, "The shape of a whirlwind is round…therefore the fire which is enclosed [within the hypothetical cometary whirlwind] ought to be like a whirlwind. And yet the first is elongated and scattered and not at all like something round-shaped." It might just be that Seneca and Epigenes are describing different aspects of comets—Seneca the coma and tail, and Epigenes, a close look at a rapidly rotating nucleus. (Before the invention of the telescope, the comet would have to come very close to the Earth for rotation to be apparent.) We therefore ask if there is some widespread ancient symbol, associated with the sky, that indicates rotation. Very tentatively, we suggest that there is one such symbol, the swastika.

This symbol of four bent arms emanating from a common center was officially adopted and made synonymous with horror by the Nazi regime in Germany. The Nazi crimes against humanity are well documented if insufficiently reflected upon; long after their passing, the Nazis are still poisoning amity among nations. But there was a time, long before Nazis, when swastikas abounded, benign symbols known to almost every culture on the planet. If you can for a moment, try to ignore the association of the Nazis with this symbol, and consider it for itself.

In 1979, for our *Cosmos* television series, we were in India to film the celebration of the cycle of the seasons, called Pongal. We were deeply moved by the generosity and kindness of the Hindu villagers of Thanjavur, in Dravidian South India. We were a little taken aback, however, when, in chalk, they began cheerfully marking swastikas on their doorsteps. It was an ancient symbol of good fortune, they explained. Indeed it is.

In the two deepest and therefore oldest levels in Troy—dating back to 3000 B.C. in the Early Bronze Age—no signs of swastikas are found; but beginning with what he called the third or burnt city, dating back to the beginning of the second millennium B.C., Heinrich Schliemann, the discoverer of Troy, found them everywhere. Hundreds of recovered artifacts, especially spindles operated in rotary motion, were festooned with swastikas. In Tang Dynasty China, public misuse of this important symbol had reached such a state

Coggia's Comet (1874 III), drawn by Chambers. Courtesy National Aeronautics and Space Administration.

A drawing of Encke's Comet in 1871. Courtesy R. A. Lyttleton.

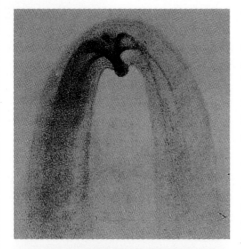

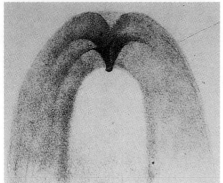

Two drawings by Wolf of Comet Daniel (1907 IV). Courtesy National Aeronautics and Space Administration.

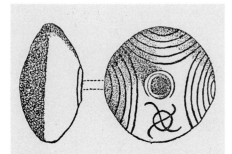

Spindle, adorned with a swastika, one of hundreds found by Schliemann at Troy. From Thomas Wilson, *The Swastika: The Earliest Known Symbol, and Its Migrations; With Observation on the Migration of Certain Industries in Prehistoric Times*, Smithsonian Institution, Washington, D.C., 1896.

that an imperial decree had to be issued forbidding the imprinting of swastikas on silk fabrics. In the west of India there are caverns which serve as Buddhist shrines; most of the rock inscriptions are preceded or followed by a swastika. The Jains—a religious community in India which, in its respect for the sanctity of all life, stands at the opposite pole from the Nazis—use the swastika "as a sign of benediction and blessing," for which reason also the Japanese once put it on their coffins. An account of Tibet in *The Times* of London in 1904 describes "a few white, straitened hovels in tiers…On the door of each is a kicking swastika in white, and over it a rude daub of ball and crescent," astronomical symbols representing Sun and Moon. On blankets, beadwork, pottery, and other artifacts, the swastika was once a common, almost typical, symbol of the indigenous peoples of North America. Thomas Wilson, the curator of the Department of Prehistoric Anthropology at the U.S. National Museum, wrote in 1896:

We know not whether it is intended as a religious symbol, a charm of blessing or good luck, or whether it is only an ornament. We do not know whether it has any hidden, mysterious, or symbolic meaning; but there it is, a prehistoric or Oriental Swastika in all its purity and simplicity, appearing in one of the mystic ceremonies of the Aborigines in the great American desert in the interior of the North American continent.

How does the same curious symbol become established in the ancient cultures of India, China, the American Southwest, Mayan Mexico, Brazil, Britain, and Turkey, among others? The swastika was in general use in Bronze Age Europe from the Arctic to the Mediterranean, spreading, in the Iron Age, to the Etruscan, Mycenaen, Trojan and Hittite civilizations. The word itself is Sanskrit:

The root word, *svasti*, literally means well-being. The *sign* swastika must have existed long before the *name* was given to it. It must have been in existence long before the Buddhist religion or the Sanskrit language…
Looking over the entire prehistoric world, we find the swastika used on small and comparatively insignificant objects…such as vases, pots, drugs, implements, tools, household goods and utensils…and infrequently on statues, altars, and the like…[It was used] in Italy on the urns in which the ashes of the dead are buried; in the Swiss lakes stamped in the pottery; in Scandinavia on the weapons, swords, et cetera, and in Scotland and Ireland on the brooches and pins; in America on the metates for grinding corn; the Brazilian women

wore it on the pottery fig leaf; the Pueblo Indian painted it on his dance rattle, while the North American Indian, at the epoch of the moundbuilding in Arkansas and Missouri, painted it in spiral form on his pottery; in Tennessee he engraved it on the shell, and in Ohio cut it in its plainest normal form out of sheets of copper...As we do not find it represented in America on aboriginal religious monuments, on ancient gods, idols, or other sacred or holy objects, we are justified in claiming that it was not here used as a religious symbol...With this preponderance in favor of the common use, it would seem that, except among the Buddhists and early Christians, and the more or less sacred ceremonies of the North American Indians, all pretense of the holy or sacred character of the swastika should be given up.

This is from Wilson's classic work on the ethnography of the swastika. Wilson was skeptical of the idea that a swastika was passed from culture to culture by diffusion:

If the sign bore among the aborigines in America the name it bore in India, Swastika, the evidence of contact and communication would be greatly strengthened. If the religion it represented in India should be found in America, the chain of evidence might be considered complete.

But this is not the case. On the other hand, Wilson argues that the swastika is by no means so simple a design that it would have arisen spontaneously the world over:

For evidence of this, I cite the fact that it is not in common use, that it is almost unknown among Christian peoples, that it is not included in any of the designs for, nor mentioned in any of the modern European or American works on decoration, nor is it known to or practiced by artists or decorators of other countries...
   The straight line, the circle, the cross, the triangle, are simple forms, easily made, and might have been invented and re-invented in every age of primitive man and in every quarter of the globe, each time being an independent invention, meaning much or little, meaning different things among different peoples or at different times among the same people; or they may have had no settled or definite meaning. But the Swastika was probably the first to be made with definite intention and a continuous or consecutive meaning, the knowledge of which passed from person to person...

This is a real puzzle: a symbol thousands of years old

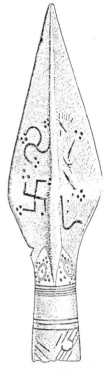

Iron age spear, recovered near Brandenburg, Germany, with symbols including a right-angled four-armed swastika (*lower left*) and an ogee three-armed swastika (*top left*). From Wilson.

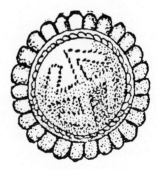

Swastika symbol on an Etruscan gold "bulla." From Wilson.

An extended ogee swastika on a tripod pottery vase from pre-Columbian Arkansas. From Wilson.

Mycenean wooden button, showing an ogee swastika flanked by two four-armed crosses. From Wilson.

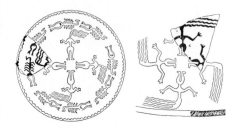

Pottery from ancient Samarra. The recovered shard, with figures shown in black, is used in an attempt to reconstruct the complete pattern. After Count Eugène Felicien Albert Goblet d'Alviella, *The Migration of Symbols*, Paris, 1891.

that neither arises spontaneously in the mind of the artist nor, primarily, passes from culture to culture. It baffled Schliemann too, who remarked, "The problem is insoluble." And perhaps it is. However, if the swastika were originally something in the skies, something that could be witnessed independently by widely separated cultures, the mystery might be solved. The symbol then would arrive in each culture from outside, and yet not be transmitted from other cultures.

In the more ancient representations of the swastika, one often sees the arms curved, not bent; this is called the ogee swastika. In thinking about the significance of the swastikas he had discovered in ancient Troy, Schliemann thought he saw an attempt to depict spin, and suggested that the direction of motion was specified by the direction of the arms, which always trailed the rotation. But as to what it was that was rotating, he offered no hypothesis.

Another dilemma running through scholarly writings on the swastika is that, on the one hand, it appears to be connected with something brilliant in the sky, and on the other hand it is clearly something separate from the Sun. To give a flavor of the often turgid scholarly debate on this aspect of the swastika, Count Goblet d'Alviella argued in 1891, as follows: The arms of the swastika "are rays in motion." The images most closely associated with it represent the Sun or the Sun gods. Sometimes the swastika alternates with representations of the Sun. From this Goblet d'Alviella deduces that the swastika means the Sun. A critical piece of evidence is a Thracian coin on which the word for day is replaced by the swastika symbol. This, he believes, is a complete identification between the swastika and "the idea of light or of the day." But, critics argue, there is no need for an additional symbol for the Sun, and the swastika in no way resembles the Sun. In some Indian coinage the swastika appears separate from but with equal prominence as the great wheel of the Sun. Impasse.

All of these difficulties seem to be resolved if there once was a bright swastika rotating in the skies of Earth, witnessed by people all over the world. Ordinarily, the notion seems so far from astronomical reality that, while it must have been briefly considered by others who have wondered about the origin of the swastika, no one proceeded further, for the simple reason that there is nothing remotely like a burning swastika now apparent in the heavens. But we need only examine the sketches and photographs of the spraying fountains in a cometary nucleus, recorded by generations of astronomers, to realize that there is here the potential for generating such a prodigy.

What we are imagining is something like this: It is early

1938

in the second millennium B.C. Perhaps Hammurabi is King in Babylon, Sesostris III rules in Egypt, or Minos in Crete. More likely it is a time not today associated with any famous personage. While all the people on Earth are going about their daily business, a rapidly spinning comet with four active streamers appears. When people look up at the comet, they are looking down on the axis of rotation. The four jets, symmetrically placed around the equator on the daylit side, generate—because of the comet's rapid rotation —curved streamers, as you can easily see in the patterns formed by a rotary garden sprinkler. For the usual representation of the swastika, observers would have seen the pinwheel spinning counterclockwise, with the arms trailing. As long as all four jets were on at once, the inhabitants of Earth would see a brilliant swastika, perhaps somewhat foreshortened, in the daylight sky.

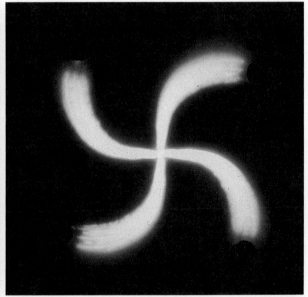

There is something anthropomorphic about the swastika. We read it as arms and legs in motion. It is one of the few symbols, not ordinarily found in nature, that is both simple and compelling. If something like it slowly materialized in your night sky amidst shrouds and fountains of dust—something self-propelled, animate, almost purposeful—you would surely find the experience noteworthy. You would speculate on its meaning, its religious significance, its portent. People would copy the symbol down so others would know about it, so that this marvel would not be forgotten. Whether you view it as an auspicious sign or as a harbinger of disaster, no one need explain to you that the thing is important.

The swastika form is not very different from the pinwheel structures seen in many comets, and brought out in short exposure photography with a large telescope. Comet Ben-

Four symmetrically placed jets in the illuminated (upper) hemisphere of a non-rotating cometary nucleus (left). If the nucleus were rapidly spinning (counterclockwise in this schematic representation), the arms would trail the rotation and something like a swastika would be produced. Diagram by Jon Lomberg/BPS.

nett 1970 II is a recent example. For Comet Bennett, at least, the color of the pinwheel was yellow, implying that the structure is in the dust. You look at these forms and perhaps you will grant at least that if enough spinning, jetting comets pass by the Earth, sooner or later there will be one that presents something like a swastika to view. But the argument alone is insufficient to convince you, or us, of a cometary origin for the symbol. For so speculative a subject, at least one piece of more direct evidence is needed.

Under these circumstances it is arresting to find, in the culture with the longest tradition of careful observations of comets, a straightforward, apparently unambiguous description of a swastika as just another comet. Such is the case of the twenty-ninth and final comet to appear in the ancient silk atlas of cometary forms that was unearthed in a Han Dynasty tomb at Mawangdui, China (Chapter 2). It dates from the third or fourth century B.C., but is an anthology of observations that must be much older. The twenty-ninth comet is called "Di-Xing," "the long-tailed pheasant star." The caption is the lengthiest of all twenty-nine, because the swastika comet alone is subject to differing interpretations. This comet is associated with change. "Appearing in spring means good harvest, in summer means drought, in autumn means flood, in winter means small battles." Of course, these auspices are fanciful, but the origin of the swastika in a cometary apparition now seems to us a real possibility. We wonder if the connection is drawn in other little-noticed artifacts of ancient cultures.

Long ago, the swastika achieved worldwide distribution, and was, unlike the usual cometary attributes, almost every-

The last seven comets shown in the Mawangdui silk from third or fourth century B.C. China. Nothing in the caption indicates that the last form, in the shape of a swastika, was considered fundamentally different from the other comets shown. Wen Wu. "Ma Wang Tui po shu 'T'ien wen ch'i hsiang tsa chan' nei jung chien shu" and "Ma Wang Tui Han ts'ao po shu chung to hui hsing t'u." Beijing: Wen wu ch'u pan she. 1978, Volume 2, pp. 1–9.

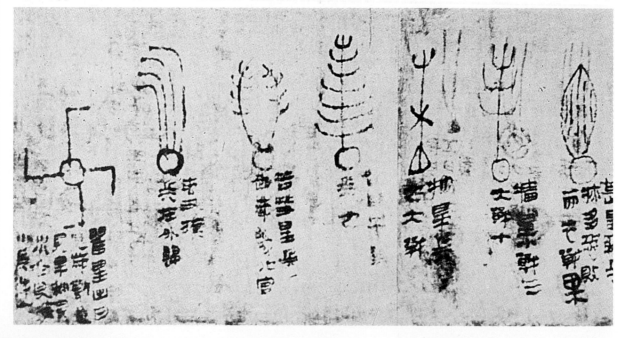

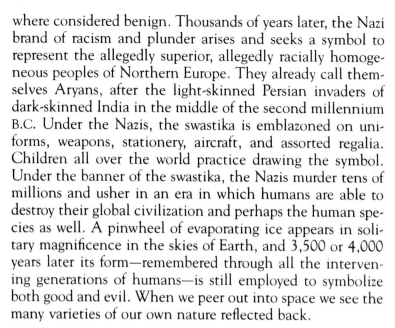

1939

where considered benign. Thousands of years later, the Nazi brand of racism and plunder arises and seeks a symbol to represent the allegedly superior, allegedly racially homogeneous peoples of Northern Europe. They already call themselves Aryans, after the light-skinned Persian invaders of dark-skinned India in the middle of the second millennium B.C. Under the Nazis, the swastika is emblazoned on uniforms, weapons, stationery, aircraft, and assorted regalia. Children all over the world practice drawing the symbol. Under the banner of the swastika, the Nazis murder tens of millions and usher in an era in which humans are able to destroy their global civilization and perhaps the human species as well. A pinwheel of evaporating ice appears in solitary magnificence in the skies of Earth, and 3,500 or 4,000 years later its form—remembered through all the intervening generations of humans—is still employed to symbolize both good and evil. When we peer out into space we see the many varieties of our own nature reflected back.

Both the form of this luminous ejection, and the direction in which it issued from the nucleus, underwent singular and capricious alterations, the different phases succeeding each other with such rapidity that on no two successive nights were the appearances alike. At one time the emitted jet was single, and confined within narrow limits of divergence from the nucleus. At others it presented a fan-shaped, or swallow-tailed form, analogous to that of a gas flame issuing from a flattened orifice; while at others again two, three, or even more jets were darted forth in different directions.

—John Herschel,
"On the 1835 Apparition
of Halley's Comet,"
Outlines of Astronomy,
London, 1858

PART II

ORIGINS AND FATES OF THE COMETS

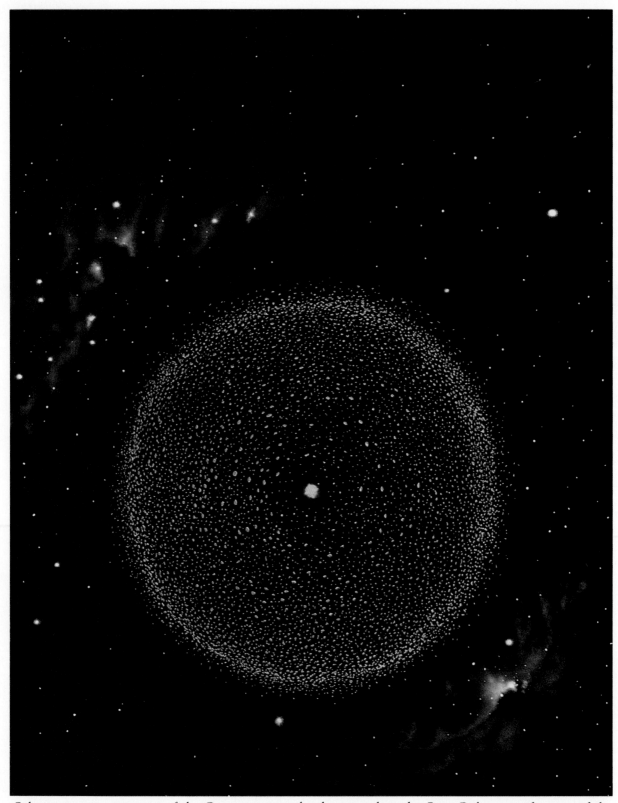

Schematic representation of the Oort cometary cloud surrounding the Sun. Only a tiny fraction of the trillion comets in the outer Oort Cloud are shown, most of them lying about one-third of the way between the Sun and the nearest star. Much farther in toward the Sun is the probably still more numerous cometary population of the inner Oort Cloud. Painting by Jon Lomberg.

Chapter XI

AT THE HEART OF A TRILLION WORLDS

How many other bodies besides these comets move in secret, never rising before the eyes of men! For god has not made all things for man.

—Seneca,
Natural Questions,
Book 7, "Comets"

We have reason to suspect that there are a great many more comets, which being at remote distances from the Sun, and being obscure and without a tail, may for that reason escape our observation.

—Edmond Halley,
Transactions of the Royal
Society of London,
Volume 24, page 882, 1706

It seems however not agreeable to the uniformity of the universe, that after a short view of the Sun [the comets] should be continually flying farther off, in that wide void beyond the planetary bounds, to creep along that dark cold region for millions of years...but that they should rather revolve about the Sun, in certain, though long periods.

—Thomas Barker,
Of the Discoveries Concerning Comets, London, 1757

THE ANCIENTS IMAGINED THE PLANETS TO BE attached to invisible machinery—transparent crystal spheres, elegantly coupled and geared. We now know that the ancients were wrong. The planets orbit the Sun guided only by the invisible hand of Newtonian gravitation. Some worlds are rock, some gas, some ice, and nowhere, from Mercury to Pluto, is there anything like a crystal sphere. But imagine ourselves leaving the solar sytem at some impossible speed, until even the orbits of the outermost planets are too small for us to see, until even the Sun is only a point of light no brighter than the brightest stars seen from Earth. Then we do encounter something like a crystal sphere, but a shattered one—a cloud of a trillion shards and fragments of ice, little worlds each the size of a city, feebly illuminated in the great dark between the stars.

We live at the heart of a trillion worlds, all of them invisible. It sounds like the teaching of some New Age sect. And we are not talking of metaphorical worlds; rather a trillion places, every one of them as distinct a world as ours is, every one gravitationally bound to the Sun, every one with a surface and an interior and on occasion even an atmosphere.

If there is a ceiling above you, step outside. Cast your eye upward. Concentrate on the smallest piece of sky you can make out. Imagine it extending in a widening wedge far out into space, to the stars. In that little patch of sky are a hundred thousand worlds or more, worlds unseen, unnamed, but in some sense known. These distant cousins of the Earth are the cometary nuclei—cold, silent, inactive, slowly tumbling in the interstellar blackness. But when they are induced to fall into our part of the solar system, they creak and rumble, begin to evaporate and jet and eventually produce the tails so admired by the inhabitants of Earth. How we know of this invisible multitude of icy worlds is one more scientific detective story that begins with Halley.

After Edmond Halley made the first inventory of cometary orbits, it was clear that many comets return infrequently, once every few centuries or even longer (see Chapter 4, page 71, bottom). At any given moment, he knew, there must be unseen long-period comets that have not lately visited the Sun. And if by chance we see comets with periods of years to centuries, perhaps there are others with periods measured in millennia or more. As one of the epigraphs to this chapter shows, Halley was prepared to believe in a large population of undiscovered comets with very long periods and high eccentricities. But he did not envision truly immense numbers of comets. When Thomas Wright drew a rosette of orbits surrounding the Sun, he was not tempted to include more than the few known comets (page 71, bottom), although he did conclude "the Comets...I

judge to be by far the most numerous Part of the Creation."

The key to the discovery of the comet cloud is the orbits of the comets we see. We must bear in mind that this constitutes only a small sample of all cometary orbits. For all we know, our sample may not even be representative of the larger population. But it is our only starting point.

An elliptical cometary orbit has a certain size. Its near point to the Sun is called its perihelion and its far point, its aphelion, terms we have been using throughout this book. The line from perihelion to aphelion, running through the Sun, is the major axis of the ellipse, and half the major axis is called the semi-major axis. The semi-major axis of the Earth's orbit is, of course, one Astronomical Unit (A.U.). Comets with small semi-major axes never leave the planetary part of the solar system, and comprise the kingdom of short-period comets. Such comets are tightly bound to the Sun's gravity; it would take a very major influence to perturb their motion significantly. But comets with large semi-major axes spend most of their time far beyond the region of the planets, and less often than once a human lifetime they make a brief foray into the inner solar system. Such long-period comets are much more loosely bound to the Sun, and are more easily perturbed. The convention is to call a comet with a period less than two hundred years a short-period comet, and one with a period longer than two hundred years a long-period comet. There is nothing magic about two hundred years; it is chosen only because this is very roughly (now a little less than) the period of modern astronomical study of the comets. So a comet like Encke's or (by this definition) Halley's are short-period comets, while one like Comet Kohoutek, which passed by the Earth in 1973 and will not return for another ten million years, is a long-period comet.

Laplace once imagined the Sun stationary in space, but surrounded by a vast population of randomly moving interstellar comets. Some would, by chance, find themselves moving very slowly with respect to the Sun; they would be attracted by its gravity, and would fall into the inner solar system. He showed (Chapter 5) that the net result in the vicinity of the Earth would be many comets on highly eccentric orbits, but bound to the Sun, and more rarely a comet on a hyperbolic orbit, dipping once into the inner solar system, never to return again. And this is just what we seem to see. Because the calculation agrees with the actual observations of comets, Laplace took it to be a confirmation of the existence of a great cloud of interstellar comets in which the Sun and planets are embedded.

But later investigators pointed out that the Sun has a proper motion of its own, at present moving at a goodly clip

A photograph of Comet Kohoutek taken on January 11, 1974. Note the extraordinary structure in the tail, especially in the lower right-hand corner. This comet had just arrived from the Oort Cloud. Courtesy the Joint Observatory for Cometary Research, Laboratory for Astronomy and Solar Physics, NASA/Goddard Space Flight Center, and New Mexico Institute of Mining and Technology.

Jan Hendrik Oort of the University of Leiden in The Netherlands. Born in 1900, Oort mapped the spiral structure of the Milky Way Galaxy, and investigated the peculiarly polarized light from the Crab Nebula, among his many noncometary accomplishments. Courtesy Yerkes Observatory.

[This] article indicates how three facts concerning the long-period comets, which hitherto were not well-understood, namely the random distribution of orbital planes and of perihelia, and the preponderance of nearly-parabolic orbits, may be considered as necessary consequences of the [stellar] perturbations acting on the comets.

—J.H. Oort,
   "The Structure of the Cloud of Comets Surrounding the Solar System, and a Hypothesis Concerning Its Origin,"
   Bulletin of the Astronomical Institute of The Netherlands,
   Volume 11, page 91, 1950

toward a point in the constellation Hercules.* When the fall of randomly-moving interstellar comets is calculated for a Sun in motion, a large number of hyperbolic comets is predicted, contrary to observation. By the late nineteenth century, Laplace's concept of a cloud of interstellar comets was firmly rejected. The resolution of this difficulty—to imagine the exterior comets loosely bound to the Sun, so that the Sun is not moving with respect to them—seems not to have been thought of until the second third of the twentieth century.

Laplace had also calculated that the short-period comets were being destroyed—by gravitational ejection from the solar system, by running into a planet now and then, or (we can now add) merely by dissipating into interplanetary space after a sufficient number of perihelion passages. If there were a vast cloud of interstellar comets, then the population of short-period comets could be resupplied by the cascade from interstellar comet to long-period comet to short-period comet, the planetary billiards we have already discussed (see Chapter 5, page 93, bottom). But if the Sun isn't sweeping up new comets from interstellar space, how is the population of old comets in the inner solar system replenished?

There are only two possibilities: Either comets are being made today somewhere in the solar system, or there is a vast repository of hidden comets that supplies a steady trickle of samples. All suggestions about how comets might be manufactured lately, in anything like sufficient numbers, have failed. That leaves only the possibility that the comets are sequestered. If they were stored nearby, we would have some sign of them. It follows that they must be stored at a great distance from the Earth (and the Sun). But where? And how many?

If we are able to see a few long-period comets, on highly eccentric orbits, plummeting into the inner solar system, might there not be much vaster numbers in slow, circular orbits, disdainful of the inner solar system, out there beyond Pluto? That might account for the nearly random orbital inclinations of the long-period comets; we can imagine them as decoupled from whatever influence confines the planets and the short-period comets to the ecliptic plane. The cloud of comets would move as Newton imagined the planets should had God not intervened at the Beginning. Such comets would be much too far from the Sun to develop comas or tails. They would be invisible from Earth.

* * *

---

*That the stars *have* a proper motion was first demonstrated by the ubiquitous Mr. Halley.

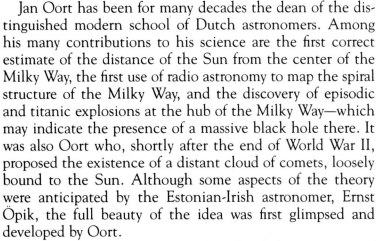

Jan Oort has been for many decades the dean of the distinguished modern school of Dutch astronomers. Among his many contributions to his science are the first correct estimate of the distance of the Sun from the center of the Milky Way, the first use of radio astronomy to map the spiral structure of the Milky Way, and the discovery of episodic and titanic explosions at the hub of the Milky Way—which may indicate the presence of a massive black hole there. It was also Oort who, shortly after the end of World War II, proposed the existence of a distant cloud of comets, loosely bound to the Sun. Although some aspects of the theory were anticipated by the Estonian-Irish astronomer, Ernst Öpik, the full beauty of the idea was first glimpsed and developed by Oort.

Just as Halley had examined the orbital characteristics of the handful of comets available to him, Oort studied nineteen long-period comets with well-determined orbits. He found a smattering of long-period comets with semi-major axes of a few thousand Astronomical Units (A.U.), and even a few tens of thousands of A.U. These are already very far from the Sun, hundreds of times further away than Pluto is. But the bulk of the comets seemed to be clustered in the vicinity of 20,000 A.U. or more. Nineteen comets is not a large sample, but it is enough. Since Oort's pioneering study in 1950, the statistics have improved, but the conclusion remains the same: Most long-period comets come to us from a region roughly 50,000 A.U. from the Sun.

Oort proposed that a vast cloud of unseen comets surrounds the Sun at these immense distances, and that all the comets we see are the deserters and refugees from that distant assemblage. Most of these comets are on fairly circular orbits, with modest eccentricity. They never enter the planetary part of the solar system, and we never see them. But occasionally a cometary nucleus leaves its fellows and plummets into the inner solar system, where it may come close enough to the Sun for us to designate it as a long-period comet; or else it might make a close pass by one or more of the major planets, and have its orbit progressively altered, so that eventually we describe it as a short-period comet.

But what induces this occasional comet, weakly held by the Sun's gravity, to enter the inner solar system? Oort calculated that the Sun, in its motion about the center of the Milky Way Galaxy, would sometimes come close enough to other stars to make a kind of gravitational flurry in the comet cloud—spilling numbers of them in all directions, including to the vicinity of the Sun. A typical comet in the Oort Cloud is circuiting the Sun at the leisurely pace of about a hundred meters per second, around 220 miles an hour. The change in speed administered by the passing star

The astronomer Ernst Julius Öpik. Öpik (1893–1985) made major contributions to our understanding of comets, asteroids, meteors, and meteorites for half a century starting in the 1920's. His work on the ablation of meteors in entering the Earth's atmosphere had unexpected later applications in the design of heat shields and nose cones of ballistic missiles and space vehicles. Reproduced from the *Irish Astronomical Journal*, Volume 10 (Special Issue), Plate I, with the permission of the editors.

Öpik had earlier investigated stellar perturbations of distant cometary orbits. His conclusions were described as follows:

"The orbital inclinations, too, are subject to change by stellar perturbations. The effect, in the long run, is to mix them up at random, and it may be that the observed indiscriminate distribution of cometary inclinations is thus to be explained."

—Henry Norris Russell,
The Solar System and Its Origin,
New York, 1935.

We may therefore take something like one hundred thousand times the Earth's distance from the Sun as a limit for the average major axis of a comet's orbit, giving a period of about ten million years. Since we see about three comets of long period per year and we may miss several, there may be (say) fifty million comets.

—Herbert Hall Turner,
   Savilian Professor of Astronomy,
   Oxford University, at a
   Friday evening discourse,
   The Royal Institution,
   London,
   Friday, February 18, 1910

This calculation ignores comets which do not enter the inner solar system. But it does indicate that many of the ideas central to the Oort Cloud hypothesis had been in the air for decades, indeed for centuries. Halley, in his 1705 paper, had noted that "the space between the Sun and the fix'd Stars is so immense that there is Room enough" for many comets of long period.

is only a few tens of centimeters a second, near the top speed your fingers can manage to walk across a tabletop. It represents a very small change in the overall speed of the comet, but it's enough to send a few of them careening down among the planets. No single gravitational impulse from a passing star causes the comets to flutter about. Rather, the accumulation of a few dozen close stellar passages has produced a jittery population of faster-moving comets, and the latest stellar encounter provides the small increment needed to drive some of them down toward the Sun or out into the interstellar medium. It's the straw that breaks the camel's back.

Even if a star were to plow straight through the comet cloud the consequences would not be spectacular. Öpik likened it to a bullet traversing a swarm of gnats: comparatively few of the gnats are scattered or destroyed; the swarm continues almost undisturbed.* And comets deep within the Oort Cloud are not ejected by stellar perturbations at all; residing closer to the Sun, they are more tightly bound to it by gravity, and cannot readily be shaken loose by a passing star, unless it comes very near the Sun.

In addition to nearby stars, today we also know that large clouds of interstellar molecules exist in our part of the Galaxy, and the solar system should plow through a few of them every billion years. Each time this happens, there will be an additional gravitational stirring within the circumsolar cometary halo, and a further flurry of comets will be sent into the inner solar system. Oort deduced a vast reservoir of comets in deep freeze and mint condition. How vast? The eighteenth-century German astronomer Johann Heinrich Lambert argued that the space around the Sun was probably filled with as many comets as could be squeezed in without frequent collision, and from this deduced that there were "at least 500 millions of comets" in the solar system. With the present size of the Oort Cloud and with the present rate of gravitational perturbation from passing stars, astronomers deduce at least a trillion cometary nuclei. The number of comets in the Oort Cloud is thus larger than the number of stars in the Milky Way Galaxy. However, this estimate is almost certainly too modest. A still larger number of comets seems likely. Recent evidence supports the idea that the Oort Cloud stretches from its periphery near 100,000 A.U. from the Sun continuously inward, reaching almost to the orbit of Pluto.

Such comets are too far out for encounters with the planets to change their orbits, and too close in for the usual pass-

---

*The star would drill a hole through the Oort Cloud about 1,000 A.U. across that would gradually be repaired.

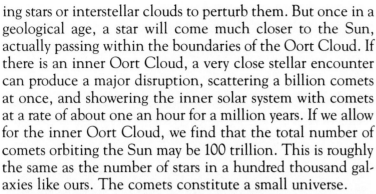

ing stars or interstellar clouds to perturb them. But once in a geological age, a star will come much closer to the Sun, actually passing within the boundaries of the Oort Cloud. If there is an inner Oort Cloud, a very close stellar encounter can produce a major disruption, scattering a billion comets at once, and showering the inner solar system with comets at a rate of about one an hour for a million years. If we allow for the inner Oort Cloud, we find that the total number of comets orbiting the Sun may be 100 trillion. This is roughly the same as the number of stars in a hundred thousand galaxies like ours. The comets constitute a small universe.

These numbers, even the mere trillion comets in the outer cloud, are so staggering that they invite disbelief. "The chief difficulty about this hypothesis," wrote a highly respected American astronomer known for his openness to new ideas, "is that it demands the existence of an enormous number of comets of large perihelion distance." Well, exactly.

What is the scale of the classical Oort Cloud? A hundred thousand A.U. is a little less than two light-years, roughly halfway between the Sun and the nearest star. If we were standing on the comet, we would be so far from the Sun that we would see it as it was almost two years ago. The typical period for a comet in the Oort Cloud to circle the Sun is a few million years. Since the age of the solar system is about 4.6 *billion* years, a typical such comet has made a thousand revolutions of the Sun. A year on the comet is a million times longer than a year on the Earth, and there the words of the Psalmist are literally correct: "For a thousand years in Thy sight are but as yesterday when it is past, and as a watch in the night."*

With a trillion or more comets in the Oort Cloud, you might think that comets are closer together there than anywhere else in the solar system—perhaps huddling together, far from the Sun, like one of Doré's illustrations of the souls of the dead. But the number of comets is more than balanced by the immensity of the space they occupy, and the typical distance from one comet to another out there is about 20 A.U., roughly the same as the distance from the Earth to the planet Uranus. The greatest concentration of comets in the solar system happens, it seems, to be in the very innermost parts, where by lucky chance we happen to live. The three or four long-period comets that each year are passing through together with all the short-period comets comprise, so far as we know, the greatest density of comets anywhere between here and the nearest star system, Alpha Centauri.

---

*365 days/year × 1000 years = 365,000 : 1, crudely 1,000,000 : 1.

Like the slow-moving, dimly illuminated comets of the Oort Cloud, the souls of the slothful huddle together far from the Sun. (From "The Multitude of the Slothful" by Gustav Doré, after Dante.)

The *Voyager 2* spacecraft was launched in 1977 on an unusual high energy trajectory so that it would reach the planet Uranus in 1986 and Neptune in 1989. If our present spacecraft could get out to the Oort Cloud, it would take us a decade or more to fly from comet to comet. But we cannot soon get to the Oort Cloud. The *Voyager* spacecraft—the fastest ever launched by the human species—took nine years to go from Earth to Uranus; it will take them 10,000 years to reach the main comet repository. The comets themselves take a few million years to fall from the solar system frontiers to the vicinity of the Earth. The Oort Cloud is very far away.

Altogether, how much do a trillion comets weigh? If every one of them is about a kilometer across, then the total mass of all the comets in the outer Oort Cloud is about the same as the present mass of the Earth. Put another way, if you took the Earth and divided it up into little kilometer-sized chunks, you would have in numbers and size (but not in composition) some approximation of the current population of the outer Oort Cloud. If, as seems likely, the typical comet is a little bigger, or if you include the inner Oort Cloud, then the total mass of the cloud will be considerably more.

The short-period comets tend to orbit the Sun in the same plane as the planets, in the ecliptic or zodiacal plane. They also tend to go around the Sun in the same direction as the other planets. The long-period comets, on the other hand, show a chaotic mix of every orbital inclination, and are as likely to be going around the Sun clockwise as counterclockwise. Newton thought that the chaos of the long-period comets was to be expected in a universe in which only gravitation was calling the shots; while the orderly regularity of the short-period comets was a sign of divine intervention at the Beginning. But Laplace showed (Chapter 5) that the orbital characteristics of long-period comets could be transformed into those of short-period comets during gravitational capture by Jupiter, and religious opinion changed. As late as 1835 a lecturer at The Royal Institution was able to conclude that the orbital inclinations and eccentricities of comets do "not depend upon physical laws, but upon the will of the Creator." However, the will of the Creator may be difficult for mere humans to recognize.

Oort argued that no matter what the original inclinations of the cometary orbits in the Oort Cloud, stellar perturbations were more than adequate to redistribute the orbital planes. Even if all the comets in the Oort Cloud were once in the same plane as the planets, by now passing stars would have randomized the inclinations—also producing an equal number of long-period comets in retrograde as in prograde revolution. Any information on the original distribution of

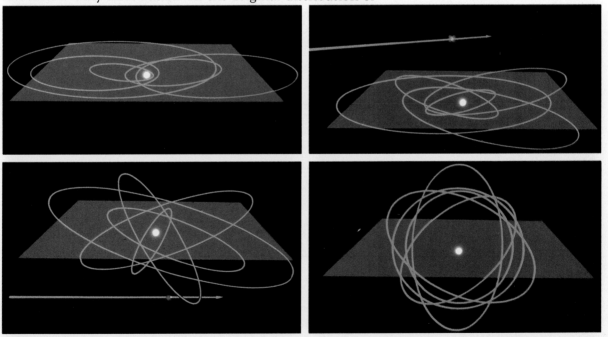

Initially ordered cometary orbits are made random by the gravitational influence of passing stars. The central luminosity is of course the Sun. The blue plane is the ecliptic or zodiacal plane in which lie the orbits of the planets and the short-period comets. Imagine (upper left) that comets in the Oort Cloud and long-period comets were originally restricted to this plane. As, through the age of the solar system (*top right, bottom left*), stars pass by, the cometary orbits are lifted out of the plane, until (*lower right*) the cometary orbits have random inclinations and are more likely to be out of the plane than in it. The gravitation influence of these passing stars serves also to propel an occasional comet from the Oort Cloud into the inner solar system. Diagram by Jon Lomberg/BPS.

cometary orbits has by now been lost through multiple stellar encounters. From order, chaos.

A typical comet in the Oort Cloud has existed for billions of years at a temperature only a few degrees above absolute zero. There are no collisions, no heating of the comet, no outgassing. It is very quiet in the Oort Cloud. There is a flux of galactic cosmic rays which slowly intrudes into the top meter or two of the comet. Each cosmic ray leaves a trail of broken chemical bonds behind it. As the fragments slowly reassemble themselves in the frigid and solid surface, new molecules are generated. If there is initial methane or carbon monoxide ice, the net result over the age of the solar system will be the generation of a considerable abundance of complex organic molecules, but only in the outer shell of the comet. If you took a very slow stop-motion film of the cometary surface (one frame every million years, say), the ices would slowly become darker and redder due to the complex organic molecules being synthesized. Suppose the comet then is nudged into the inner solar system. During a single perihelion passage, all of the processed ices would be vaporized to space—the work of a billion years undone in a month. The underlying ices revealed after the blow-off of the top meter will be close to their pristine form—pure white ice, if that is how the comet formed—or with a reddish primordial taint from the organic molecules of the interstellar medium that went into the comet in the first place. In any case, under the thin surface layer worked by cosmic rays is material virtually untouched since the beginning of the solar system.

Why is the outer boundary of the Oort Cloud a hundred thousand Astronomical Units away? Does the population of comets in the Oort Cloud fall off slowly thereafter, perhaps bumping into the Oort Cloud of some other star? We might imagine the two cometary halos interdigitating, interpenetrating—the individual comets mingling, but widely separated—while remaining loyal to the stars around which they were born. Eventually, they pass on. But we now know that the Oort Cloud cannot extend very far beyond a hundred thousand A.U. The Soviet astrophysicist G. A. Chebotarev has shown that the massive center of the Milky Way Galaxy, thirty thousand light years from the Earth and Sun, is adequate to loosen the feeble gravitational grip of the Sun on any comet more distant than about two hundred thousand A.U. A small part of the mass of the center of the Galaxy is probably provided by a black hole that resides there. Without the black hole, the Oort Cloud would be slightly bigger.

Thus the Oort Cloud connects familiar events in this backwater of the inner solar system not only with the nearby stars but also with the center of the Galaxy, so far

A postage stamp of the United Nations connecting a comet and a galaxy.

away that, through the telescope, we see it as it was 30,000 years ago. The comets that rush into our part of the solar system are from a skittish population made excitable by the passage of stars and nebulae. If the solar system were isolated from the rest of the Galaxy, we would never know that comets exist—because then the passing stars and interstellar clouds would be unavailable to shake the Oort Cloud occasionally and send some comets fluttering into the inner solar system.

And the number of comets that arrive down here (as well as the external boundaries of the Oort Cloud from which they come) may be determined in small measure by a black hole at the galactic center—an object undreamt of only a few decades ago. The comets are unexpectedly and profoundly connected with the Milky Way, a conclusion appropriate enough for Jan Oort, who perhaps more than any other person in the twentieth century has revolutionized our knowledge of the Galaxy.

The scope of Oort's idea is remarkable. To account for the handful of new comets that appear in our skies each year, a vast mind-numbing multitude of invisible comets living far beyond the orbit of Pluto is postulated. The idea explains what we know about comets in an elegant way that no other theory even approaches. The trillions of comets are now widely accepted by astronomers all over the world and called, properly, the Oort Cloud. Many scientific papers are written each year about the Oort Cloud, its properties, its origin, its evolution. Yet there is not yet a shred of direct observational evidence for its existence. We are not yet able to poke around in the Oort Cloud. No spacecraft has voyaged to count the comets there. It will be a while before any do. There is one recent measurement that might conceivably be relevant, the discovery of distant wispy aggregations of matter by the *Infrared Astronomy Satellite;* the contention that this is the structure of the Oort Cloud is, however, wildly controversial.

But with the refinement of our scientific instruments, and the development of space missions to go far beyond Pluto, our chances of observing Oort Cloud comets will improve. There will be some day in the future of our species —provided we are not so foolish as to destroy ourselves first —when we will directly measure the population of the Oort Cloud, designate and characterize each of the large comets there, plot their orbits into the future, and perhaps make plans for their utilization. We do not know how long it will be until some representation of the Oort Cloud like that in the frontispiece to this chapter will be accumulated from real data. We wish the astronomers of that distant epoch well, and in our mind's eye share with them the joy that they will surely take in those great discoveries to come.

An early view of the connection between comets and the Milky Way Galaxy. This cartoon by Olaf Gulbransson shows the chief of police of Berlin—who had lately suppressed public demonstrations with excessive force—admonishing Halley's Comet for traveling through the Milky Way. The Milky Way ("Milk Street" in German), he scolds, is no place for a demonstration. From *Simplicissimus,* April 4, 1910. Courtesy Ruth S. Frietag, Library of Congress.

A snapshot from the origin of the solar system, almost five billion years ago. The flattened disk surrounding the young Sun is the solar nebula, in which enormous numbers of kilometer-sized objects of snow and rock are condensing. Some of them (only a few are shown here) develop tails and can certainly be described as comets. It is from the solar nebula that all the planets, moons, asteroids, and comets in the solar system today are thought to derive. Painting by Jon Lomberg.

Chapter XII

MEMENTOS OF CREATION

Out of whose womb came the ice?

—Job 38:29

It often happens that comets arise. These...are not any of the stars that were made in the beginning, but are formed at the same time by Divine command and again dissolved.

—John of Damascus,
*Exact Exposition of
the Orthodox Faith,*
eighth century

HUMANS WOKE UP ONE MORNING, LOOKED AROUND and discovered that our star, the Sun, has a retinue of other worlds. Not one of them is just like the Earth. The inner, nearby, so-called terrestrial planets (Mercury, Venus, Earth, and Mars) travel on nearly circular orbits in the same direction around the Sun as the Sun's own rotation.* They have thin atmospheres. Their silicate mantles conceal metal cores.

Beyond them are the asteroids, thousands of small airless irregular objects—some rocky, some metallic, and some rich in dark, complex organic matter. The biggest are hundreds of kilometers across, but many more are a kilometer in size or smaller. They also move in prograde motion around the Sun, closely confined to the plane within which all the other planets perform their circumnavigations. Most of the asteroids are in nearly circular orbits between the paths of Mars and Jupiter. A few have more elliptical orbits that carry them inside the orbits of Mars, the Earth, Venus, or even Mercury.

Farther out are the giant gas planets, Jupiter, Saturn, Uranus, and Neptune, made mostly of hydrogen and helium, but with smaller amounts of water, ammonia, and methane. They vary in mass from about 15 to about 300 Earths, and appear to have small interior cores of rock and metal. Orbiting these so-called jovian planets are dozens of moons—each constructed from some different proportion of rock, ice, and organics.

In this realm and far beyond are trillions of icy comets, most of them in somewhat elliptical orbits enormously distant from the planets; half are in prograde revolution, half in retrograde. Their orbits are at random inclinations. A much

Rotation and revolution in the solar system. One of the regularities of planetary motion is that the planets tend to rotate on their axes and revolve about the Sun in the same direction as the Sun spins (center)—suggesting that the Sun and planets all condensed out of the same flattened spinning cloud of gas and dust. The rotation of Venus and Uranus do not follow this regularity, and must be due to other, perhaps catastrophic, later events. Diagram by Jon Lomberg/BPS.

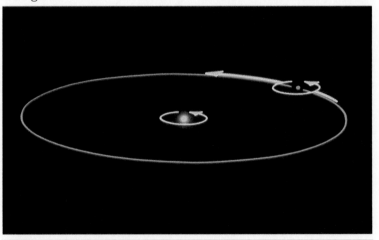

*This direction of rotary motion—counterclockwise if you were looking down on the solar system from high above the North Pole—is called direct or prograde. Motion in the opposite direction is called retrograde.

1946

smaller number of comets have highly eccentric orbits that take them near the inner terrestrial planets, where still fewer short-period comets can be found. Almost all of them are orbiting in a prograde direction and fairly close to the ecliptic plane. Now where did this profusion of worlds come from? Why the regularities in their motions?

On first hearing where babies come from, many of us express a certain skepticism. Often what is being described does not correspond to the listeners' own observations. Compared to storks, angels, and cabbage patches, the machinery seems unwieldy and improbable. But no matter how implausible the story may seem at first, there is a reasonable consensus among knowledgeable individuals on the matter. Direct experience eventually converts even the most committed skeptic.

In the study of cometary or planetary origins, exotic processes over immense distances and in remote epochs are routinely invoked. If we are skeptical of an astronomical theory of our origins, we cannot poll the opinions of reliable astronomers on nearby worlds. There is no one to tell us the cosmic version of the birds and the bees. We must figure it out for ourselves. Astronomers have a potential advantage, though: if they are curious about the stars, they can examine billions of them. Even a very unlikely process can be wit-

The rings of Saturn as observed by the *Voyager 1* spacecraft. Saturn is just out of the picture in the upper right-hand corner. The minor color differences between these many flat rings have been greatly enhanced by computer processing. The large dark gap in the rings, about three-quarters of the apparent distance to the periphery in this picture, is the Cassini Division. For Kant and Laplace, the rings of Saturn served as a model for the structure of the solar nebula that surrounded the Sun at the time the planets were being formed. Courtesy National Aeronautics and Space Administration.

Rings of other planets. Saturn has an elegant ring system composed of enormous numbers of fine icy particles orbiting the planet in its equatorial plane (see page 208). The rings of Saturn have been known since the eighteenth century. But recently it has been discovered that other giant planets also have equatorial ring systems. On this page are two false color images of the rings of Jupiter, taken by the *Voyager 2* spacecraft and contrast-enhanced by Mark Showalter of Cornell University. On the opposite page is the earliest unambiguous image (although still of very poor resolution) of the rings of the planet Uranus, taken in the infrared, where the rings are relatively bright and the planet is relatively dark; it was taken at the 200-inch Hale telescope at Palomar Observatory on May 26, 1983, by Philip Nicholson of Cornell University and Keith Matthews of the California Institute of Technology. Courtesy Philip Nicholson.

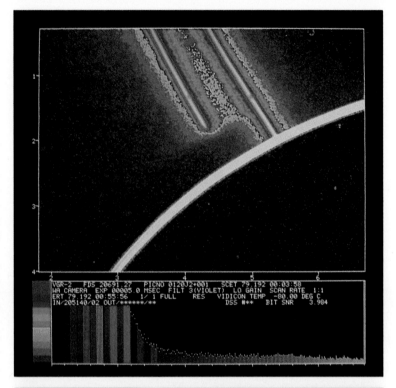

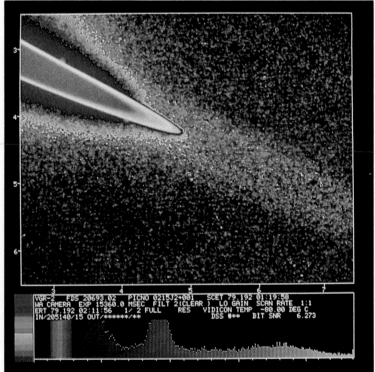

nessed if enough cases are examined. Failing that, astronomers are reduced to the fundamentals—what is already known about the universe, and the principles of physics.

Something like the modern view of the origin of the solar system was first proposed by two extraordinary thinkers

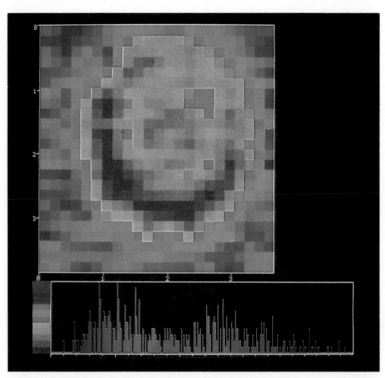

whom we have met already in these pages—Immanuel Kant and Pierre Simon, Marquis de Laplace. They had a distinct advantage over their predecessors: In the eighteenth century, observational astronomy was making remarkable progress. Kant and Laplace were intrigued by the structure of the rings of Saturn as had been revealed through the discoveries of Galileo, Huygens, and their successors. Here was a planet with a flat disk of particles surrounding it in its equatorial plane. Did the Sun once have a much larger ring system, from which the planets somehow condensed?

Kant had accepted and further developed Thomas Wright's insight and believed that the Milky Way is a thin plate of stars, one of which is the Sun. Astronomers were discovering strange flattened luminous forms (see page 77) in the night sky—spiral nebulae they were later called. (Nebula, plural nebulae, is a Latin word meaning cloud.) Nature seemed to have a propensity for flattening extended systems whether they are made of dust or stars.

The Kant-Laplace hypothesis for the origin of the solar system involves the interplay of rotation and gravitation. Imagine some irregular cloud of interstellar matter made of gas and dust and destined to form the solar system. All such clouds known today exhibit slow rotation. If the cloud is sufficiently massive, the random molecular motions are overwhelmed by self-gravity—the mutual attraction of the atoms and grains in the cloud for one another. The cloud then begins to contract, the distant provinces falling inward; the density of the cloud increases as a fixed amount

However arbitrary the system of the planets may be, there exists between them some very remarkable relations, which may throw light on their origin; considering them with attention, we are astonished to see all the planets move around the Sun from west to east, and nearly in the same plane, all the satellites moving around their respective planets in the same direction, and nearly in the same plane with the planets. Lastly, the Sun, the planets, and those satellites in which a motion of rotation have been observed, turn on their own axis, in the same direction, and nearly in the same plane as their motion…A phenomenon so extraordinary is not the effect of chance, it indicates an universal cause…

> —Laplace, "On the Regularity of the Solar System," *System of the World,* Part 1, Chapter 6, 1799

To this he added two more regularities, the near circularity of the orbits of the planets, and the great eccentricity and random inclinations of the orbits of the (long-period) comets.

of matter squeezes itself into progressively smaller volumes. As it contracts, the cloud spins faster, for the same reason that a pirouetting ice skater does as she brings her arms in. (The experiment can also be done with a small person, seated on a rotating piano stool, holding a brick in each outstretched hand, and then rapidly drawing them in. This demonstration must be performed with caution.) The principle of physics involved is called the conservation of angular momentum, and can be derived from Newton's laws of motion.

But as the gas and dust and occasional condensations that make up the cloud spin faster around their common axis of rotation, they experience an increasing reluctance to continue falling inward, sometimes called centrifugal force (centrifugal meaning fleeing the center). A pail of water on a rope whirled sufficiently fast around your head does not spill—at least not until you stop whirling. The centrifugal force balances gravity.

There is a category of amusement park ride in which a hollow cylinder is made to spin rapidly, with a crowd of laughing, screaming people—poised somewhere between terror and delight—glued by centrifugal force to the rotating interior surface. When the cylinder stops spinning, the people come tumbling down off the walls.

The contracting cloud also will experience centrifugal force, which will slow it down and eventually stop the contraction—but only in the plane of rotation. If you are standing on a small lump of matter falling toward the center of the cloud but along the axis of rotation, rather than in the equatorial plane, you do not feel any centrifugal force. The result is that matter in the equatorial plane stops collapsing, while matter along the axis continues to fall in. As a result, an initially irregular cloud in time becomes a flattened disk. The further the disk collapses, the more rapidly it rotates, and the denser it becomes at the very center. The collapse stops, or at least slows, when the disk is spinning so fast that matter spews off at the periphery.

The Kant-Laplace hypothesis proposes that, long ago, an irregular, rotating interstellar cloud collapsed in this manner, with the central condensation forming the Sun. There is no doubt today that interstellar matter compressed to the density and temperature of the Sun will initiate thermonuclear reactions and begin to shine like a star. But it was a daring hypothesis for the eighteenth century. Other, smaller nearby condensations, Kant and Laplace proposed, formed the planets, each sweeping out a wide swath of adjacent debris as it grew in size. The result would be a regular spacing of the newly-formed planets, something like the layout of the solar system today. Still smaller condensations near the planets would form their moons. The general idea behind the Kant-Laplace hypothesis is more important than the precise details: The solar system, they proposed, *evolved* —from a very different primordial state and with no outside intervention, natural or supernatural.

Because the word nebula means cloud, and out of analogy with the spiral nebulae (which are, of course, of much larger, galactic dimensions), the contracting cloud that formed the Sun and the planets is traditionally called the solar nebula. Today we know a much larger variety of flat rotating clouds around the nearer stars. They are called accretion disks.

Laplace suggested that during the formation of the solar system, the Sun's atmosphere once extended far out into space, perhaps in consequence of an enormous explosion in the Sun, like Tycho's supernova of 1572, produced by a star in the constellation Cassiopeia. Or perhaps it was the residuum of the original solar nebula. Laplace's interstellar comets were, he imagined, falling in toward the Sun. The material in the solar nebula slowed comets down in the inner solar system, altered their orbits and induced them to impact the Sun. The drag of the solar nebula cleaned the inner solar system of comets with nearly circular orbits, but left the comets at much greater distances unaffected. Through gravitational perturbations by the jovian planets, an occasional comet is induced to visit the inner solar system. The idea is remarkable in several respects. It indicates a kind of natural selection in the physical world well before Darwin; it proposes that there were once many more objects in the solar system than there are now; and it hypothesizes a large repository of comets beyond the most distant planet known.

Why then were the planets not similarly disturbed, and induced to collide with the Sun? Laplace proposed that the planets had formed by successive condensations in the early solar nebula. A tube of empty space, centered around the orbit of each new planet, was formed as the planet, growing at the expense of adjacent material, swept its surroundings clean of nebular debris. Perhaps he toyed with the idea that many dark breaks should exist in the rings of Saturn if there are moons among the rings. However, he urged caution in accepting his hypothesis which he offered "with that distrust which every thing ought to inspire that is not the result of observation or calculation." Probably because of his flirtation with a possible interstellar origin of comets, it seems not to have occurred to him that comets as well as planets might condense out of the solar nebula.

That the rotation and revolution of the satellites are in the same direction as the rotation of their planets; that the planets rotate in the same sense that they revolve; and

1948

The hypothesis of Kant and Laplace is seen to be one of the happiest ideas in science, which at first astounds us, and then connects us in all directions with other discoveries...

—"On the Origin of the Planetary System," lecture delivered by H. Helmholtz in Heidelberg and Cologne, 1871, published in *Popular Lectures on Scientific Subjects* by H. Helmholtz, New York, 1881

What may be another solar system in the late stages of formation is shown in this picture of the star Beta Pictoris, masked at the telescope (and at the center of the picture) in order to bring out the fainter disk of debris (the yellow and pink diagonal feature). This faint disk, seen nearly edge-on, may be no more than a few hundred million years old. The picture was taken by Bradford A. Smith (University of Arizona) and Richard J. Terrile (Jet Propulsion Laboratory) at the 2.5-meter telescope of the Las Campanas Observatory, Chile. Courtesy Bradford A. Smith and Richard J. Terrile.

that the orbits of the planets are close to circular, while the comets have highly eccentric orbits, all followed naturally if everything (including or excluding the comets) had condensed out of the same rotating and collapsing cloud.

For both Kant and Laplace, the nebular hypothesis explained the regularities of the solar system as the end result of an evolution of worlds. Both believed that other stars were surrounded by planetary systems evolved from their own accretion disks. In the last few years, groundbased and spaceborne observations have confirmed that many nearby stars are surrounded by accretion disks. The initial discovery was made by a space observatory called IRAS, the *Infrared Astronomy Satellite*, a joint Anglo-Dutch-American

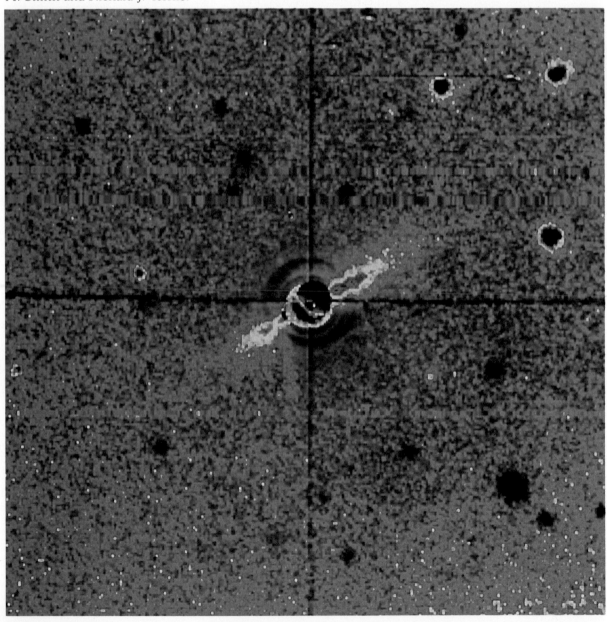

endeavor. Vega is one of the brightest stars in the night sky, only twenty-six light-years distant, and it was a real surprise to discover that this well-studied star is surrounded by a previously unsuspected disk of debris. It showed up as an extended source of infrared radiation centered on Vega. Now, Vega is a star considerably younger than the Sun. To find an accretion disk around Vega strongly suggests that most, perhaps even all, ordinary stars are surrounded by such a disk during and immediately after their time of formation. Something eventually tidies up the disk—perhaps a combination of radiation pressure, the stellar wind, and planetary formation. But it takes time. And in that time, additional bodies may be condensing out of the nebula.

IRAS also provided infrared evidence of an accretion disk around a star called Beta Pictoris, among many others. Soon after, Bradford Smith of the University of Arizona and Richard Terrile of the Jet Propulsion Laboratory attached a special highly sensitive camera, developed for a forthcoming space observatory, to a groundbased telescope, and were able to photograph the Beta Pictoris accretion disk in ordinary visible light. The disk extends at least 400 Astronomical Units from the central star (here blocked out, so its radiation will not overwhelm and wash out the much more feeble light reflected off the disk). If this were a picture of the Sun in its early history, the accretion disk would extend much further from the Sun than does the orbit of the farthest known planet (some 30 to 40 A.U. out). Smith and Terrile deduce a relative absence of debris in the interior of the disk, and suggest that this region has already been swept up by the condensation of planets—that are, however, much too small to be seen directly. Many other accretion disks around adolescent stars have been sighted recently. Accretion disks have also been found around infant stars formed only a million years ago.

Thus, it now seems that the Kant-Laplace hypothesis is in its fundamentals verified, and by a technology that would have delighted both of them. The Sun, the planets, and their moons all condensed out of the same rotating and collapsing disk of gas and dust. This is why all the planets revolve in the same plane in which the Sun rotates. Newton's view that the regularity of the planetary motions is direct evidence of divine intervention has been superseded by another more evolutionary view—still determined, to be sure, by the laws of nature, which, if we wish, we may still attribute to a god or gods. But when queried—by Napoleon, of all people—about why his account of the origin and history of the solar system made no mention of God whatever, Laplace replied, "Sire, I have no need of that hypothesis."

Vast collections of interstellar gas and dust: the Eagle Nebula, also known as M-16, 5,500 light-years away, and the Eta Carinae Nebula (opposite page), beyond the Southern Hemisphere constellation Carina, 4,200 light-years distant. Behind the bright foreground stars we see a vast concentration of interstellar matter so opaque that the stars behind are invisible. In such dark spherical and filamentous clouds solar nebulas and new planetary systems form—as ours did almost 5 billion years ago. The red color is mainly due to hydrogen gas in these interstellar clouds. Courtesy National Optical Astronomy Observatories.

Let us now follow a modern rendition of the Kant-Laplace hypothesis, in which we pay special attention to the origin and evolution of the comets. From direct spectroscopic evidence, we know the interstellar gas to be composed mainly of hydrogen and helium, although it is rich in many other materials, including complex organic molecules (Chapter 8, pages 150–151). Besides the gas, the other chief constituent of interstellar space is an enormous number of motes of dust. One of them, placed on a table before you, would be entirely invisible. They are, typically, a tenth of a micron across. But concentrate enormous numbers of them over hundreds or thousands of light-years, and you can have enough dust to blot out the stars behind them. The chemistry of the grains can also be inferred. Most seem to be made of ices, silicates, and organics—very roughly in equal proportions. Since this mix of gas and grains makes up the interstellar clouds everywhere in the Milky Way, it must also have constituted the early collapsing solar nebula. Since interstellar space ordinarily holds much more gas than grains, this should have been true for the solar nebula as well.

As the nebula contracts and its density increases, collisions of grains with one another become more frequent. In part because of the organic and icy content of these grains, when they collide they tend to stick. Big grains annex smaller ones. But all this does not go on in the dark. The primitive Sun has begun brightly shining. In the outer parts of the disk, it is still sufficiently cold that exotic ices such as methane or carbon monoxide are perfectly stable in the growing condensations of matter. But in the very inner solar

1949

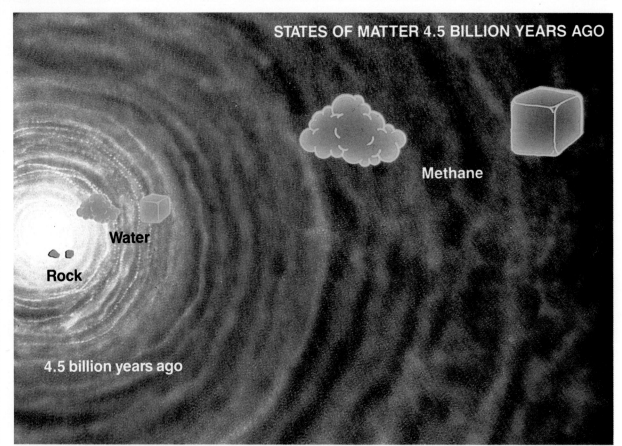

STATES OF MATTER 4.5 BILLION YEARS AGO

Methane

Water

Rock

4.5 billion years ago

Schematic illustration of condensation in the early solar nebula. We are looking down on the nebula, its bright hot interior shown at left. The farther from the forming Sun, the cooler it is. Three different materials are shown: methane (CH_4) in green; water (H_2O) in blue; and silicates (SiO_2) in orange. The cubes indicate solids, the clouds vapors. In the outermost portions of the solar nebula, methane condensed as a solid; further in, it was present only as a gas. Water is solid as ice well into the interior of the solar system, and silicates survived as solids almost to the surface of the Sun. Thus rocky planets were formed in the interior of the solar system and icy bodies farther out. Diagram by Jon Lomberg/BPS.

system, it is too hot even for water ice. There the ices on the grains evaporate and dissipate, and what survives is made mainly of silicates. You have to carry a rock very close to the Sun, only a few million kilometers away, for it to boil. As a result of all this, the chemistry in the inner solar system must have been very different from the chemistry in the outer solar system—silicates predominating inside and ices with a few percent of organics outside.

According to several calculations, a vast number of kilometer-sized objects should have accumulated throughout the nebula—silicate-rich ones on the inside, ice-rich ones on the outside. These objects should have been generated, not primarily through grain-by-grain collisions, but by a fundamental gravitational instability in the solar nebula, in which objects a few kilometers in size were quickly and preferentially formed.

Both dust and gas gravitationally collapsed to form the disk. But it takes a great deal of gravity to hold on to so lightweight and thus fast-moving a molecule as hydrogen. In the middle part of the nebula, the kilometer-sized lumps collided and grew into still larger objects, until a few aggregations of matter were able to retain the cold gas around them. This was the evolutionary line to the jovian planets. The original accretion core is smothered in a vast sphere of

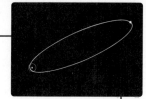

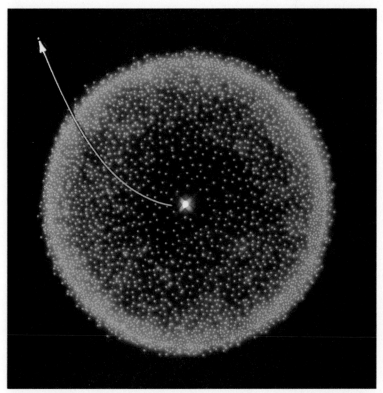

In the early history of the solar system, the Oort Cloud is formed at least in part by ejection of comets from the vicinity of the giant planets. (The solar nebula and the giant planets are much too small and far away in this picture to be seen.) Cometary bodies in the vicinity of Uranus and Neptune were gravitationally ejected to help form the Oort Cloud, shown here in blue. Cometary bodies in the vicinity of Jupiter and Saturn were ejected from the solar system altogether (arrow). Diagram by Jon Lomberg/BPS.

gas. In the warmer inner solar system the grains, divested of their ice, grew more slowly, and the temperatures were higher—both effects making it more difficult for the gas to be captured by the growing rocky spheres. This was the evolutionary line to the terrestrial planets.

Big objects would sweep up smaller ones on adjacent orbits. Because the relative velocities were low, the two bodies would tend to collide softly and merge. Eventually, a few large objects were produced, in orbits that never intersect. These became the planets. There is a kind of collisional natural selection at work here. You start out with a large number of growing objects in chaotic orbits, but through a process of collision and growth and only occasionally the shattering of worlds, the solar system becomes regularized, simplified. The number of worlds steadily declines, from trillions to thousands to dozens. If you look at the planets today you find them decorously spaced, their orbits by and large almost perfectly circular; except for the case of Pluto,* planets give each other wide berth. Those early bodies on highly eccentric orbits were in danger; very soon they would collide with a world or be ejected from the solar system. Eventually, the only planets left were those that had by chance developed on orbits that quarantined

---

*And for this reason some scientists have supposed that Pluto is an escaped satellite from Neptune, returning periodically to the scene of the crime.

It appears reasonable to suppose that during the phase of the formation of the major planets there were besides the planets many *small* condensations, which may have had comet-like structures…many of these stray bodies must ultimately have been absorbed by the planets (or proto-planets), but it would be unavoidable that during this period of conglomeration a number of the small condensations suffered large perturbations, bringing them into orbits of considerable eccentricity. The perturbations at subsequent perihelion passages would then start a process of diffusion…in the outward direction…The minute stellar perturbations required to bring the perihelion outside the region of the major planets do not…contribute appreciably to the escape of comets. Their main effect is to…"catch" them semipermanently in the large cloud surrounding the solar system…Once the stellar perturbations had deviated the orbit so that they no longer pass through the inner part of the solar system, evaporation would practically cease, and the comets could easily retain their volatile components up to the present time.

—J.H. Oort, "Empirical Data on the Origin of Comets," Chapter 20, in G.P. Kuiper and B.M. Middlehurst, eds., *The Solar System*, Volume 4, Chicago, 1965

them from their neighbors. It is just as well for us that they did; frequent world-shattering collisions are probably not good for the development of life.

The planets so formed would be orbiting the Sun in the sorts of orbits we recognize for the planets today. While no one has been able to prove that exactly nine planets should form—and not, say, six or forty-three, the entire question of the ultimate number of planets being a matter of collision statistics—the general picture is very successful, and explains not only the orbits, but also the overall chemical differences between the terrestrial and the jovian planets that we observe today.

If you picture the collapsing disk of gas and dust flattening and spinning faster, the grain collisions generating larger and larger objects, the eventual formation of kilometer-sized objects that collide and grow still further, a question may occur to you: What happened to all those kilometer-sized objects? Are there any left? Were they all swept up as the growing planets ran into them, or might some of them still exist somewhere, unchanged from the epoch in which the solar system was formed?

When we calculate the fate of that original population of small worlds, we discover that gravitational interactions with the newly-finished jovian planets would have ejected multitudes of kilometer-sized worlds into the outermost gravitational frontier of the solar system, like an automatic pitching machine throwing baseballs into the bleachers once a minute for a hundred million years. This is how the Oort Cloud was generated. There is a population of primitive bodies that four and a half billion years ago were sequestered so far from the Sun that no vaporization, no collisions, nothing at all could transform them. They are the stuff from which the solar system was formed, and they are waiting for us in the Oort Cloud. Even a single comet newly arrived from the solar system frontiers is the answer to an astronomer's dream.

In the 1960's, V. S. Safronov, a Soviet specialist in the early history of the solar system, and in 1981, J. A. Fernandez, a young Uruguayan astronomer, and W. H. Ip, in Germany, showed that if primitive cometary bodies (those kilometer-sized objects) were formed in the vicinity of Jupiter and Saturn, gravitational perturbations by these massive planets would eject them out of the solar system altogether. But if these protocomets were born in the vicinity of the less massive planets, Uranus and Neptune, their gravitational influence would tend to eject the cometary bodies into the Oort Cloud, but not out of the solar system. So if these primitive icy and rocky worlds had condensed throughout the solar system, most would have been used up

1951

in making planets and in being ejected into interstellar space. But trillions at least would have been relocated to the Oort Cloud.

If the protocomets had been formed in the vicinity of Jupiter, exotic ices would not have survived; and, if formed still closer to the Sun, even ordinary water ice would not be retained. Thus, two independent considerations—making the primitive comets out of the right stuff, and ejecting them into the right orbits—point to an origin in the rough vicinity of Uranus and Neptune.

Comets, it seems, were formed ultimately out of interstellar grains within the solar nebula, just a little before the moons and planets formed, some 4.6 billion years ago. Many comets collided with each other, forming larger bodies, and sacrificing themselves so the planets would be made. Our planet also seems to have been formed from such objects, poor in ice, rich in rock. Many other comets were gravitationally ejected from the solar system altogether as, sooner or later, they made close passes by the jovian planets, and especially Jupiter. But the calculations show very clearly that a substantial population of the original comets must have been ejected to the far reaches of the solar system, where the random gravitational shuffling of passing stars would have forced them into more circular, randomly inclined orbits. Not all would have been ejected out to the very periphery of the solar system, and the calculations predict a substantial population of comets on near-circular orbits from hundreds to tens of thousands of Astronomical Units out—a population of comets fairly impervious to gravitational disturbances by passing stars. Comets may also have formed at these distances in the accretion disk of the solar nebula. It is therefore possible that a typical denizen of the inner Oort Cloud has never been seen by astronomers on Earth. It is entirely plausible that much bigger comets than those several kilometers across were ejected into the Oort Cloud. But there are far fewer of these, and much more rarely will we see one redirected into our small but well-lit volume of space.

If this currently popular picture is correct, a typical short-period comet is an aggregate of interstellar matter condensed during the origin of the solar system almost five billion years ago, ejected by the newly-formed planets, Uranus or Neptune, to the solar system frontiers, its orbit there circularized by gravitational encounters with passing stars. A few billion years later, the cumulative gravitational influence of further stars and interstellar clouds drives the comet back into the planetary part of the solar system, where close planetary encounters—this time especially with Jupiter—reduce the large elliptical orbit into the more mod-

Evolution of comets and the solar system. On this page, within an interstellar cloud of gas and dust (*top left*) a solar nebula (*top right*) forms and evolves. The collision of kilometer-sized condensations and subsequent collisions and ejections eventually produces a solar system with a small number of planets (*bottom*) and a large number of comets (*not shown*). Painting by Kazuaki Iwasaki.

The evolution of the residual small debris—objects 100 kilometers in size and smaller—is shown schematically on the facing page. A cometary nucleus (a) transported to the inner solar system becomes a comet with a developed tail (d). Some comets collide with moons and planets, producing craters (f), which are also produced by rocky asteroidal objects that have survived their own collisional history (c). Other cometary fates include ejection back into the Oort Cloud or into interstellar space (b); evaporation of the topmost layers of ice, converting the cometary nucleus into an apparently rocky asteroid (e); and nearly complete vaporization of ices, so that the comet falls to pieces and becomes a meteor shower (g). Cometary debris falling on the Earth may have made some contribution to the origin of life (h). Opposite page, (a) Michael Carroll; (b) Jon Lomberg; (c) Don Dixon; (d) Dennis di Cicco and International Halley Watch; (e) Don Dixon; (f) photograph by Michael Collins, courtesy National Aeronautics and Space Administration; (g) Don Dixon; (h) Jon Lomberg.

1952

Iapetus, one of the outer moons of Saturn, as photographed by *Voyager 2*. The Earth aside, as we go farther from the Sun evidence for organic matter in the solar system seems to increase. The bright material on the surface of Iapetus is known to be mainly water ice; the dark material is thought to be a stain of complex organic matter, the origin of which is still under debate. Courtesy National Aeronautics and Space Administration.

est dimensions of a short-period comet. The homecoming has been long delayed, and the solar system has changed considerably in the interim.

Like everything else of which we have evidence, comets are born, live for a time, and then die—or at least disappear. After a short-period comet wends its long way from Uranus to the Oort Cloud to Jupiter, what happens next? Each time the comet passes through the inner solar system, it runs a gauntlet of risks. Eventually, the odds catch up. For some, every perihelion passage shrinks the comet by a meter, until in the end there is hardly anything left. Other comets collide with something in their path, transmogrify themselves into a different world, or set out for the interstellar void. These several fates have, it turns out, deep consequences for the planets today and, it seems likely, for ourselves. We trace these connections in the following chapters. A graphic summary of the evolution of the solar system and the birth and deaths of the comets is provided on pages 218 and 219.

Because of these various ways for a comet to die, there would after a while be no short-period comets at all—if new comets were not being recruited into the inner solar system by Jupiter's gravity. As on Earth, the places of those recently departed are taken by a new, exuberant, but relatively inexperienced generation.

The comets are way stations in the evolution of planets.

They have seen much. As remnants of the forming solar system they can tell us much. Both the comets and the planets are formed of interstellar materials. The difference is that the planets have been enormously reworked, physically and chemically, since the beginning of the solar system, while the comets of the Oort Cloud remain comparatively unscathed by the ravages of time. This is the principal motivation for the dawning age of spacecraft exploration of comets. When we study the comets, we study our own beginnings.

1953

Meteor radiant in Leo. In a meteor shower, most of the meteors (represented here as arrows, indicating the direction of their fall through the Earth's atmosphere) seem to originate in a particular spot in the sky—here within the Constellation Leo, whose stars are indicated in blue. The shower occurs when the Earth, at that moment carried by its orbit around the Sun in the direction of the Constellation Leo, is plowing through a cloud of meteoric debris. Other meteor showers have radiants in different parts of the sky and occur on different days of the year. Diagram by Jon Lomberg/BPS.

Chapter XIII

THE GHOSTS OF COMETS PAST

Oft you shall see the stars, when wind is near,
Shoot headlong from the sky, and through the night
Leave in their wake long whitening seas of flame.

—Virgil, *Georgics*, Book 1

...Through the tranquil and pure evening skies, a sudden fire shoots from
time to time, moving the eyes which
were steady, and seems to be a star
which changes place, save that from
the region whence it was kindled
nothing is lost, and it lasts a short
while.

—Dante Alighieri,
Paradise, Canto XV

Daimachus, in his *Treatise on Religion*...says that...for seventy-five days
continually, there was seen in the
heavens a vast fiery body, as if it had
been a flaming cloud, not resting, but
carried about with several intricate
and broken movements, so that the
flaming pieces, which were broken off
by this commotion and running
about, were carried in all directions,
shining as falling stars do. But when it
afterwards came down to the ground
in this district, and the people of the
place recovering from their fear and
astonishment came together, there
was no fire to be seen, neither any sign
of it; there was only a stone lying, big
indeed, but which bore no proportion, to speak of, to that fiery compass. It is manifest that Daimachus
needs to have indulgent hearers.

—Plutarch
(ca. 46 to ca. 120 A.D.),
Lysander

YOU HEAR ABOUT THEM BEFORE YOU EVER SEE THEM.
"Did you see the falling stars last night?" the adults ask one
another. You wonder how old you'll have to be before they'll
let you stay up late and see for yourself. Falling stars. The
words conjure up something tragic—a star, high and proud
for years and then, for some secret transgression, humbled
before our eyes. Cosmic justice.

When you're older, ten say, and you finally get to see your
first falling star, you're pleased; it's like fireworks. An "ooh"
or an "ah" might even escape your lips. You try to remember
back, and see whether any star that used to be up there is
now missing. But it's hard to do. There are so many faint
stars. Even so, with stars falling every night, you wonder why
there are any stars left at all.

The phrase "falling star" has a model of the universe built
into it: Stars can be loosened from their attachments to the
firmament and plummet to Earth. Stars, therefore, must be
little things. You see the point of light streaking from the
horizon toward the zenith, brightening and then fading.
Where did it go? It seems more natural when it falls the
other way, from zenith toward horizon. Then you might be
tempted to look for a fallen star—by taking a brisk walk over
to where the trail of light seemed to end. In the next county,
maybe. At age ten, what do you imagine you might find?
Something with five points, wrapped in silver paper, glisten-
ing in the snow? Perhaps.

"Falling star" and "shooting star" are of course not scientific terms. The proper word is meteor. A meteor is an object
that as it falls through the Earth's atmosphere produces a
trail of light. It looks something like a magnesium flare.
Unlike the comets with which they are sometimes confused, meteors, of course, do streak across the sky. They are
very tiny. If one falls in solitary splendor, it is called a sporadic meteor. If it is a member of a group of meteors all falling on the same night from the same part of the sky, it is a
part of a meteor shower. The brightest meteors are called
fireballs, and the brightest fireballs can outshine the Moon
or even the Sun. Its brilliant head is teardrop-shaped and
accompanied by a streak of light and scattered sparks. After
a daylight fireball falls, a trail of dark smoke is sometimes
seen. By definition, a meteor does not strike the ground. If
you follow the streak toward the horizon, rush over to the
next county and indeed recover a rock newly fallen from the
sky, it is not a meteor, but a meteorite. The suffix suggests
that meteorites come from meteors, and thus are smaller
than meteors—which, in general, is untrue. That big hole in
the ground in Arizona is called Meteor Crater—but the
object that dug the crater was much too big to be a meteor.
It was a meteorite. Meteorites, though, are pieces of other

1954

worlds. It is a reasonable guess that meteors are also.

"The meteors whisper greenly overhead" is a lovely and evocative line by the American writer Loren Eiseley. In fact, meteors whisper only to themselves. They streak too silently through the upper air to be heard down here. Like comets, and children in Victorian dramas, they are seen but never heard. Meteorites—fragments chipped off the asteroids or the Moon or Mars or extinct comets—*can* be heard; they and the fireballs produce on occasion a sonic boom or a deep rumbling roar, the only sounds made by another celestial body that Earthbound ears unaided have ever heard.

Prescientific cultures held meteors—like comets—to be portents of something or other, usually impending evil. More rarely, other explanations are proffered. In West Africa, there are traditions which hold meteors and meteorites to be a kind of solar excrement. Like the notion of a meteor as a star falling to Earth, this teaching of the Atakpame people has more than a grain of truth to it, as we shall see. The Herero call them "buzzing stones," which doubtless reflects some direct experience with a meteorite fall. In other traditions, the meteors are the souls of the dead returning to Earth to be reborn; or a thunder ax; or the heralds of Mbomvei, the supreme being. The Jukun hold that a meteor is a gift of food carried from one star to another—an extraterrestrial take-out service. According to the Kamba, a meteor is a kind of royal ensign, signifying that the beings who live on the stars are this day visiting the Earth. In Islamic Africa, a shooting star used to be described as a dagger thrown by angels to thwart those genies who aspire to ascend to heaven. But in general—in Africa and throughout the world—meteors, like comets, were considered portents of pestilence, disaster, witchcraft, and death. Perhaps because meteors are so much more com-

The night was fine; the moon was absent. The meteors were distinguished not only by their enormous multitude, but by their intrinsic magnificence. I shall never forget that night....For the next two or three hours, we witnessed a spectacle which can never fade from my memory. The shooting stars gradually increased in number until sometimes several were seen at once. Sometimes they swept over our heads, sometimes to the right, sometimes to the left, but they all diverged from the East. As the night wore on, the constellation Leo ascended above the horizon, and then the remarkable character of the shower was disclosed. All the tracks of the meteors radiated from Leo. Sometimes a meteor appeared to come almost directly towards us, and then its path was so foreshortened that it had hardly any appreciable length, and looked like an ordinary fixed star swelling into brilliancy and then as rapidly vanishing. Occasionally luminous trains would linger on for many minutes after the meteor had flashed across, but the great majority of the trains in this shower were evanescent. It would be impossible to say how many thousands of meteors were seen, each one of which was bright enough to have elicited a note of admiration on any ordinary night.

—Account of the Leonid meteor shower of November 13 and 14, 1866. Robert Ball, *The Story of the Heavens*, London, 1900

A meteor falls to Earth. From Amédée Guillemin, *The Heavens*, Paris, 1868.

mon than comets—if you're patient, you can see them fall on any clear dark night—the evils attributed to meteors tend to be more humdrum than those for which the comets are held accountable.

From early times, the Chinese (of course) kept meticulous records of meteor showers, with careful attention to color. The earliest known description, "Stars fell like a shower," is in the *Ch'un Ch'iu*, the *Spring and Autumn Annals*, describing an event that happened on March 23, 687 B.C. In ancient and medieval Europe, careful record-keeping on meteors was virtually unknown, but the occasional bright fireball was considered noteworthy. For example, in the year 1000—which had been widely advertised to be the date of the end of the world—a contemporary account relates:

> The heavens opened, and a kind of flaming torch fell upon the Earth, leaving behind a long track of light like a path of a flash of lightning. Its brightness was so great that it frightened not only those who were in the fields, but even those who were in their houses. As this

An eyewitness rendition of the Great Leonid Meteor Shower of November 13, 1833. After Fletcher Watson, *Between the Planets*, Harvard University Press, 1956.

1955

opening in the sky slowly closed, men saw with horror the figure of a dragon, whose feet were blue and whose head seemed to grow larger and larger.

Modern scientific interest in meteors was prodded by the following account by the German scientist Alexander von Humboldt in Camana, Venezuela, on the night of November 11, 1799:

From half after two in the morning, the most extraordinary luminous meteors were seen in the direction of the East. M. Bonpland, who had risen to enjoy the freshness of the air, perceived them first. Thousands of bolides and falling stars succeeded each other during the space of four hours...From the first appearance of the phenomenon, there was not in the firmament a space equal in extent to three diameters of the Moon, which was not filled every instant with bolides and falling stars...All these meteors left luminous traces from five to ten degrees in length...the phosphorescence of these traces, or luminous bands, lasted seven or eight seconds. Many of the falling stars had a very distinct nucleus, as large as the disk of Jupiter, from which darted sparks of vivid light.... The light of these meteors was white, and not reddish ...The phenomenon ceased by degrees after four o'clock, and the bolides and falling stars became less frequent; but we still distinguished some to North-East by their whitish light, and the rapidity of their movement, a quarter of an hour after sunrise.

Humboldt found that many observers, including some in Europe, had seen the same marvel that night, and concluded that a meteor shower was something that happened over a large area of the Earth and high up in the atmosphere. But this straightforward conclusion posed numerous problems:

Whatever may be the origin of these luminous meteors, it is difficult to conceive an instantaneous inflammation taking place in a region where there is less air than in the vacuum of our air pumps...Does the periodical recurrence of this great [meteor shower] phenomenon depend upon the state of the atmosphere? Or upon something which the atmosphere receives from without, while the Earth advances in the ecliptic? Of all this we are still as ignorant as mankind were in the days of Anaxagoras.

It is indeed a remarkable fact that meteor showers recur nearly on the same date of every year—on August 11, say, or December 14—although it might take a day or two to reach

A meteor shower as depicted in Camille Flammarion's *Astronomie Populaire*, Paris, 1880. The date is listed as November 27, 1872; if 27 is a misprint for 17, this would be the Leonid Shower.

maximum intensity and a day or two to fall off. You can go out on a clear night, and count the numbers of bright meteors. With an intense shower you might have to count dozens every second. With a pedestrian sort of shower, you might have to wait a minute between bright meteors. You note where the meteors appear to be coming from. They are not randomly distributed over the sky, but instead are concentrated toward a particular place in a particular constellation (frontispiece, this chapter). This focus, from which the meteors appear to be radiating, is called the radiant. As the constellation rises and sets, the radiant moves with it. So the meteor shower is characterized by the constellation from which it seems to emerge. The Leonid shower, around November 17, pours out of the constellation Leo; the Perseids, around August 11, out of the constellation Perseus, and so on. Now how, the nineteenth-century astronomers asked themselves, could meteor showers know what day of the year it was, and how did they manage this conjurer's trick of pouring out of a tiny spot of sky, which rises and sets with the stars?

In a meteor shower, the trails all diverge from the same point in space. This is a circumstance familiar to those who must drive in snowstorms at night. As the automobile plunges through the falling snowflakes, they speed by on all sides, apparently diverging from a fixed point just ahead*— or to the side if a steady wind is blowing. Similarly, a meteor

---

*This would also be true for hypothetical interstellar comets attracted by the Sun's gravity; they would most often seem to be arriving from the direction in space toward which the Sun is moving, in the constellation Hercules. The fact that comets do not show a preference for Hercules is one of several powerful arguments against Laplace's proposal that the long-period comets come to us directly from beyond the solar system.

1956

shower occurs when the Earth, in its swift annual motion about the Sun, passes through a swarm of fine particles (see the illustrations on page 230).

What is happening was described vividly at the turn of the last century by the British astronomer Robert Ball:

> Let us imagine a swarm of small objects roaming through space. Think of a shoal of herrings in the ocean, extending over many square miles, and containing countless myriads of individuals…The shoal of shooting stars is perhaps much more numerous than the herrings…The shooting stars are, however, not very close together; they are, on an average, probably some few miles apart. The actual bulk of the shoal is therefore prodigious; and its dimensions are to be measured by hundreds of thousands of miles.*

To the question of how there come to be shoals of meteors passing by the Earth, an answer was supplied when meteor showers were discovered to be associated with comets—especially defunct comets such as Comet Biela/Gambart (Chapter 5). There is now good evidence that fine particles abound in the dust tails of comets, and in the coma. Those too large to be removed by radiation pressure and the solar wind continue to move around the Sun as separate microplanets, sharing essentially the same orbit as the parent comet. Because of jetting and the successive shedding of comas, some particles will be moving a little faster and some a little slower than the cometary nucleus. As a result, the particles have slightly different periods around the Sun. A particle somewhat slower than the rest will trail a little bit after one perihelion passage, more after two, and so on. Eventually, the particles would be spread out over the entire length of the orbit, dispersing a little in lateral dimensions as well. Also, small particles feel radiation forces to which big particles are immune, and the gravity of nearby planets also serves to spread the stream of particles. A swarm of small meteors can orbit the Sun shoulder to

---

*Ball reasoned something like this: Suppose the peak intensity of a meteor shower lasts about an hour. If we knew how fast the Earth is moving through the swarm, we could then calculate the distance from the front of the swarm to the back. The Earth is 150 million kilometers from the Sun (93 million miles) and so its entire circular orbit is $2\pi \times 1.5 \times 10^8$ kilometers around. It takes 365 days for the Earth to make this circuit. So the Earth travels at about $2\pi \times 1.5 \times 10^8$ kilometers/365 days = 2.5 million kilometers per day. This is the speed of the Earth around the Sun. Thus, if the meteor shower lasts a day, the swarm must be 2.5 million kilometers across. Since generally the peak intensity lasts only a few hours, the inner core of a typical meteor swarm is a few hundred thousand kilometers across.

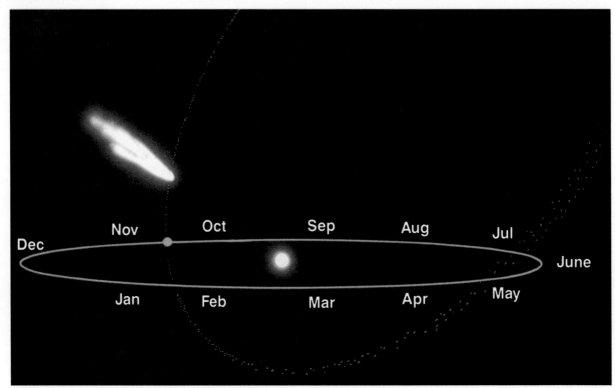

A still active comet, through successive perihelion passages, has begun to break up, and fine cometary debris now litters its entire orbit. The particles have also spread out from the width of the original cometary orbit. The comet itself passes well behind the Earth's orbit in this diagram, but the debris intercepts the Earth's orbit. Thus, on a given date of each year, the Earth runs into the cometary debris, producing a meteor shower.

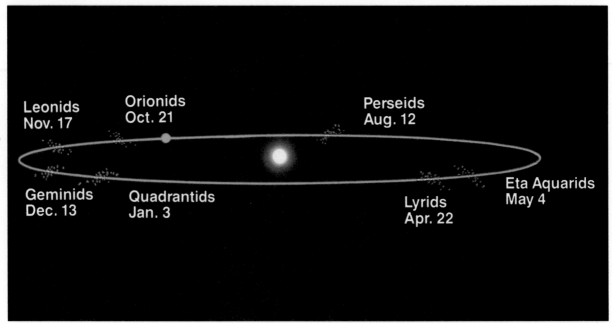

The Earth does not pass through most meteor streams left by decaying comets. In the few cases where the debris in the cometary orbit intersects the Earth's orbit, meteor showers occur. Here small segments of the meteor-strewn orbits of several comets are shown where they intercept the orbit of the Earth. Each intersection, for the debris from a different comet, corresponds to a particular date in the year. The Earth itself is shown in blue. Diagrams by Jon Lomberg/BPS.

1957

shoulder in the same orbit with virtually no collisions, like skydivers falling together from an airplane. As a comet slowly dissipates, it fills its orbit with debris.

In most cases, the shoals of meteors in a cometary orbit are entirely invisible to astronomers on Earth. But occasionally, by chance, the orbit of the comet intersects the orbit of the Earth. Since the Earth is in a specific segment of its orbit on every day of the calendar, the resulting meteor shower must occur on a specific day of the annual calendar (page 230, bottom). The main meteor streams today are the Perseids, the Leonids, the Orionids, and the Geminids. Also, the Taurids, associated with the periodic Comet Encke. There are two meteor streams associated with Comet Halley (see Appendix 3). Every comet as it dies litters its orbit with debris. Some orbits cross the Earth's. Meteor showers are the ghosts of comets past and passing.

Ball continued,

> The meteors cannot choose their own track, like the shoal of herrings, for they are compelled to follow the route which is prescribed to them by the Sun. Each one pursues its own ellipse in complete independence of its neighbors…We never see them till the Earth catches them. Every 33 years the Earth makes a haul of these meteors just as successfully as the fishermen among the herrings, and in much the same way, for while the fisherman spreads his net in which the fishes meet their doom, so the Earth has an atmosphere wherein the meteors perish. We are told that there is no fear of the [supply of] herrings becoming exhausted, for those the fishermen catch are as nothing compared to the profusion in which they abound in ocean. We may say the same with regard to the meteors.

Just as Tycho Brahe was able to determine that comets passed far beyond the Moon from observations of parallax (Chapter 2), so can photographic observations of the same meteor, made by separated cameras on the Earth, determine how high up the meteor trail is. The typical answer is somewhere in the vicinity of 100 kilometers (60 miles). At this altitude the atmospheric pressure is 0.00003 percent that at the surface of the Earth; Humboldt was entirely right to wonder how so brilliant a trail could be made by passing through air so thin.

Imagine you're holding a small piece of comet (perhaps only a grain of dust) so it is hovering far above the Earth, and then you let it go. Of course it speeds up as it falls. By the time it reaches the Earth's upper atmosphere it will be moving at escape velocity, about 11 kilometers (or 7 miles) a second. Ordinarily a cometary fragment will be traveling very fast relative to the Earth before it is attracted by the

Earth's gravity, and will therefore hit it at a higher speed. A particle on a highly eccentric orbit, moving in a retrograde direction so it collides with the dawn hemisphere, can be traveling as fast as 72 kilometers (or 45 miles) per second. By contrast, the typical muzzle velocity of a rifle bullet is about one kilometer per second. When a meteor enters the Earth's atmosphere it is heated to incandescence by friction with the thin air at an altitude of around 100 kilometers. Spectroscopy of meteors shows spectral lines of iron, magnesium, silicon, and a range of other elements that make up ordinary rocks. Organics and even some ices may be in them before they enter the Earth's atmosphere, but at least the stony component is still there as the meteors burn and die.

Big particles the size of your fist, or larger, streak through the Earth's atmosphere, heat up by friction with the air, and char, melt, or burn off a thin crust. This process protects the interior of the meteorite just as the ablation shield on a spacecraft protects astronauts during reentry. The remainder of the object survives passage through the Earth's atmosphere and reaches the ground, where, when they are recovered, they are called meteorites.

Very small particles are able to radiate their heat away quickly—because their area is so large compared to their mass—and so they never melt. They simply slow down at around 100 kilometers altitude, where they contribute to the rare "noctilucent" (literally, bright at night) clouds that reflect sunlight back to the nighttime Earth. Gently, they fall for years through a barrage of bombarding air molecules which tend to keep them suspended. Eventually, they enter the circulation of the lower atmosphere and are carried down to the Earth's surface. They are called micrometeorites.

Particles of intermediate size are too small to survive the charring of even a thin crust, and too large to radiate all their frictional heat away and float gently down. They burn up entirely during entry. These are the meteors.

With radar techniques and a network of fast cameras, it is possible to measure how the meteor decelerates when it enters the Earth's atmosphere, and how it flares. Mass and density information is derived. Typical visual meteors are millimeter sized—no bigger than a small pea. A fireball as bright as the brightest star typically weighs less than a hundred grams (about an ounce). A porous object presents more area than does a denser object with the same mass, and so it decelerates differently. In this way, densities are determined for the meteors of different showers. For example, the Geminids are meteors with the density of ordinary terrestrial materials, around one gram for every cubic centimeter. Most meteors, however, have much lower densities. Comet

A swarm of cometary debris falling into the Earth's atmosphere. Very small particles float down as a fine mist, large chunks survive entry and arrive slightly scorched at the Earth's surface, and particles of intermediate size burn up as meteors. Painting by Don Dixon.

Giacobini-Zinner is the presumed source of the Draconid (or Giacobinid) meteor shower—once formidable, now not very impressive—approximately on the evening of each October 9. The density of the Draconids is very low, possibly as small as 0.01 grams per cubic centimeter. To maintain such fragile structures these meteoroids could not have been violently ejected from their parent body. So there seem to be at least two populations of objects that enter the Earth's atmosphere—one very much like the meteorites that are recovered, and the other, very fragile porous structures that are unlike macroscopic objects made on Earth. There is probably a continuous range of bodies of intermediate density as well.

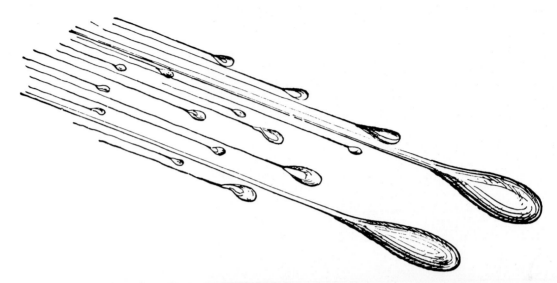

Meteors melting and burning up as they enter the high atmosphere of the Earth. From H. W. Warren, *Recreations in Astronomy*, 1879.

With modern photographic and radar techniques it is possible to calculate the speed and direction from which a meteor comes, and then to extract from this data the meteor's orbit. Sporadic meteors—those unconnected with meteor showers—tend to lie in the ecliptic plane and go around the Sun in the same direction that the planets and the short-period comets do. The shower meteors, on the other hand, have much larger eccentricities and orbital inclinations, although some certainly lie in the ecliptic plane. Here again, a kind of natural selection has occurred. Short-period comets with small inclinations produce meteor streams that tend to be disrupted by Jupiter's gravity, and fragments in a range of orbits are produced—some of which become sporadic meteors. But comets with large orbital inclinations tend to avoid Jupiter, so the meteor streams they produce tend to remain intact for much longer. Of more than 40,000 meteor trails studied, not a single one has an orbit that originated beyond the solar system.

Only three meteorites have ever been recovered that are fireball remnants: Lost City, Pribram, and Innisfree, each named after the locale near which it was recovered. They tend to be ordinary stony meteorites that derive from the asteroid belt, interior to Jupiter's orbit.

Those bright meteors that arrive from beyond Jupiter have been given the stirring name of transjovian fireballs. As determined from their entry characteristics, they are as fragile as the most delicate meteor known. If a sizable piece of such material were gently placed on the table before you it would collapse under its own weight. It is possible that the spaces in these silicate dust balls were originally on the parent comet, filled with ices and organics.

* * *

1958

A lithograph by Honoré Daumier, entitled "Comet of 1847." However, the woman's urgency suggests that the visitor is streaking across the sky, in which case it would be not a comet but a meteor. Since we do not know beforehand the precise position of each meteor, it is much more sensible to use a very wide angle telescope rather than the spyglass variety depicted here. From the collection of D. K. Yeomans.

On May 5, 1960, the Soviet Premier, Nikita Khrushchev, made a brief announcement that an American airplane, four days earlier, had violated Soviet airspace and had been shot down. A little later on May 5, the newly formed U.S. National Aeronautics and Space Administration issued a related bulletin that revealed, for the first time to many, that there was a new kind of aircraft called the U-2. It could fly very high. A research plane of this sort, in the course of studying "meteorological conditions at high altitudes," NASA said, had been missing over the "mountainous and rugged" area of Lake Van, Turkey. Perhaps it had accidentally strayed over the border into the U.S.S.R. The airplane was described variously as a "flying test bed," and a "flying weather laboratory." It had been used by NASA, among other purposes, to determine "the concentration of certain elements in the atmosphere." State Department spokesman Lincoln White said "there was absolutely no deliberate attempt to violate Soviet airspace, and never has been." Three days later, Mr. Khrushchev announced that the U-2 airplane had been shot down near Sverdlovsk, more than 2,000 kilometers from Lake Van. Its pilot, Francis Gary Powers, and some of his photographic equipment had been recovered in one piece. Powers admitted to working for the U.S. Central Intelligence Agency on one of a series of daring espionage overflights of the Soviet Union. The plane, it turned out, had no equipment for analyzing the atmosphere, but Khrushchev displayed some of the equipment that Powers acknowledged bringing with him—including a pistol with silencer; a poison capsule to swallow if captured; 7,500 rubles in Soviet currency; French, West German, and Italian money; three watches; and "seven gold rings for ladies." Khrushchev asked,

The Lockheed U-2 aircraft, designed for surveillance overflights for the Central Intelligence Agency, and now rededicated to scientific research by the National Aeronautics and Space Administration, flies into the stratosphere to capture cometary debris. Courtesy Ames Research Center, National Aeronautics and Space Administration.

Why was all this necessary in the upper layers of the atmosphere? Or maybe the pilot was to have flown still higher to Mars, and was going to lead the martian ladies astray. You see how thoroughly American pilots are equipped before setting off on a flight to take samples of air in the upper layers of the atmosphere.

The American response, making no reference to the NASA cover story of only a few days earlier, much less Mr. White's assurances, talked about the importance of obtaining data about Soviet military capability because of the closed nature of Soviet society. The incident is historically important for a number of reasons, including its scuttling of the scheduled summit conference between Khrushchev and the American President, Dwight Eisenhower. It also threatened the integrity of the spanking new National Aeronautics and Space Administration, which at Eisenhower's explicit request was to have been dedicated to peaceful research in science and technology.

The U-2 was designed for photographic reconnaissance from an altitude too high to shoot down conveniently. But as Soviet surface-to-air missiles improved, U-2's were mainly supplanted for intelligence work by reconnaissance satellites. Once they became obsolete for espionage, U-2's began being used for science in earnest. It is therefore a mildly ironic footnote to history that, many years later, the U-2 was the key to fundamental—even trailblazing—discoveries in what might properly be called upper atmospheric research.

But it is better described as, for the first time in human history, capturing pieces of a comet and bringing them home for examination. The driving force behind this program has been Donald Brownlee of the University of Washington in Seattle.

A U-2 takes off from NASA's Ames Research Center at the Moffett Field Naval Air Station near Mountain View, California. The aircraft has immense wings for its size; it looks a little ungainly, as if it were a cross between a glider and a jet aircraft. Attached to a pylon mounted on the wing are covered collecting surfaces of sticky silicone grease. The surfaces are not opened to the airstream until the U-2 reaches an altitude of some 20 kilometers. The aircraft flies more or less at random, because it cannot know where in the stratosphere there might be concentrations of fine cometary or meteoritic dust. For every hour that it flies, ramming the sticky plate into the air in front of it, it collects about one big particle (more than ten microns across, still invisible to the naked eye), and many smaller ones. The plate is then automatically covered, the airplane swoops down out of the almost black sky and the day's catch of stratospheric dust is examined through the microscope back on Earth.

It rarely happens that a small particle lifted a little off the Earth's surface one day finds itself in the stratosphere. Such particles tend to be rained out or carried out before they reach high altitudes. There is a natural barrier to circulation between the lower and the upper atmosphere. (This restriction is removed in a nuclear war, but that is another story.) So as long as humans are somewhat restrained about polluting the atmosphere with fine particles, and we are not looking just after a major volcanic explosion, the stratosphere

A sticky collecting surface attached to the wing of a U-2 aircraft to gather cometary debris from stratospheric altitudes. Courtesy Ames Research Center, National Aeronautics and Space Administration.

should act as a useful natural catchment region for extraterrestrial particles falling to the Earth from space.

You take the sticky plate off the wing of the U-2 and put it under the microscope. You count the number of particles of various sizes, you photograph them, you try to do some chemical analysis—although this is difficult because there are so few big particles. For the same reason, you try to avoid destructive testing. There are many other scientists in line to examine this material after you. Between examinations, the particles are stored at the old Lunar Curatorial Facility at NASA's Johnson Spaceflight Center in Houston, where the rocks returned from the Moon by Apollo astronauts are also kept. There is, at this institution, the only person in the world with the official title "Curator of Cosmic Dust."

One kind of dust you find is very simple—roughly spherical particles of pure aluminum oxide, collected at 20 kilometers altitude. How did they get there? These particles have been generated by solid fuel rockets—largely of American, Soviet, and French manufacture—as they accelerate through the stratosphere on their way to more distant places. The population of aluminum oxide particles in the stratosphere is growing, but it still represents only a distraction. The sticky plates have captured far stranger particles.

When the most abundant kind of stratospheric dust particle is examined under extremely high magnification (opposite), it shows itself to be an irregular aggregate of still tinier particles, individually about a tenth of a micron across: a hundred thousand of them side by side would be as long as the nail of your little finger. There is no known industrial or biological process that makes particles like this. Even if there were, there is no way for such particles to be transported to the stratosphere in such numbers. A range of physical and chemical tests—cosmic ray tracks, say, or nickel/iron ratios—all point in the same direction: these particles have originated on another world. Since the principal source of fine particles in our region of space is the comets, it seems likely that we have before us the stuff of comets.

We are not sure why the particles stick together. Perhaps there was once (or still is) an adhesive material on their surfaces. You can see at a glance that this is a fragile structure. It will collapse if you put much weight on it. If these aggregates were ever at depth inside some small world, they would have been crushed—unless the spaces had been filled by some other, more volatile, material, now gone. This is another reason why the stratospheric dust particles plausibly arise from the tails of comets. The ice has now all evaporated, and even some of the organics have probably been lost.

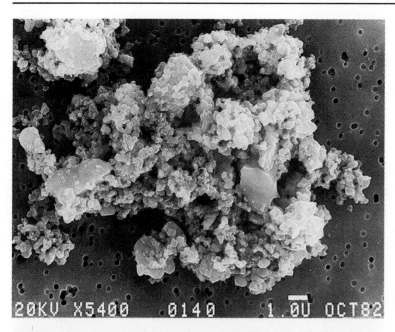

1960

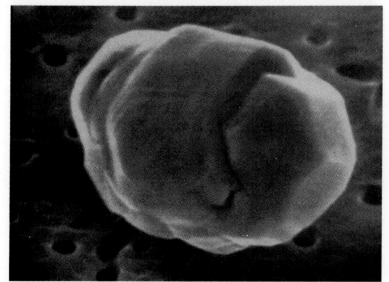

Probable cometary debris collected at stratospheric altitudes and examined under a scanning electron microscope. At top we see many fine particles stuck together in a morphology resembling a cluster of grapes. The magnification is 5400 times and the small horizontal bar is one micron (a ten-thousandth of a centimeter) across. At bottom is a close-up of an individual submicron particle which may just conceivably be a relatively unprocessed interstellar grain. Courtesy Don Brownlee and Maya Wheelock, University of Washington, and National Aeronautics and Space Administration.

Look at these tiny grains again. They were, probably, once on the surface of a cometary nucleus. If you were able to view close up an undisturbed piece from the interior of a cometary nucleus, you would probably see such clusters of small, dark particles, but with all the voids and surrounds filled with ice. As time passes in the inner solar system, the ice evaporates, the particles are dragged off into space, and some of them by merest accident eventually enter the Earth's atmosphere (see page 240).

Ninety percent of the fluffy particles of probable cometary origin collected in this way are aggregates, as we see in the illustration on the top of page 239, of tiny submicroscopic dark grains, an intimate mix of silicates and other minerals with complex organics. These individual grains

1960

are, very roughly, the same size as the dust that fills the spaces between the stars. It seems just barely possible that we have before us a sample of the original material from which the solar system formed. There is not yet enough of it available to be certain. Much larger scale collection programs by the U-2 and its successors are needed—both to accumulate enough material to do a variety of physical and chemical tests, and to search for rarer particles that may have something surprising to tell us. Ultimately, we will want to compare these particles with particles actually collected from the nucleus or coma of a comet. But for now, it seems at least possible that we are looking for the first time at the bricks and mortar that built the planets.

Schematic diagram of the evolution of a cometary surface, in extreme close-up. Far from the Sun, the surface is an intimate mixture of ice (shown here in blue) and rocky and organic material (shown here in brown). As the comet approaches the Sun, the temperatures increase until (*top left*) the ices begin to vaporize. Soon (*top right*) most of the surface ices have been lost to space until (*lower left*) only rocky and organic materials are left. Fragments of this composition break off from the comet; when one enters the Earth's atmosphere (*bottom right*) it becomes a meteor or, if quite large, a fireball. Paintings by Kim Poor.

A comet impacts the luminous surface of the Sun. The dark features are sunspots with locally enhanced magnetic fields. There are no solids or liquids anywhere in the Sun; everything is gaseous. Painting by Anne Norcia.

1961

Chapter XIV

SCATTERED FIRES AND SHATTERED WORLDS

Often, when the Sun has set, scattered fires are seen not far from it.

—Seneca,
Natural Questions,
Book 7, "Comets"

The angels all were singing out of tune,
  And hoarse with having little else to do,
Excepting to wind up the sun and moon,
  Or curb a runaway young star or two,
Or wild colt of a comet, which too soon
  Broke out of bounds o'er the ethereal blue,
Splitting some planet with its beautiful tail,
As boats are sometimes by a wanton whale.

—Lord Byron,
  The Vision of Judgment,
  1822

GEORGE-LOUIS LECLERC, COMTE DE BUFFON, A naturalist who died on the eve of the French Revolution, was one of the first scientists to attempt, from the record in the rocks, to reconstruct the history of the Earth as a succession of geological epochs. He also proposed that in ages past a massive body that he called a comet struck the Sun, launching out into space great blobs of fiery matter which cooled, condensed, and formed the planets and their moons. Laplace soon showed that this reconstruction of events could not account for the orbital regularities in the solar system; but it is the first mention in science of cometary matter hitting the Sun, and the first modern attempt to describe how at least some comets die. In this chapter, we describe four of the many fates of comets: to hit the Sun directly; to disintegrate, with the remains spiraling into the Sun; to impact the moons and planets; and to be transformed into another world.

In the nineteenth century—when no one knew that there was such a thing as the nucleus of an atom, much less a science of nuclear physics—it was widely held that meteors make the Sun shine. Meteors must be falling into the Sun, it was argued, surrendering their energy of motion, and thereby heating further the fiery surface—which, in consequence, obligingly radiated light and heat to the needy inhabitants of the Earth. But meteors are the debris and sometimes the shrouds of dead and dying comets; so if this view were correct, comets would be responsible for life on Earth. In fact, meteors provide pitifully little of the Sun's light, and it is hydrogen fusion that makes the Sun shine, although comets and their debris regularly strike the Sun, the Moon, and the planets—and may in several senses be responsible for life on Earth.

* * *

The Great September Comet (1882 II) came fairly near to hitting the Sun; it is called—one of the more romantic names in cometary science—a sungrazer. Before perihelion passage, it showed a single nucleus. Afterward, as it headed out past the planets, it revealed itself to have split into four separate nuclei, which gradually drew away from one another. The Sun tugs on the near side of the comet a little more strongly than on the far side; also, the sunward side feels more heat than the night side. With a very fragile structure and a close passage these unequal stresses might be enough to break it into two or more pieces. The computed periods for the return of these components are all between five hundred and nine hundred years in the future, but separated by intervals of about a century. The four fragments—

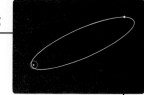

1962

A comet is discovered near the Sun during a total solar eclipse. Tidal disruption is tearing the comet's nucleus apart. Painting by Anne Norcia.

each of considerable size—have been predicted (with considerable uncertainty) to return in the years 2546, 2651, 2757, and 2841. The generations of observers in those remote times will see four comets, spaced about a century apart, pouring out of the same small region of the sky, heading for the Sun.

The Comet of 1882 is itself one of the so-called Kreutz family of sungrazing comets that episodically arrive from the same place in the sky. The great sungrazing comets of 1668, 1843, 1880, and 1887 (the last-named, like the ghostly horseman seen by Ichabod Crane, apparently headless) are all part of this family, as are such more recent visitors as 1963 V and 1970 VI. It is natural to wonder whether they might all be fragments of a still larger ancestral comet that, in another age, came too close to the Sun and was torn apart by the solar tides.

The closer a comet passes to the Sun, the stronger is this disruptive tidal force. Comet Ikeya-Seki (1965 VIII) was also a daylight sungrazer of the Kreutz family and split into two pieces just after perihelion. The Great Comet of December 1680, studied by Newton and Halley, passed less than

Comet West. A set of four different exposures taken on each of three different nights in 1976. On the long time exposures the outer coma is brought out, but details in the inner coma are obscured. On the shortest exposure, the splitting of the cometary nucleus is evident. As the comet split, the components moved apart, each varying in brightness from night to night, and each exhibiting a separate tail. Photographs taken at the 154-centimeter instrument, Mount Lemmon Observatory, by Stephen M. Larson. Courtesy Lunar and Planetary Laboratory, University of Arizona.

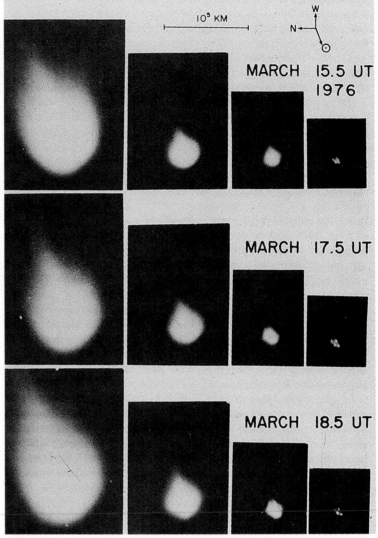

MARCH 15.5 UT 1976

MARCH 17.5 UT

MARCH 18.5 UT

100,000 kilometers from the Sun's surface, considerably closer than the Moon is to the Earth;* it did not split during perihelion passage. On the other hand, Comet West (1976 VI), coming no closer than thirty million kilometers to the Sun, split into four fragments that slowly receded from each other at more than their mutual escape velocity. Thus the gravitational tides of the Sun or unequal heating cannot be the sole causes of the splitting of comets. We still do not know why comets split.

Splitting and jetting may be connected (see Chapter 7, page 129). At the moment Comet West split, the individual fragments brightened noticeably, and propelled large quantities of dust into space in the first of some dozen bursts.

*This is very close. If you had been standing on the Comet of 1680—or a little above it so you could see out—you would have discovered that the Sun filled half your sky. In more than one respect, you would have been inside the Sun.

1963

!

Scientific papers tend to be low key, and the appearance of an exclamation point—except in mathematical papers, where it stands for something very different—is rare. Here is one of the exceptions. Brian Marsden of the Center for Astrophysics in Cambridge, Massachusetts, is the director of the International Astronomical Union's office that alerts astronomers all over the world to the discovery of a new comet, and much else. Marsden himself has made a number of important contributions to understanding the comets. In a scientific paper* published in the 1960's, he is discussing the sungrazing comets, and finds that there are two members, 1882 II and 1965 VIII, that look as if they had split apart near aphelion. But aphelion is well beyond the orbit of Neptune, disconcertingly far away:

> Although most of the comets observed to split have done so for no obvious reason, one really does require an explanation when the velocity of separation is some 20% of the velocity of the comet itself! A collision with some asteroidal object at 200 A.U. from the sun, and 100 A.U. above the ecliptic plane, even though it would only have to happen once, is scarcely worthy of serious consideration.

> A violent explosion 200 Astronomical Units from the Sun, virtually in the interstellar night, is hard to understand. Collisions there should be extremely rare. The problem is left unsolved.

*B.G. Marsden, "The Sungrazing Comet Group," *Astronomical Journal*, Volume 72, page 1170, 1967.

Eighty percent of comets that split do so when they are far from the Sun. Comet Wirtanen fragmented in 1957 a little inside the orbit of Saturn. Comet Biela/Gambart is a similar case. Such splitting may be due to the vaporization of exotic ices, or to collisions with otherwise undetected interplanetary boulders. These sungrazers have all passed within about half a million kilometers of the Sun's surface, penetrating through the thin, hot gas of the solar corona where the temperature hovers around a million degrees. But the comet spends so little time in the solar corona, and the tremendously hot gas there is so thin, that it is probably the Sun's gravity and not so much its heat that is responsible for the fragmentation of the sungrazers.

If some comets come so close to hitting the Sun, should not others fall in and impact it head on? Occasionally comets approach the Sun, are lost in its glare, and are never seen again; but perhaps they have only been tidally disrupted into many small pieces by the Sun's gravity. However,

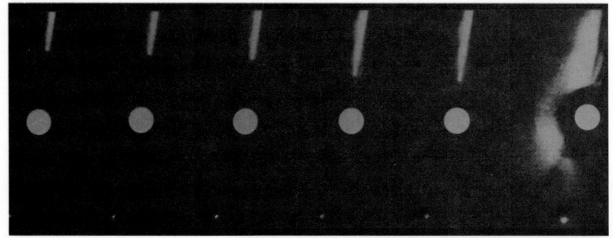

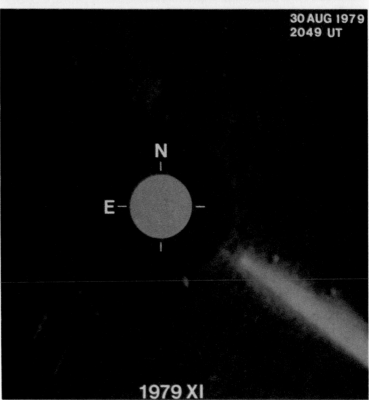

Frames from a movie shot in space: Comet Howard-Koomen-Michels 1979 XI impacts the Sun on August 30, 1979. The "disturbance in the solar corona" may in fact be the comet tail continuing intact for about 24 hours. Courtesy Naval Research Laboratory, Washington. D.C.

through an entirely chance discovery, we can now be reasonably sure that comets do sometimes hit the Sun, as Buffon, on a larger scale, had long ago proposed.

The U.S. Naval Research Laboratory flies a video camera and telescope piggyback on an Air Force satellite, to monitor the changing activity of the solar corona. Perhaps it is an alarm system for solar flares and other storms of charged particles from the Sun which may damage instruments and astronauts in space (see Chapter 9). A small opaque disk is positioned in front of the telescope so it blocks out the Sun; the fainter, extremely hot solar corona can now be photographed. In late August 1979 this instrument made an

1963

extraordinary discovery, which, however, remained unreported for two and a half years, "owing to delays in release of the data tapes to experimenters." It obtained, by accident, a time lapse motion picture of a comet on an impact trajectory with the Sun (see opposite page). The comet, called Howard-Koomn-Michels (1979 XI), turns out also to be a probable member of the Kreutz family of sungrazing comets. The comet was headed around the Sun, but its perihelion distance was too near. The head never emerged on the other side of the Sun, and must have vaporized and fragmented into its consitituent small grains. But the tail of this now decapitated comet continued intact for at least a day, before being dissipated and blown back from the Sun. Imagine the comet streaking through the solar corona, outgassing madly, tumbling erratically, fragmenting, grains vaporizing in a staccato of small flashes, until, its substance spent, all that remains in the vicinity of the Sun is a cloud of atoms—mostly H and O—stirred in the white-hot hurricane that constitutes its outer, cooler layers.

We do not know how many other sungrazers have suffered a similar fate. Comet Howard-Koomn-Michels had not previously been known to exist. It seems unlikely that more than one comet hits the Sun each year; but even this would correspond to about ten times more mass falling into the Sun every year from intact comets than from interplanetary dust. Over the age of the solar system, this might be enough to make the dying Sun of billions of years from now shine on for one more decade.

* * *

Find a place, if you can, far from the city, far from atmospheric pollution, far from artifical light. Pick a clear and moonless night, preferably in springtime in early evening. Look near where the Sun has set. Direct sunlight is now blocked by the western horizon, and you can make out a more feeble light. If you live in northern midlatitudes, you'll see a faint luminous band rising upward and to the left from

Photograph of the zodiacal light extending almost 50 degrees above the horizon. This is a time exposure (note that the stars are shown not as points but as short trails) taken about one hour after sunset. Courtesy William K. Hartmann, Planetary Sciences Institute, Tucson, Arizona.

A longer time exposure of the zodiacal light (note longer star trails; the north celestial pole about which the sky seems to revolve is just off the upper right-hand corner of the picture). To the naked eye, of course, the zodiacal light is much fainter. From *Moons and Planets*, 2nd ed., by William K. Hartmann. © 1983 by Wadsworth, Inc. Used by permission of William K. Hartmann and the publisher.

the place where the Sun has set. It is not the Milky Way, and is only a little harder to find. Seneca called it "scattered fires," and that is what it looks like, distant flames in, or somehow reflected off, the sky.

If you observe in very clear air, especially at tropical latitudes, you can detect this band of light sweeping completely around the sky, through the zodiacal constellations among which the planets move; and for this reason, it is called the zodiacal light. At night, just opposite where the Sun is on the other side of the Earth, is a patch of light, brighter than its surroundings, called the Gegenschein, German for counterglow. From the geometry, it is clear that the Earth—along with the other planets—is immersed in a flat ring of material that envelops the Sun, and that reflects its light back to Earth.

This further example of disk-like structures in astronomy did not escape the notice of Immanuel Kant. While Kant had more than one view on the matter, his description of the nature and origin of the zodiacal cloud is memorable:

> The Sun is surrounded with a subtle vapour which surrounds it in the plane of its equator, and is spread out to only a small breadth on both sides but extending to a great distance. We cannot be certain whether…it comes into contact with the surface of the Sun…or, like Saturn's ring, is everywhere removed to a distance from it. Whichever of these views is correct, there remains resemblance enough to bring this phenomenon into comparison with Saturn's ring…This necklace of the Sun…may have been formed out of the universal primitive matter whose particles, when it was floating around in the highest regions of the solar world, only sank to the Sun after the completely finished formation of the whole system.

Kant seems to be saying that the zodiacal light is reflected off a flattened disk of small particles deriving from the solar nebula, and only lately arrived in the inner solar system. If this reading is correct, once again Kant proves to be a century or two ahead of his time.

When you view the zodiacal light, you are observing the remains of comets (and, to a lesser extent, fine debris from the collisions of asteroids between the orbits of Mars and Jupiter). There isn't much of the stuff. The total mass of all the particles in the zodiacal cloud is only some hundred billion tons, the mass of a single comet a few kilometers across. This cloud is made of fine grains of dark dust, mixes of silicates and organics, typically tens of microns across and apparently very like the particles recovered by aircraft in the stratosphere (Chapter 13). If you grind up a small comet into

And if there are several streams of meteors, which come across that little line in space which constitutes the Earth's orbit, what untold multitudes of them must be within the whole length and breadth of the solar system! Perhaps it may even turn out that the mysterious zodiacal light which attends the Sun, is due to countless hordes of these little bodies flying in all directions through the space that lies within the Earth's orbit.

—G. Johnstone Stoney, "The Story of the November Meteors," Friday-evening discourse, The Royal Institution, February 14, 1879

The zodiacal light as observed in Japan by a Mr. Jones, from Amédée Guillemin, *The Heavens*, Paris, 1868.

ten-micron pieces, and ignore the ice, you produce a trillion trillion particles. Spread them through the inner solar system and they can reflect a fair amount of light.

But we should not imagine interplanetary space to be full of small zodiacal particles. Hold your thumb in front of your nose and move it toward the Sun. If you were able to continue traveling, thumb first, all the way into the Sun, how many zodiacal particles would you encounter? The answer seems to be only a few. A single comet, during a single perihelion passage, can provide a hundredth of a percent of all the particles in the zodiacal cloud.

These particles are now measured routinely by interplanetary spacecraft. For example, between 1972 and 1974, the *HEOS-2* Earth satellite of the European Space Agency searched for dust between 5,000 and 244,000 kilometers from the Earth, and found itself flying through occasional swarms of fine particles, previously unknown. The swarms must recently have formed from a larger parent body (thought to be Comet Kohoutek), because the lifetime of a coherent cloud of grains is very brief. Possibly a large, extremely fluffy body fragmented near the Earth. But, in the large, the zodiacal cloud is uniform, homogeneous, without lumps or strands or holes—except when a planet plows through it, and temporarily leaves a kind of tunnel behind it. Spacecraft on their way to some distant planet regularly bump into the stuff. But there's so little of it, and the individual particles are so far apart (a kilometer separation is typical) that they do not impede interplanetary flight.

However, the paucity of this material is puzzling. In the last hundred thousand years or so, all the zodiacal particles now in the inner solar system should have been provided by the dust tails of comets. The solar system is 4.5 billion years old. So why isn't there much more zodiacal dust? You might very well expect 4.5 billion/100,000 = 45,000 times more zodiacal dust than there is; in which case, the zodiacal light would be brighter than the planets and the stars. Some biblical fundamentalists have even used this puzzle to argue for a solar system no more than 100,000—and hopefully (in their view) less than 10,000—years old, in a brave but hapless attempt to reconcile a literal reading of the Book of Genesis with the findings of modern science.

The great mass of fine particles released from comets over the eons is missing. Where have they all gone? They have, it turns out, been devoured by the Sun. Small particles around a few tenths of a micron across are lost from comet tails and from the inner solar system altogether, accelerated out into the depths of space by the pressure of sunlight. Such particles seem to have been detected by *Pioneers* 8 and 9, among other interplanetary spacecraft. Larger particles also feel the radiation pressure, although it is not enough to drive them significantly outward. But it does serve to lessen the Sun's gravity, making the particle a little lightfooted. There is another, contrary influence, first described by the British physicist J. H. Poynting:* The particle is hot, he said, and in consequence, emits radiation to space. Of course, it also reflects sunlight. But because of its motion, it

> crowds forward on its own waves emitted in front, and draws away from those emitted behind, so that there is increase of pressure in front and a decrease behind. Thus, there is a force resisting the motion.

---

*From Poynting's Friday-evening discourse at The Royal Institution, "Some Astronomical Consequences of the Pressure of Light," May 11, 1906.

1965

The zodiacal light as seen in Europe by a M. Heis, from Amédée Guillemin, *The Heavens*, Paris, 1868.

Since now the particle isn't moving as fast as is necessary to balance the Sun's gravity, it falls a little inward, where it is still hotter, reflects more sunlight, moves faster, and generates still more resistance to its motion. Slowly the particle spirals into the Sun. Poynting calculated that a tiny particle of rock with a radius of ten microns would reach the Sun in less than a hundred thousand years. Much larger particles are too big to be pushed around by sunlight, and do not spiral in. This fiery fate of small particles orbiting the Sun is now known as the Poynting-Robertson Effect. (The American physicist H.P. Robertson later made the most general formulation of the phenomenon.)

Poynting went on to describe how particles of various sizes will eventually move in orbits

> so different that they may not appear to belong to the same system. In the course of time they should all end in the Sun. Perhaps the Zodiacal light is due to the dust of long dead comets…It appears just possible that Saturn's rings may be cometary matter which the planet has captured, and on which these actions have been at play for so long that the orbits have become circular.

So a typical zodiacal particle spends about a hundred thousand years before it is gobbled up by the Sun, and essentially all the particles in the zodiacal cloud are replenished in such a period. Thus, the great majority of the particles you see in the zodiacal light left their comets of origin long before recorded history; but almost none of them before human beings first evolved. All told, about ten tons of interplanetary particles spiral into the Sun every second, 300 million tons a year.

In its annual voyage around the Sun, the Earth runs into particles of the zodiacal cloud, mainly in the dawn hemi-

The View from the Golf Course

Even the impact of a smallish fragment of a comet (or asteroid) with the Earth can have significant effects far away. The day after the Tunguska explosion in Siberia, on June 30, 1908, the following letter was written to the *Times* of London. It was published two days later:

Sir,—Struck with the unusual brightness of the heavens, the band of golfers staying here strolled towards the links at 11 o'clock last evening in order that they might obtain an uninterrupted view of the phenomenon. Looking northwards across the sea they found that the sky had the appearance of a dying sunset of exquisite beauty. This not only lasted but actually grew both in extent and intensity till 2.30 this morning, when driving clouds from the East obliterated the gorgeous colouring. I myself was aroused from sleep at 1.15, and so strong was the light at this hour that I could read a book by it in my chamber quite comfortably. At 1.45 the whole sky, N. and N.E., was a delicate salmon pink, and the birds began their matutinal song. No doubt others will have noticed this phenomenon, but as Brancaster holds an almost unique position in facing north to the sea, we who are staying here had the best possible view of it.

—Yours faithfully,
Holcombe Ingleby

Dormy House Club, Brancaster, July 1.

sphere; more rarely, faster moving debris catches up with the Earth in its twilight hemisphere. The total being accumulated, all over our planet, is about a thousand tons of dust a day. The number of fine particles collected by stratospheric aircraft—U-2's and others—is about what we would expect for particles in the zodiacal cloud captured by the moving Earth. If this cometary dust had fallen on the Earth over its entire history at the same rate that it does today, and if nothing destroyed it after landfall, there would be a dark, powdery layer about a meter thick everywhere on Earth. (If a single large comet were pulverized, and all its debris were spread smoothly over the Earth, a layer about a centimeter thick would result.)

These tiny grains have had an epic history. Intermingled with specks of ice, they were for eons floating in the gas between the stars; then caught up in a contracting, spinning interstellar whirlpool that eventually formed the solar system; growing to form cometary nuclei that were promptly ejected into cold storage at the outskirts of the solar system; then, plunging toward the Sun, propelled out from the comet as the ices evaporate; orbiting the Sun as individual microplanets; and, some of them, finally spiraling in until— during their passage through the solar corona—they are turned into puffs of gas.

The constituent atoms—silicon, oxygen, iron, aluminum, carbon, hydrogen, and the rest—are then diffused through the upper atmosphere of the Sun, eventually being gathered into its internal circulation, and carried down deep into its interior. Some of these atoms will be transported to the very core of the Sun, and engage in the thermonuclear alchemy that makes our star shine. But they constitute the most minor of contributions. Every now and then, though, an occasional light beam, an odd photon—perhaps illuminating for an instant the way of a gnat—originates in the belt of distant comets.

*   *   *

You take even a casual glance at most of the worlds in the solar system, and you discover that they're full of holes. There aren't many analogues that come readily to mind for this sort of battered surface—a fact that says something about the Earth. People once compared the cratered surface of the Moon to Swiss or Emmenthaler cheese, but that doesn't quite evoke its real appearance—craters upon craters upon craters, down to the smallest crater you can see. Some of these worlds look more like a rain-spattered beach. Drop marbles or BB's of different sizes at random in plaster of paris, and let the surface set. The craters will be nicely circular, with rims and ramparts, and occasionally a central mountain peak. Some craters will overlap others. The plaster of paris begins to look very much like the surface of a world.

Some of these impact craters on the Moon, say, are made by asteroids, but most—especially in the outer solar system —are made by comets. Rarely, a world will overtake a comet. More often the comet will overtake the world, or crash into it head-on. Like the Sun, the moons and planets collect their share of cometary impacts. The Sun, being made of gas, can retain no impact craters. But worlds with uneroded ancient surfaces remind us at a glance of how many comets have, over the eons, died in fatal collisions.

On June 30, 1908, something fell out of the sky in Siberia and, at an altitude of eight kilometers, exploded and knocked down a forest. The blast was more powerful than that of the highest yield nuclear weapon in the current arsenals. No impact crater was ever found. The responsible agent is thought by some to be a good-sized but still fragile piece of Comet Encke. Others think it was an asteroid, on

A comet or asteroid strikes the Moon forming Crater Tycho (named after Tycho Brahe, Chapter 2). This crater was formed in the lunar highlands after many of the other craters there; note craters on top of craters. Painting by Don Dixon.

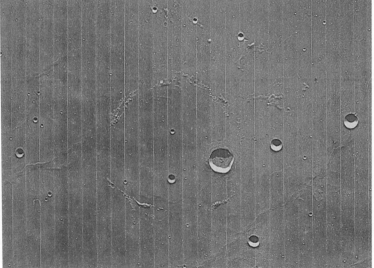

Photograph of the sparsely cratered lunar lowlands. The largest sharp crater in this image is Flamsteed (named after Britain's first Astronomer Royal, Chapter 3). Visible beneath Crater Flamsteed is an old, almost entirely lava-flooded ancient crater from the earliest history of the Moon. *Lunar Orbiter IV* photograph in Oceanus Procellarum. Courtesy National Aeronautics and Space Administration.

the grounds that a comet would be too fragile to travel so deep into the Earth's atmosphere before fragmenting. Unless the impacting object is fairly delicate, it will make an impact crater as it strikes the surface. Because the falling body is traveling very fast, the hole it excavates will be larger than the falling body. If no subsequent geological event obscures the surface, every crater corresponds to an impact, and almost every impact corresponds to a crater. The surface

of a moon or planet is therefore a diary of assault and battery. If you know how to decipher that record, you can uncover the catastrophes of ages past.

Take, for example, the Earth's Moon. The side that permanently faces the Earth and that we know with the naked eye has two kinds of terrain—dark, smooth lowlands and

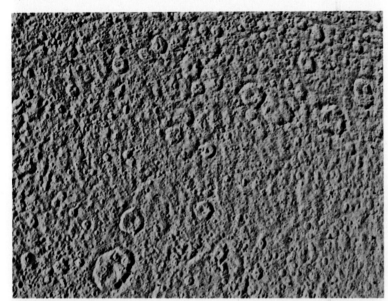

Saturation cratering on the surface of Saturn's moon Rhea. Most of these large holes in the ground have been made, over billions of years, by cometary impact. The color in this image is greatly exaggerated by computer processing. *Voyager 1* photograph courtesy National Aeronautics and Space Administration.

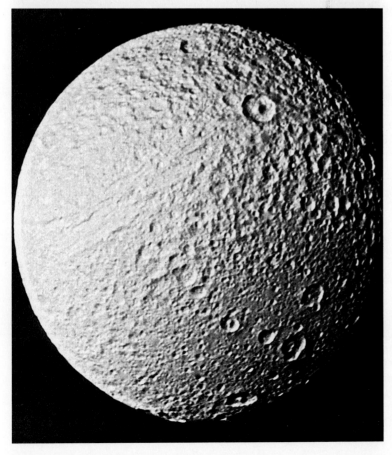

Voyager 1 view of the Saturnian moon Tethys. Because the moons of Saturn are in the outer solar system, the abundant cratering on them is produced almost entirely by comets. Courtesy National Aeronautics and Space Administration.

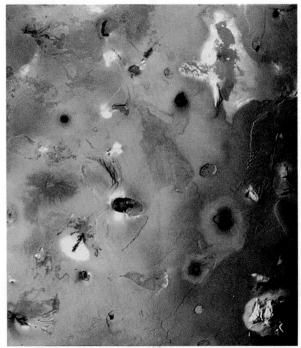

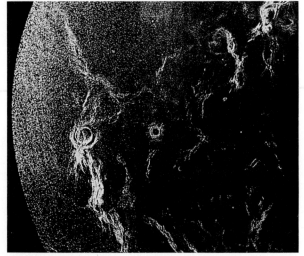

Images of the innermost large moon of Jupiter, Io, taken by *Voyager* spacecraft in 1979 and reproduced in false color. There are many dark spots and apparent holes in the surface (top left, Pele crater). But when examined more closely (top right, Ra Patera, and bottom left, Naasaw), every one of them turns out to be volcanic in origin. There are no known impact craters on Io. Eight or nine active volcanoes were observed during the two *Voyager* flybys. Cometary impact craters are certainly being generated on the surface of Io, but the volcanic flows are eroding them away very rapidly. Courtesy National Aeronautics and Space Administration.

Bottom right, a radar image of the surface of Venus obtained with the largest radar telescope on Earth, the Arecibo Observatory, Arecibo, Puerto Rico. The smaller crater at the center is Lise Meitner, about 60 kilometers across. There is still debate about whether such craters are of impact or of volcanic origin, but the relative sparseness of craters on Venus shows that the surface is continually being modified—probably by volcanism, as on Io. Courtesy Donald Campbell, National Astronomy and Ionosphere Center.

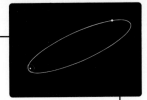

1967

bright, rough highlands (see page 256). Both highlands and lowlands are cratered, but there are many more craters in the highlands than in the lowlands. Because American astronauts and Soviet robots have brought back samples from nine different locales on the Moon, we know something of the composition and—through radioactive dating—the ages of various provinces of the lunar nearside.

The dark lowlands are made of lava that, 3.3 to 3.9 billion years ago, gushed out of the then-hot lunar interior, wiping out any previous craters. Thus, for the lowlands, the earliest master record has simply been lost. The sparse number of lowland craters is just what we might expect if the inner solar system had always been as full of comets and asteroids as it is today. But the highlands have far too many craters for that. Here craters have been superimposed on preexisting craters—so that again a record of the earliest times has been lost. One way or another, these worlds hide the evidence of their very beginnings, although the record of later catastrophes is clear.

From the story told by the dense cratering in the lunar highlands, it is clear that, in the first few hundred million years of lunar history, there were enormously more impacting objects around—comets and asteroids—than there are in interplanetary space today. The same story is told on world after world as our reconnaissance spacecraft fly out among the planets. On Saturn's moon Rhea (see page 257), for example, there has been no geological activity for billions of years, and we are presented with a world cratered to saturation, from pole to pole. In the beginning, interplanetary space must have been full of boulders and icebergs that crashed into the forming worlds. Indeed, this is *how* the worlds formed, in violent collisions.

Meteor Crater, Arizona. This crater is 1.2 kilometers in diameter and was probably produced 15,000 to 40,000 years ago when a lump of iron 25 meters across impacted the Earth at a speed of 15 kilometers per second. The energy released was roughly equivalent to that of a 4-megaton nuclear explosion.

Not all the worlds retain the craters of eons past. On some, as in the lunar lowlands, the traces have been eradicated. Something fills the craters in, or rubs them out, or covers them over. On Venus, there are recent lava flows; on Mars, great sand drifts; on Jupiter's moon Io, a surface rich in recently frozen sulfur; and on Enceladus, a satellite of Saturn made almost entirely of ice, something has melted the surface—the cratering record on that world was literally writ on water. On Io, the craters may be wiped out in centuries; on Venus, it may take a billion years. But stretching through the entire solar system is, written on top of one another, the chronicles of ancient impacts and, fairly often, of more recent geological processes.

Something similar is true on Earth, where even in arid terrain running water rather quickly destroys the craters—so that unless they are very recent, they are very hard to find. So-called Meteor Crater Arizona is only about 25,000 years old (see preceding page). Most terrestrial impact craters are much older, and these are generally large or in geologically inactive regions.

Sometimes a collision scar will reveal something of the underlying planetary surface. On Mars, for example, there are craters with scalloped flow patterns around them, indicating the presence of subsurface water ice that has been momentarily liquified in the collision; surface debris is carried outward until the water freezes. The impact of a large comet might bring water or an atmosphere to worlds that are nearly airless.

When a comet strikes a gas planet like Jupiter, it sweeps through the upper atmosphere, encounters more and more resistance as it plunges deeper, and somewhere below the visible clouds breaks up, its material eventually circulating over much of the planet (see opposite). The stuff of comets is mixed with the air of Jupiter. But it does not seem likely that any known feature in the clouds of Jupiter can be attributed to a recent cometary collision.

Craters have shapes. Some are perfectly bowl-shaped; others are flat, shallow, gently sloped. The shape of the crater does not much depend on the velocity of the impacting object, provided it hits the ground hard enough. The result is the same as detonating a large explosion at the point of impact. The crater shape does depend on how soft the surface is, and on the fragility of the infalling material.

American astronauts on the lunar surface photographed craters too small to see with the unaided eye. Samples returned to Earth have also revealed a plethora of microcraters. Some must be due to asteroidal dust swept up by the Moon; others to fine particles generated on and sprayed over the Moon by large impacts. We also expect some fraction of

1968

A comet streaks into the jovian atmosphere, penetrating below the clouds. Because of its size, and the orbits of the Jupiter family of comets, the largest planet receives comparatively many such impacts. Painting by Don Dixon.

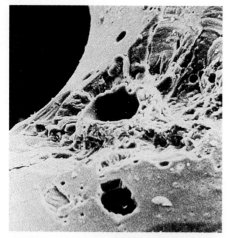

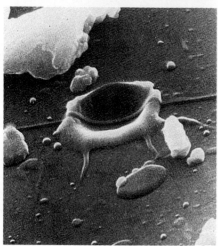

Microcraters in rock samples returned from the Moon. Both pictures were taken through the microscope. Top, from *Apollo 12*, the biggest crater is 30 microns across. Bottom, from *Apollo 15*, the big crater with the raised lips is less than 2 microns across. Courtesy D. S. McKay and Johnson Space Center, National Aeronautics and Space Administration.

the lunar microcraters to arise from cometary dust, excavated puff by puff. The bowl-shaped microcraters seem to be due to cratering by fine, rocky particles—from asteroids or cometary silicates. But there is also a population of microcraters that are extremely shallow for their size, more depressions than bowls. The shallow microcraters can only be understood if they were excavated by low density grains impacting the Moon, and these can only be cometary in origin. Microscopic bits of cometary fluff are very gently softening and abrading the face of the Man in the Moon.

And by collision roulette, larger comets must also hit the Moon from time to time. A lunar soil sample called 61221 shows evidence of the molecules H_2O, CH_4, CO_2, CO, HCN, H_2, and N_2. Volatiles present in this sample—especially HCN—and absent in other lunar samples are suggestive of a recent cometary impact. In a collision, comets can also impress a pattern of magnetism around the impact area, and this may account for some otherwise puzzling magnetic anomalies found on the Moon.

* * *

Sooner or later, comets in the inner solar system run out of gas. Typically, the amount of water ice that disappears in a single perihelion passage is a few meters (about 10 feet). Each time the comet passes by the Sun, it shrinks. Of course, the comet is not pure ice, but an intimate mix of ice and dust. Huge quantities of dust are blown off in the great jet fountains that sometimes play within the coma, and dust is also lifted more sedately by evaporating ice. So a comet made mainly of ice will lose ice and dust every perihelion until the comet is altogether gone. Nothing is left but a powdery contribution to the zodiacal cloud and, perhaps, an occasional sporadic meteor in the skies of Earth.

But now imagine that the comet has more dust than ice. After the first perihelion passage, there is a layer of dust left on the surface that geologists would call a lag deposit. Some of the dust has been carried away by the evaporating ice, but not all. The next time the comet approaches the Sun, the dust serves as insulation for the underlying icy dirt. It is now harder for it to heat up. And if it does heat up, it is harder for the underlying vapor to escape—the way is barred by the layer of dust. The pent-up pressures of subsurface gases may blow off the overlying mantle of dust, and then the process begins again. After a number of perihelion passages, so large a lag deposit may be created that no further ice can be lost to space. No longer does the comet form a coma or a tail. It shuts down, closes up shop, becoming a small, dark lump of matter in the inner solar system.

1969

Hypothetical evolution of a short-period comet into an asteroid. An intermittently active comet comes close to the Sun and a small amount of remaining surface ice vaporizes *(top left)*. The comet then fragments *(bottom left)*, exposing previously sequestered ice that also vaporizes. Eventually *(top right)* an apparently rocky, ice-free body is left which is described as an asteroid even if it has an interior still rich in ice. Paintings by Don Dixon.

Something similar will occur if an icy comet with a rocky core periodically enters the inner solar system. Such a comet would have to be very large—at least tens of kilometers across and perhaps hundreds—for it to have melted, and vertically segregated materials of different properties, the denser rock on the inside, the lighter ices on the outside. Thus, there are at least two ways for comets to evolve into small bodies with surfaces of silicates and organics. But there *is* a category of such small, dark objects, some of them in quite eccentric orbits, called the Earth-approaching asteroids. The entrancing suggestion that comets might make a metamorphosis from an icy world to a rocky one was first made by Ernst Öpik (Chapter 11).

If the metamorphosis really occurs, then there might be a few comets caught in transition—in, you might say, their last gasps. They would sputter as they approach the Sun, the final bare patches of ice vaporizing and the lag deposit enforcing quiescence everywhere else. There are such comets, and, if you observe assiduously, you can sometimes catch them in their quiescent phase. Some comets show no decline in brightness even after ten or more perihelion passages, but others tend, after a while, to fade steadily into inconspicuousness. Take Schwassmann-Wachmann 1, for instance, although it may be very far from extinct. There is no question that it outgases like a comet: the diffuse coma is at least on occasion clearly seen, and even the emission lines of CO^+ have been observed. Schwassmann-Wachmann 1 seems to show more coma activity when it is closer to the Sun than when it is further away. But when it is quiescent, it has the brightness and color of an RD-type asteroid—so-called because it is red and dark from the complex organic molecules it contains. A number of active comets, observed far from the Sun before substantial comas form, also resemble RD asteroids—including Halley's Comet, seen in 1985 when it was near the orbit of Jupiter. Other intermittently active comets—Arend-Rigaux, for example, or Neujmin—when quiescent resemble S-type asteroids, grayer, made mainly of silicates and metals, but still quite dark. It seems clear that when such comets complete their outgassing, they will be indistinguishable from asteroids.

Asteroids collide with one another and occasionally a fragment is broken off and finds its way to the Earth's surface, where it often ends its days in a museum, catalogued as a meteorite. There are many different kinds of meteorites. Some have never been heated very much in four and a half billion years, contain complex organic matter, and somewhat resemble the cometary fluff that is collected in the stratosphere. Some of these carbonaceous meteorites could conceivably be of cometary origin, arising from collisions

1969

with asteroids or other causes of cometary fragmentation. Indeed, the color and darkness of RD asteroids is like that of the organic-rich clays found in carbonaceous meteorites.

Other meteorites, though, show very little organic content, and are made mainly of stone, or even metal. Some stony meteorites might arise from collisions with asteroids like the one Comet Arend-Rigaux is on its way to becoming. The huge iron meteorites proudly displayed in museums throughout the world tell a story of great heat, melting, and coalescence of molten iron droplets into a large mass before the whole thing cooled. In their history, the iron meteorites have been so strongly heated and reworked that they cannot possibly be unaltered, primordial cometary stuff. The same is true, although to a lesser extent, of the stony meteorites. But the carbonaceous meteorites are much closer to the primitive matter from which the solar system formed.

Many of the asteroids originated roughly where they are right now—mainly between the orbits of Mars and Jupiter. They may be the remnants of the rocky (and metallic) bodies that, along with some cometary matter, accreted to form the Earth and the other terrestrial planets. They must be the source of many meteorites. But interspersed among them, traveling incognito, cunningly disguised, are comets. They've put on an overcoat of dust, or stripped down to some hidden core. Unless you look quite closely, you might never notice that there are blue-jay eggs in the robin's nest. A few asteroids even have cometary sorts of orbits—highly elliptical, carrying them near Mars, or the Earth, but especially near Jupiter. One such asteroid is called Hidalgo, of the RD class. Perhaps because of the rocket effect (Chapter 6), Comet Encke is slowly walking its orbit inward. There is abundant evidence that Encke is falling to pieces—including the Taurid and Beta Taurid meteor showers which it trails.

One of these suspect asteroids, Phaeton (1983 TB), was discovered by the *Infrared Astronomy Satellite* (Chapter 12) in 1983, and shares its orbit with the meteor stream responsible for the Geminid meteor shower every December 14. It is extremely unlikely that 1983 TB would find itself exactly in the Geminid meteor stream by accident, and it is equally improbable that it is an asteroid that just before we started observing it was fragmented by collision. It is probably a piece of an extinct comet. But the Geminid meteors are known to be stony, not fluffy, material. As with comets like Arend-Rigaux, there is a hint here that comets can metamorphose into true rocky asteroids. Thus, when we find a crater on Earth with evidence that the impacting body was made of rock, it may nevertheless have been generated by an extinct comet.

Ice and iron cannot be welded.
—Robert Louis Stevenson,
Weir of Hemiston,
1896

Asteroid Oljato may be another comet with an identity crisis. Its highly eccentric orbit, stretching from Jupiter to Venus, is comet-like. It reflects light at visible and at radar wavelengths in an entirely anomalous way; no other object in the solar system looks like Oljato. And it seems to be associated with one of the meteor showers. Oljato has also been detected in an unusual way: In orbit about the nearest planet is an American spacecraft called *Pioneer Venus*, which, among its other duties, is busily recording day-by-day the strength of the local magnetic field. There is a steady background field, and occasional anomalies. About a quarter of the magnetic disturbances that have been detected occurred when Oljato passed very close (within 1.3 million kilometers) to Venus and its attendant spacecraft. Asteroids are not expected to have detectable magnetic fields. But if Oljato is a slightly active comet, it would generate a thin coma. The coma would be quickly ionized by solar ultraviolet light and would in turn compress the magnetic field carried by the local solar wind. As this compression passed by Venus, an intensification of the local magnetic field might be detected. Oljato's strange spectrum suggests it is something rare, but several other pieces of evidence strongly imply it is a comet at death's door.

Adonis is another suspect asteroid, on a very elliptical comet-like orbit that crosses the Earth's orbit. Its radar echo has peculiar properties that may be explicable if its surface layers are highly porous—as if the ice has all been vaporized, leaving only a loosely bonded matrix of dust.

Every year, on average, there is a new recruit to the ranks of the short-period comets, dragooned from longer period orbits by Jupiter's gravity. Aided by sunlight, gravitational tides and rotation, many of these comets fall to pieces and vanish altogether. Others are ejected from the solar system or run into worlds. But a steady population is apparently metamorphosed into asteroids, with orbits ranging from near-circular to highly elliptical. Over the age of the solar system, the number of comets thus disguised must become quite large.

Many asteroids that cross the Earth's orbit are probably extinct comets. Because they tend to be small and dim, they are hard to find. There are fewer than a hundred known objects that cross the orbit of Mars or Earth, and this is after a decade of intense searching. These Earth-approaching asteroids are of particular interest to us, because they represent something with serious and calculable consequences: the high-speed collision of an object kilometers across with the Earth would represent a major catastrophe, of a sort that must have happened from time to time during the history of the Earth. It is a statistical inevitability. And while we

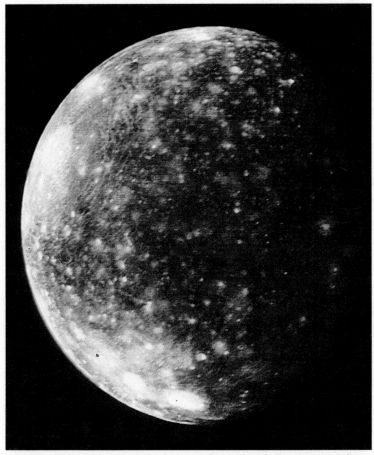

Voyager 2 view of Callisto, the outermost of the four big moons of Jupiter. Something like half of these craters have been made by comets. Courtesy National Aeronautics and Space Administration.

may describe the Earth as having been hit by an asteroid, we recognize that the falling body may have been a comet in disguise.

* * *

From the observed population of comets today, and the cratering record, it is even possible to estimate how many comets or asteroids and of which kind have excavated the craters on a given world. Eugene M. Shoemaker is the founder of the U.S. Geological Survey's Branch of Astrogeology; one of the scientists behind the earliest serious work in America to explore the Moon; responsible for the discovery of many near-Earth asteroids; and the world expert on Meteor Crater, Arizona. Shoemaker combines down-here geology with up-there astronomy. With his collaborator, Ruth Wolfe, he has estimated that something like a quarter of the craters seen on the big moons of Jupiter are produced by the long-period comets. Perhaps another quarter are due to active short-period comets—ones that are still producing copious dust tails, outbursts, brightness changes and the like. Fully half the craters are dug out by non-cometary bodies, including extinct short-period comets

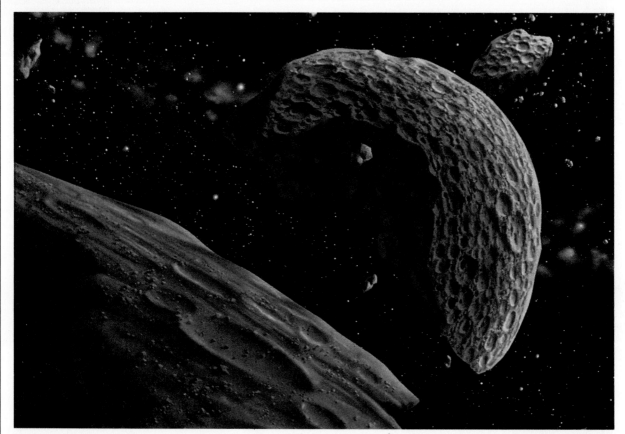

The fragments of Mimas, hypothetically shattered after a cometary impact. Mimas is the innermost of the moons of Saturn that were discovered from the Earth. *Voyager* spacecraft photography reveals it to have a crater so large that had it been struck by a somewhat larger comet it would have shattered. Comets can destroy small worlds. Painting by Kim Poor.

disguised as asteroids. For the Earth, about a third of recent craters are due to the long-period comets, not very different from the story on Jupiter's moons. But the remainder of terrestrial craters are due to Earth-approaching asteroids, at least some of which are extinct comets. Here, according to Shoemaker and Wolfe, the known short-period comets make little contribution to cratering.* Because their aphelia are near to Jupiter through the circumstances of their capture, the short-period comets much more often run into the moons of Jupiter than they do the Earth.

When a big comet or asteroid strikes a small world, which sometimes happens, the world may be destroyed, smashed to bits. This almost happened on Phobos, a moon of Mars (Chapter 7, page 131), and to Mimas, a moon of Saturn. If the object that produced the big crater in Mimas had been somewhat larger, the result might have been what is shown above. Large comets were abundant in the early history of the solar system, and as they were being incorporated into the growing moons and planets, catastrophic collisions happened often.

On the Moon and Mars and the large satellites of Jupiter,

---

*On this point, experts differ; some believe that active short-period comets may be quite important for terrestrial cratering.

there is evidence of enormous ancient collisions that produced not craters, but basins hundreds or even thousands of kilometers across. In one promising theory, the Moon itself was born from a massive impact of some failed small planet with the Earth, the ejected fragments from the collision coalescing to make our natural satellite. The scar has healed long before now, but if that collision had been somewhat more energetic, the Earth might have been destroyed altogether, and we would not be here, attempting to decipher the evidence of our origins.

The spacing between the worlds, their masses, and even their survival depended on how many primitive bodies were moving on what orbits in the early solar nebula; that is, on the chance concatenation of a multitude of events, some probable, others less so. Start the solar system over again, let only random factors operate, and there will be a slightly different collection of rocky and icy bodies formed, each a few kilometers across; another sequence of collisions building new worlds; an altered number of planets, with somewhat different masses and orbital positions. But in general the rocky terrestrial planets will be on the inside, and the gas giants still on the outside. Plus icy and organic moons, and the comets in a great cloud beyond. There may be a vast multitude of planetary systems in the Galaxy that look roughly like our own, but with significant differences from type to type and system to system, depending on the complex collisional roulette that shaped their origins.

*   *   *

Comets rework the surfaces of worlds, carry volatiles to parched planets, dig up buried treasures for the planetary scientists, and leave a record of their history throughout the solar system: they are busy, useful, obliging bodies. But comets also make worlds and destroy them. They remind us of the time some 4.5 billion years ago when the solar system was made of nothing but comets and their rocky equivalents, hundreds of trillions of them or more, swarms of bodies merging and smashing, careening from one district of the solar system to another, a whirlwind of small worlds that eventually led, after the impetuosities of youth had subsided, to the present staid machinery of the solar system. The occasional passing comet is, these days, one of the few reminders of our violent and chaotic origins.

A large cometary nucleus, 10 kilometers across, is just about to strike the Earth—65 million years ago in this schematic representation. The resulting explosion is thought to have propelled huge clouds of pulverized ocean bottom all over the Earth, producing a time of cold and dark that resulted in the extinction of the dinosaurs and many other species. Nuclear Winter is a similar effect. Painting by Don Davis.

Chapter XV

THE WRATH OF HEAVEN: 1. THE GREAT DYING

A comet never descends all the way to the lowest regions of the atmosphere, and does not approach the ground.

—Seneca,
Natural Questions,
Book 7, "Comets"

Comets are such beings as have been on account of their merits raised to heaven, whose period of dwelling in heaven has elapsed and who are then redescending to Earth.

—Al Biruni, eleventh century

THE GROUND IS GROWING. SILT SETTLES ONTO THE stream bed, dust falls from the air, and the ground is a little higher. A pretty typical rate of growth is a micron a year—a film at most one microscopic particle thick, an almost imperceptible patina settling annually over the surface.* A micron is almost nothing. But over geological time, it adds up. In 10,000 years, at this rate, you build up a centimeter of the stuff, and in a million years, a meter. If you could find a place where, over a good part of the age of the Earth, material was only added, and never removed, you might discover cliffs kilometers high in which an undisturbed record of the history of our planet is preserved.

Providentially, there *are* such places; one of the best known is the Grand Canyon of North America, where the sedimentary column has been exposed by the slow but persistent erosion of the Colorado River, and the uplift of the Colorado Plateau. You see a lovely pattern in the rocks, a sequence of muted pastel layers. In adjacent cliff faces, there are almost identical patterns. Each layer is an epoch. Every boundary between layers corresponds to some major change in the environment. If the sedimentary columns show the same pattern everywhere, the environmental changes must have been worldwide. As you glance from the top to the bottom of such a cliff, you see the history of the world. You have only to step up and read.

Without doing any digging at all, you might see something bright gleaming in the sunlight. You examine it more closely and realize it's a fossil, perhaps a dinosaur's knee, or the jaw of an armored fish. With a microscope, you easily discover fossils of little creatures, perhaps single-celled plankton that floated in some primeval sea. You look elsewhere, at other sedimentary columns, and the remarkable thing is that everything goes together. This particular thick, reddish deposit tends, in many places, to be underneath these two thin gray layers; it always has a similar mix of fossils. The rocks corresponding to a given epoch may differ from site to site, but the abundant fossils are always the same. You find an ammonite shell (see illustration above), and it confirms that the vast desert landscape of which this cliff is part was once the bottom of an ocean. A few meters above or below, and the fossils look very different; you realize that if you wait long enough, the landscape will change dramatically.

The rocks are trying to tell us something. These fossils are the only remains of creatures that once strode or slithered or

An ammonite from the Cretaceous seas. Because they made abundant fossils the shapes of ammonite shells are known very well. Drawings on pages 272 through 275 by Maren Leyla Cooke.

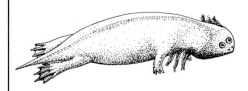

A labyrinthodont, an amphibian from the late Triassic.

---

*This material must come from somewhere, and meteoric dust generally falls thousands of times too slowly to be the source. In fact, the ground grows a little in one place, because it is eroded a little somewhere else on Earth.

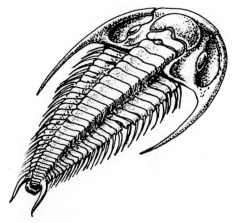

1972

swam or just took root, beings that arose and died out long before there were any humans to take note of them. You take a particular fossil and, all over the world, there is a layer—corresponding to some epoch in the past—below which that organism had not yet come into being. Generally, there is a layer above, in which that fossil no longer appears; but the fossils of new organisms are there found for the first time. The lesson is clear: Great numbers of species have become extinct, providing opportunities for other species to evolve and fill the vacated ecological niches. Many of these species were too small to be seen without a microscope, although some were as big as an office building. You count up all the beings that leave fossils and you discover that the vast majority by far of all the species that ever lived are now extinct. Extinction is the rule. Survival is the exception.

And then you notice that the different fossil forms do not succeed each other at a uniform pace. Instead, the rocks proclaim long periods in which the kinds of life change very little, punctuated by short intervals of cataclysm, turnover, the wholesale loss of many kinds of creature, sometimes all over the Earth at once—followed astonishingly quickly by many new forms, clearly evolved from the survivors of the catastrophe. The clearest boundaries in the sedimentary column are the ready markers of fearsome, planet-wide disasters. You have a planet where for tens of millions of years everything is fine. It's enough to give even inveterate pessimists a sense of security. But then, maybe when you'd least expect it, tumultuous changes occur—changes so striking that they are instantly apparent to the eye and mind of observers like us, who have come into being hundreds of millions of years later.

A trilobite from the Cambrian Period. The last trilobites died out in the Permian.

There is a steady background rate of extinction, of course. In our time, because of human activities, species become extinct every year—from land reclamation or hunting or industrial pollution or cutting down the tropical forests. The passenger pigeon, whose migratory flights once darkened the skies of North America, became extinct in the twentieth century, the great auk in the nineteenth, the dodo in the seventeenth. The last mastodon died perhaps as recently as 2,000 B.C., the giant armadillo around 5,800 B.C., and the giant ground sloth (it ate treetops), around 6,500 B.C. We humans were probably responsible for the demise of most of these creatures, but such extinctions are part of the natural ebb and flow of life, and are caused by minor changes in the physical or biological environment—in this case a newly capable predator. What we are concerned with here, however, is something quite different: massive extinctions of dozens or hundreds of families of life simultaneously, all over the Earth.

The extinction of what biologists call a "family" of plants

One of the first ocean corals, the rugose coral, gone with the Permian extinctions.

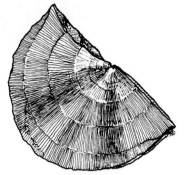

A sessile shelled animal from the late Ordovician.

or animals represents a major loss. Think of the variety of dogs in the world, from Chihuahuas to Great Danes. They are all one species, because they are all interfertile. Biologists distinguish a broader category called a genus (plural, genera) that here embraces not only dogs but wolves and jackals as well. This genus, Canis, is in turn part of a much larger category, a family, that includes foxes; together they all are the Canidae. A family is a major group of beings. The human family, the family of the hominids, includes most of those primates who walked around on two feet and tried to puzzle things out over the last few million years, even though we might not recognize many of them as human were we to encounter them on the street. The loss of a family is the hacking off of a limb of the tree of life.

Mass Extinctions			
Geological Period in which Extinctions Occur	Millions of Years ago	*Percent Marine Extinctions for:	
		Families	Genera
Late Ordovician	435	27	57
Late Devonian	365	19	50
Late Permian	245	57	83
Late Triassic	220	23	48
Late Cretaceous	65	17	50
Late Eocene	35	2	16

*Rough estimate of percent of all families and genera of marine animals with hard parts (so we have fossil evidence of their existence) rendered extinct; numbers are given to nearest 5 million years (Column 2). Extinctions of land families and genera are thought to be roughly comparable. Data courtesy J. John Sepkoski, Jr., University of Chicago. As described in Chapter 16, there is active debate about some of these numbers; the Permian and Cretaceous mass extinctions are the most reliably determined. Other mass extinctions used to derive the disputed periodicities described in Chapter 16 are not tabulated here.

In the table above, six of the most spectacular known interruptions in the history of the Earth are listed, along with an indication of how massive these extinctions were. It is with a sense of shock that we look at the last columns of this table. It shows, for each of the six catastrophes, the percent of all families and genera of known marine organisms that were lost. At the Ordovician and Devonian boundaries, fully half of the genera became extinct, less for the Eocene event, almost as much for the Triassic. In the Permian catastrophe more than half of all the families of life on Earth were lost, more than three-quarters of the genera, and more than 90 percent of all species. One of the most recent of these planetary disasters was at the end of the Cretaceous Period, some 65 million years ago, when almost

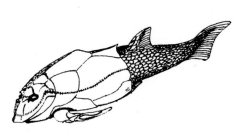

An armored fish, none of which survived the Devonian Period.

1973

one out of five families, half the genera and three-quarters of the species disappeared.

Some of the creatures that were lost are pictured on these pages—a brachiopod that attached itself to the ocean bottom, extinct by the Ordovician; a fish in armor, extinct at the end of the Devonian; a trilobite—they hunted in herds on the ocean bottoms and had existed for almost 300 million years—extinct in the Permian; a kind of coral, also gone in the Permian; an ammonite, a relative of the octopus, extinct in the Cretaceous; and Triceratops, one of the last of the dinosaurs, with three very nasty horns and clanking bone armor, also a Cretaceous casualty. You take one look at Triceratops, and you can tell that times were a little dicey toward the end of the Cretaceous.

Scientists look closely at the Cretaceous extinctions because they concern us particularly. A critical turning point in the evolutionary path to humans can be glimpsed among all those reptile bones. The Cretaceous catastrophe wiped out every family, every genus, and every species of dinosaur, and they were as varied and successful as the mammals are today. It is as if all the mammals were to roll over, their appendages stiffly in the air—every shrew and whale of us, and every person. All the flying and swimming reptiles died as well, and more than a hundred families of beings that live in the oceans. It was a catastrophe enormously beyond anything humans have ever known, at least so far.

The first mammals appeared at about the same time as the first dinosaurs. The dinosaurs were the lords of the Earth, the largest creatures, the most powerful, the ones who would capture your attention on any Cretaceous landscape. The mammals, our ancestors, were then tiny, furtive, scampering creatures, cautious, mouse-like, spending much of their time keeping out of the way of the thundering reptiles. A few dozen little mammals would have made a meager lunch for a middling-sized carnivorous dinosaur. For over 100 million years our ancestors were at an apparent evolutionary cul de sac, living at the margins and in the shadows of a world dominated by dinosaurs. If you surveyed that late Cretaceous landscape—in which the trees and flowers looked pretty much as they do today, but where the dominant animals were all reptiles—you would not have bet much on the chances of our ancestors.

But the dinosaurs went and died, every last one of them, and left the planet to the mammals. Mammals were not the only ones to survive, of course: snakes and salamanders were in good shape, and fish and insects and crocodiles, and many land plants. But the mammals soon predominated. Tentatively at first, and then with an increasing exuberance, the mammals evolved, grew, diversified, taking advantage of

Brontosaurus, a dinosaur as big as a house that flourished in North America in the late Jurassic. The Brontosaurus was made extinct long before the Cretaceous, and is an example of "background" extinctions: species die out at a steady pace over the whole history of the Earth, in addition to which there are short periods of mass extinction.

Dimetrodon, a sail-finned dinosaur, one of the casualties of the Cretaceous extinctions.

Triceratops, the last and largest of the horned dinosaurs, extinct at the end of the Cretaceous. It was bigger than an elephant.

the demise of the competition, filling untenanted ecological niches. We owe our very existence—every one of us bigger than a mouse, anyway—to the extinction of the dinosaurs. So perhaps we humans may be excused for having a special interest in this question: Why did the dinosaurs, along with most of the other species on Earth, abruptly die some 65 million years ago?

* * *

In Book 4, Chapter 4 of Laplace's *System of the World*, the great French scientist at once criticized as superstition the worldwide human fear of comets, and advanced as sober scientific hypothesis a good and practical reason for comet dread:

> In those times of ignorance, mankind were far from thinking that the only mode of questioning nature is by calculation and observation: According as phenomena succeeded with regularity or without apparent order, they were supposed to depend either on final causes or on chance; and whenever any happened which seemed out of the natural order, they were considered as so many signs of the wrath of heaven.
>
> But these imaginary causes have successively given way to the progress of knowledge, and will totally disappear in the presence of sound philosophy, which sees nothing in them, but expressions of the ignorance of the truth.
>
> To the terrors which the apparition of comets then inspired succeeded the fear, that of the great number which traverse the planetary system in all directions, one of them might overturn the earth.

Laplace himself, we remember, had carefully studied the orbital evolution responsible for the near impact with the Earth of Comet Lexell in 1770 (Chapter 5).

> They pass so rapidly by us, that the effects of their [gravitational] attraction are not to be apprehended. It is only by striking the earth that they can produce any disastrous effect. But this circumstance, though possible, is so little probable in the course of a century, and it would require such an extraordinary combination of circumstances for two bodies, so small in comparison with the immense space they move in, to strike each other, that no reasonable apprehension can be entertained of such an event.

So is the fear of a cometary collision with the Earth in the same category as superstitious dread of comets? Not at all. Laplace, who was one of the inventors of modern probability theory, explains:

> Nevertheless, the small probability of this circumstance may, by accumulating during a long succession of ages, become very great. It is easy to represent the effect of such a shock upon the Earth: the axis and motions of rotation changed, the waters abandoning their previous position, to precipitate themselves towards the new equator; the greater part of men and animals drowned in a universal deluge, or destroyed by the violence of the shock given to the terrestrial globe; whole species destroyed; all the monuments of human

1974

Something like the Laplacian view of the consequences of a comet hitting the Earth. The caption of this picture from *Pearson's Magazine* of December 1908 reads: "If a large comet approached within measurable distance of the Earth, the doom of our world would be sealed. Such tremendous heat would be engendered that everything would spring into spontaneous combustion. The hardest rocks would become molten, and no living things would remain upon the Earth's surface. Buildings and human beings would be scorched to cinders in a second."

industry reversed; such are the disasters which a shock of a comet would produce.

The expectation that comets—through universal deluge or other means—are in fact harbingers of disaster has been a respectable speculation for the whole scientific history of the subject, dating back to Edmond Halley himself (who proposed that the biblical Flood was produced by "the causal Choc [shock] of a Comet.") The brand of mischief that comets are said to bring—flood, darkness, fire, rending the Earth asunder—changes with time and astronomical fashion. But the association of comets with catastrophe remains curiously steady through the generations.

Laplace connected this awful vision of a comet striking the Earth with an apparent paradox in timescales. Human history was only a few thousand years old. But Laplace knew —from arguments such as those of Halley on the salt content of sea water—that the Earth was much older. Laplace, the cosmic evolutionist, had not a glimmering of biological evolution. Charles Darwin's *The Origin of Species* was sixty years in the future. Laplace could not imagine the world existing for long periods before there were humans. So why weren't human history and human civilization much older?

> We see…then why the ocean has abandoned the highest mountains, on which it has left incontestable marks of its former abode: we see why the animals and plants of the south may have existed in the climates of the north, where their relics and impressions are still to be found: lastly, it explains the short period of the existence of the moral world, whose earliest monuments do not go much farther back than 5,000 years. The human race reduced to a small number of individuals, in the most deplorable state, occupied only with the immediate care for their subsistence, must necessarily have lost the remembrance of all sciences and of every art; and when the progress of civilization has again created new wants, every thing was to be done again, as if mankind had been just placed upon the earth.

There is here not only the unmistakable suggestion of global cometary disasters, but, also, cometary extinctions; and even a hint that such catastrophes have occurred through the whole previous history of the Earth. These ideas are being taken up again in our time.*

---

*Nevertheless, the advance of scientific knowledge has overtaken many of Laplace's arguments here. We know that humans were on Earth for at least a million years before we invented written history and "civilization." The fossil evidence that some mountainous terrain was once beneath the oceans puzzled many others,

1975

After Laplace, cometary catastrophism became almost fashionable. Some authors imagined debris from the comets —the clay in Donnelly's *Ragnarok*, for example—distributed over much of the Earth; others imagined the effects of the collision to be felt only in limited areas. Occasionally, consequences were imagined that were even more dire than those Laplace had depicted. In 1893 the French writer Camille Flammarion—who seemed to have had a penchant far beyond that supported by the facts for frightening people with comets—wrote a science fiction story called *The End of the World*:

> Like a great celestial projectile the solid nucleus of the comet pierced the egg-shell crust of the Earth and buried itself in the semimolten interior. The comet tore its way on like a shot piercing the boiler of a battleship. The Earth was immediately converted into a planetary volcano. Oceans were spilled like thimbles of water... continents were twisted and torn like paper.

This is rather overdoing it. But if comets are a kilometer or more across and traveling at great speed in the same part of the solar system as the Earth, it follows that sooner or later a largish comet will indeed strike the Earth, with consequences that we can be sure would be catastrophic, even if we were unable to trace exactly what would happen. This is the basis of Laplace's argument. It is easy to calculate—it can be worked out on the back of the proverbial envelope— roughly how long the Earth abides between successive collisions with sizable comets. In the first decade of the twentieth century W. H. Pickering of Harvard calculated that a fair-sized cometary nucleus should strike the Earth once every forty million years. But the dozens of collisions he imagined the Earth to have suffered during its history seemed to have done little damage, a conclusion he deduced from the fact that life is all around us.

The leading astronomy textbook in America after the end of World War II* included this paragraph:

> It is probable that the Earth has undergone many collisions with comets during geological time. It may readily be computed that a small, rapidly moving body which approaches the Sun within one Astronomical

> She first beholds the raging comet rise,
> Knows whom it threatens, and what lands destroys.
> —Juvenal,
> *The Sixth Satire.*
> Translated by John Dryden.

including Petrarch. The explanation, first advanced by Leonardo da Vinci and now firmly established by modern geology, is not a universal deluge which covered the mountaintops, but rather the slow rise of the mountaintops from the ocean floor.

Astronomy I. The Solar System by H.N. Russell, R.S. Dugan and J.Q. Stewart, rev. ed. (Boston: Ginn and Co., 1945).

The record in the sediments. This is the sharp boundary in the roadside rocks at Gubbio, Italy, corresponding to the end of both the Cretaceous Period and the Mesozoic Age of geological time. At bottom right are the Cretaceous rocks, clearly shown as white limestone when broken up a little with a hammer. The color of this chalk layer is due to the fossils of microorganisms that lived in the late Cretaceous seas. The reddish-brown rocks, top left, are from the Tertiary Period, in which no dinosaur fossils can be found and where mammals are in the ascendent. A 500-lire coin, about the size of a quarter, is shown at top for scale. Snaking through this picture along the diagonal, separating the Cretaceous from the Tertiary rocks, is a layer of gray clay. It corresponds exactly to the time that the last dinosaurs appear in the fossil record, and is rich in iridium. Figure courtesy Walter Alvarez, University of California at Berkeley.

Unit stands about one chance in four hundred million of hitting the earth. As about five comets come within this distance every year, the nucleus of a comet should hit the earth, on the average, once in about eighty million years.

But because the authors had accepted the sand-bank hypothesis of comets (Chapter 6), they did not believe in massive cometary nuclei. So they concluded that collisions of comets with the Earth "would probably fall very short of producing wholesale destruction of terrestrial life." Throughout the twentieth century scientists have intermittently connected biological extinction with cometary impacts, although the prevailing wisdom held that a catastrophe on such a scale could not be visited by so minor a body.

* * *

About halfway from Florence to Rome on the Autostrada, you make a left turn, drive through Perugia, and continue an equal distance up into the Apennines. Eventually, you come to a little village called Gubbio; it dates back to the Middle Ages. There, at the side of the road, is the sequential banding of a wonderfully preserved sedimentary column

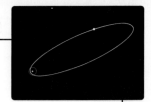

(see opposite); it antedates by millions of years any of those erected by the Caesars. If you go up to it, you can see a thin layer of pink and gray sitting above a bright white bed of rock. It marks the end of the Cretaceous Period.

You cut out a little piece and bring it back to the laboratory. The white rock is limestone, and under the microscope you can see a matrix of calcite plates and shells, manufactured by the microscopic plants and animals that lived in the warm seas. The chalk of the white cliffs of Dover, in Britain, was manufactured by such calcite-secreting oceanic microorganisms. They became extinct in the Cretaceous disaster. In fact, this is where the Cretaceous gets its name: *creta* is Latin for chalk. At Gubbio, you can clearly see that these oceanic limestones are punctuated by the gray/pink layer; it is about a centimeter thick, and made of clay. The clay and the limestone both must have accumulated as individual grains, settling through the quiet waters of the ocean tens of millions of years ago. Just above the clay, the fossils in the limestone are very different from those just below. Similar layers are found at the Cretaceous boundary all over the world. You can tell at a glance that the clay marks a catastrophe.

The kind of microscopic pollen from flowering plants changes abruptly above the clay layer—the land equivalent of the oceanic extinctions of microorganisms. Below the Cretaceous clay you find the fossils of dinosaurs who prowled the Earth for 160 million years or more. Above it there are no dinosaur fossils at all, but the remains of mammals are abundant.* This gray boundary marks a catastrophe that ended the Cretaceous Period—and with it much of life on Earth. What is in that clay? What caused the Great Dying?

Gold and platinum are valued in part because they are rare. But when you look at the spectrum of the Sun and the stars, or examine a meteorite newly fallen to Earth, you find that there is relatively much more of these precious metals up in the sky. Not that meteorites have lodes of gold or veins of platinum; but, compared to some plentiful element such as silicon, the Earth seems anomalous, strangely impoverished in precious metals. Now in molten rock, gold and platinum tend to move with iron. And the iron that was once uniformly mixed through the bodies that formed the Earth is now mainly concentrated in the liquid core of our planet, 3,000 kilometers beneath our feet. It is a good bet that most of the gold and platinum migrated there along

---

*As for humans, we have been here for 1 percent the time the dinosaurs ruled. There is no trace of us anywhere in the sedimentary column, except at the very top.

The Wedding Ring Anomaly

Of the dozens of iridium concentrations that have been found at the Cretaceous boundary, one turns out to be spurious. In ordinary rock from the crust of the Earth, the iridium content is less than a tenth of a part per billion. Yet even this is enough to be measured by neutron activation analysis. The abundance at Gubbio was much more—six parts per billion—and elsewhere still larger. But the iridium in one laboratory specimen of Montana Cretaceous clay turned out to be

> due to the platinum wedding or engagement ring worn by a technician who had prepared the samples for analysis. Platinum used for jewelry contains about 10 percent iridium…If a platinum ring loses 10 percent of its mass in 30 years, the average loss per minute, if it all deposits on a sample, is about [a hundred times] higher than our sensitivity of measurement.*

The Alvarezes and their colleagues conclude that a few seconds' exposure to a platinum wedding ring is enough to produce a spurious signal in the analysis. The more sensitive the instrument, the more careful you must be. Technicians henceforth must wear gloves.

*Walter Alvarez, Frank Asaro, Helen V. Michel, Luis W. Alvarez, "Iridium Anomaly Approximately Synchronous with Terminal Eocene Extinctions," *Science*, Volume 216, p. 886, 1982.

with the iron, when the newly formed Earth was partly molten. This is true of other elements as well, elements less well known than gold and platinum—in particular, the elements iridium, osmium, and rhodium. So if iridium were present in sufficient quantities in a particular layer of the sedimentary column, it might constitute a telltale sign of some sort of extraterrestrial intervention in earthly affairs.

In the late 1970's, a group of scientists at the University of California at Berkeley began to wonder about the clay at the Cretaceous boundary. The most prominent members of the team were Luis Alvarez, a Nobel Laureate nuclear physicist, and his son, Walter Alvarez, a geologist, both professors at Berkeley. The elder Alvarez proposed using a technique called neutron activation analysis, which could measure extremely small quantities of, among other things, iridium. The Alvarez team had the presence of mind to examine the iridium content in, above, and below the clay layer at Gubbio that marks the end of the Cretaceous. They measured the abundances of twenty-eight chemical elements through this portion of the sedimentary column, and found something astonishing. Twenty-seven of the elements showed no major changes in abundance in and out of the layer. But there was 30 times more iridium in the clay than in adjacent sediments. Similar results have now been obtained from all over the world. In Haiti, there is roughly 300 times more iridium at the Cretaceous boundary than in adjacent layers; in New Zealand 120 times more; on the shores of the Caspian Sea, 70 times; in Texas, 43 times; and in the deep ocean of the Northern Pacific, 330 times.

A worldwide iridium-rich layer marking the end of the Cretaceous looks very much like direct evidence that a large cosmic body struck the Earth 65 million years ago. You can even calculate how big the body had to be to distribute this much iridium over the Earth. The answer turns out to be about ten kilometers across, a fairly typical size for a cometary nucleus or an asteroid. Of the four comets whose diameters had been measured by radar techniques through the middle 1980's, two of them were of this magnitude. Since many Earth-crossing asteroids seem to be extinct comets (Chapter 14), the chances are apparently better than even that the Cretaceous catastrophe was triggered by a comet that hit the Earth.* If so, that layer of clay is rich in comet stuff—minerals mainly; the ices should long ago have been melted and evaporated.

From the thickness of the clay layer, it follows that the iridium was not deposited instantaneously, but over 10,000 or even 100,000 years—much longer than the timescale of a single impact. But (Chapter 16) many comets may have been involved; and particles excavated by a single impact and ejected into near-Earth orbit may have continued to fall back to Earth for long periods of time. Occasionally, volcanoes can produce anomalously high concentrations of iridium, but material from the Cretaceous boundary shows alteration of the form and chemistry of minerals that are consistent only with an enormous shock—which can be provided by a cometary impact, but not by a volcanic eruption.

This story sounds familiar. We have heard something like it before—catastrophes caused by a comet raining down a worldwide layer of clay, the very thesis propounded by Ignatius Donnelly in *Ragnarok* (Chapter 10). The catastrophes are different, the timescales are different, but the idea is unmistakably similar. The clay layers of Donnelly were not chiefly at the Cretaceous boundary, Donnelly knew nothing of iridium, and the Alvarezes were not inspired by Donnelly. If enough borderline science is written—and

1976

It does seem possible and even probable that a comet collision with the Earth destroyed the dinosaurs and initiated the Tertiary division of geologic time…It will most probably be millions of years before the next collision occurs.

—Harold C. Urey,
"Cometary Collisions and
Geological Periods,"
Nature, Volume 242,
page 32, 1973

The star which rules thy destiny
Was ruled, ere earth began, by me:
It was a world as fresh and fair
As e'er revolved round sun in air;
Its course was free and regular,
Space bosom'd not a lovelier star.
The hour arrived—and it became
A wandering mass of shapeless flame,
A pathless comet, and a curse,
The menace of the universe…

What wouldst thou, Child of Clay, with me?…

The comets herald through the crackling skies;
And planets turn to ashes at his wrath.

—Lord Byron,
Manfred, I, i, 118 and II, iv, 381
(1816)

*While, in the following pages we do not everywhere add the words "or asteroid" after "comet," we stress that there is still a significant probability that the Cretaceous impact was caused by an asteroid not of cometary origin.

enough certainly is—there are bound to be some lucky guesses, of which *Ragnarok* contains a few. But unlike the work of the Alvarez group, Donnelly's arguments are insufficient. Today the iridium speaks for itself.

If an object ten kilometers across hit the Earth at cometary velocities, it would carve out a huge crater, more than two hundred kilometers in diameter. This would be true whether the comet struck on land or ocean, because the depth of the ocean is considerably less than the size of the comet (see Chapter 7, page 130). The resulting debris—some mix of pulverized comet and pulverized earth—would be thrown high; indeed, much of it would be ejected well above the atmosphere, into space. The debris, therefore, should have been transported all over the Earth, as is observed. The clouds of fine particles, ejected to escape velocity, would travel around the Sun in orbits that repeatedly intersected the orbit of the Earth; it is possible therefore that a steady rain of fine debris would have fallen for tens of thousands of years, until the cleansing of the inner solar system by the Poynting-Robertson Effect (Chapter 14) had been completed.

A crater two hundred kilometers across is too prominent to miss. Where is it? Only three are known. Two of them are much too old, being formed more than 600 million years ago. The remaining, Popigay Crater in Siberia, seems to be some 30 million years old, and therefore too young. But the Earth is two-thirds ocean, and the Cretaceous crater was most likely formed in the ocean depths. Such a crater could

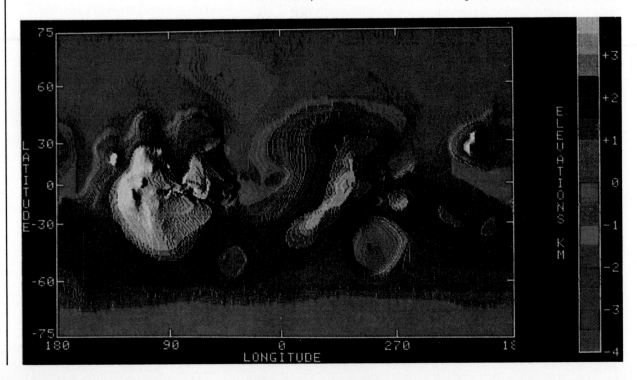

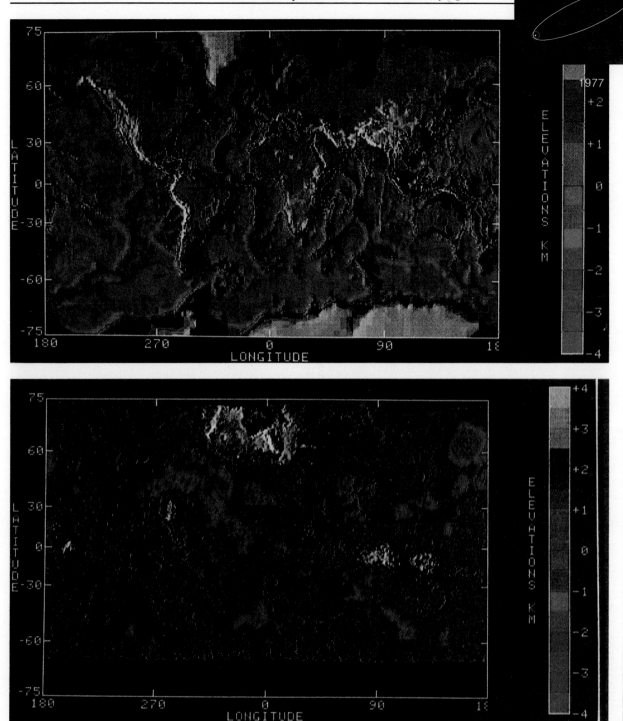

Topographic maps of three neighboring worlds—Mars (*bottom*), Earth (*top*), and Venus (opposite). On Mars there are thousands of impact craters too small to appear at this resolution, and very large craters as well—the two largest, shown here, are Hellas (near 300° longitude) and Argyre, both at southern mid-latitudes. In the absence of oceans and a thick atmosphere, Mars preserves large impact craters from a billion years ago or more. The Earth, on the other hand, has such efficient erosion due to its atmosphere and surface water, as well as plate tectonics, that there are relatively few impact craters preserved, and only a handful dating back to the early history of the planet. Venus, with a dense atmosphere and a high surface temperature (although no recent oceans), seems more like the Earth than Mars in its retention of large, ancient impact scars. These maps, based partly on spacecraft data, were prepared by Michael Kobric, Jet Propulsion Laboratory, National Aeronautics and Space Administration, and the U.S. Geological Survey.

A cometary impact with the Earth in the late Cretaceous raises an immense cloud of fine particles that are blown by the prevailing winds. No familiar continents can be discerned. (Because of plate tectonics, the continents of the Earth were very different 65 million years ago.) Omitted for clarity in this painting is the large amount of impact debris that would immediately be thrown high into the atmosphere and fill the air all over the planet. By blocking sunlight, this global obscuration is thought to have cooled and darkened the Earth. Painting by Jon Lomberg.

1978

be almost anywhere in the deep sea today, without anyone the wiser; there is, in the unclassified literature at least, no comprehensive topographic mapping of the deep sea floor. Perhaps the evidence of a vast crater remnant sits in some Soviet or American Hydrographic Office, available only, on a need-to-know basis, to the officers who command nuclear submarines. But even if a complete topographic map of the sea floor were available, we know that the Cretaceous ocean bottoms have largely been destroyed over the intervening ages by the movement of continent-sized plates back into the Earth's interior. Somewhere on the Earth, there was once a wound the size of Belgium or Corsica or Swaziland. Like a smallpox virus, the comet left a small disfigurement as a reminder of surviving a near-fatal disease. But now even the pockmark may be gone, and the thin layer of clay that

Premonition

The consequences of a collision with [a small earth-crossing asteroid] are unimaginable; the repercussions would be felt the world over. In dissipating the energy equivalent of half-a-trillion tons of TNT, 100 million tons of the earth's crust would be thrust into the atmosphere and would pollute the earth's environment for years to come. A crater 15 miles in diameter and perhaps three to five miles deep would mark the impact point, while shock waves, pressure changes, and thermal disturbances would cause earthquakes, hurricanes, and heat waves of incalculable magnitude. Should [the asteroid] plunge into the ocean a thousand miles east of Bermuda, for example, the resulting tidal wave, propagating at 400 to 500 miles per hour, would wash away the resort islands, swamp most of Florida, and lash Boston—1500 miles away, with a 200-foot wall of water…The energy involved is the equivalent of 500,000 megatons of TNT—two orders of magnitude above that involved in the largest recorded earthquake, and four or five orders of magnitude more than Krakatoa…If the strike occurred in midocean, tsunamis in the 100-foot category would cause worldwide damage. If the strike occurred on land, the blast wave would level trees and buildings within a radius of several hundred miles, and some 10^8 tons of soil and rock dust would be thrown into the stratosphere, where for several decades it would act to reduce the solar radiation ordinarily received at earth's surface and threaten the triggering of an ice age.

—MIT Student Project in
Systems Engineering,
Project Icarus,
MIT Press,
Cambridge, Massachusetts, 1968

This paragraph represents an important premonition of the Cretaceous impact, and of Nuclear Winter.

A triceratops wanders forlornly over the frozen and darkened late Cretaceous landscape in this illustration of the probable consequences of a cometary impact 65 million years ago. Painting by Don Davis.

fell from the sky may be our sole souvenir, apart from the break in the fossil record itself, of the great Cretaceous catastrophe.

But how can a cometary impact kill hundred-ton land-dwelling dinosaurs as well as microscopic algae in an ocean halfway around the world? A number of suggestions have been made: Perhaps cometary cyanides (Chapter 8) poisoned everybody, or toxic metals, or acid rain. But the probable prime mechanism was suggested by the Alvarezes themselves: If you lift a one-centimeter-thick layer of fine particles off the Earth and up into the stratosphere, the individual particles will take a year or more to fall out. And since a centimeter thickness of clay covering all the Earth is opaque, it must also be opaque when it is distributed through the upper atmosphere, slowly falling. Sunlight would not have penetrated this cloud of iridium-rich cometary clay. For months or even years, the Earth would have been darkened and cooled.

The average temperature of the Earth during the Cretaceous was some 10 degrees Centigrade warmer than it is today; by current standards, the Earth was then a tropical planet, with many lifeforms unprepared for bitter cold. Plants and animals in the tropics today, where the temperatures never fall below freezing, have few defenses against a major freeze. Calculations show that immediately after the Cretaceous impact, it would have gotten very cold all over the Earth—perhaps tens of degrees below freezing. Also, for months, the amount of light that reached the Earth's surface would have been too low for most plants to photosynthesize, and even too low for animals to see. Through seeds, spores and the like, land plants might have survived many years of cold and dark. But microscopic plants in the oceans, with no food reserves, would have died quickly—and the entire oceanic food chain which depends on those plankton would have collapsed soon after. The plight of the dinosaurs, stumbling across a frozen, somber, devastated landscape, unable to find either food or warmth, can readily be imagined. But small, warm-blooded burrowing mammals had a much better chance of surviving.

The Cretaceous extinctions thus seem to have a cause very similar to the possible consequence of modern warfare known as Nuclear Winter. Through the dust excavated by nuclear groundbursts, and smoke from the burning of "strategic targets" in and around cities, we humans can generate our own climatic catastrophe, perhaps adequate to bring about massive extinctions in our age as in the Cretaceous. The principal difference is that the dinosaurs did not contrive their own extinction.

It seems likely that had not a comet or asteroid hit the Earth 65 million years ago (or subsequently), the dinosaurs would still be here, and we would not. We would be merely one of the countless unrealized possibilities in the genes and chromosomes of the other beings of Earth—as we can be again if we do not come to our senses.

1978

The dust lanes in the plane of the Milky Way galaxy and its bulging core are shown in the distance, as the Sun and the Earth approach a giant interstellar cloud. Painting by Jon Lomberg.

Chapter XVI

THE WRATH OF HEAVEN: 2.A MODERN MYTH?

Believing that every object and every event in the universe is arranged and directed by an Omnipotent Contriver, we must admit that when the Almighty formed the wondrous plan of creation, "foreseeing the end from the beginning," he arranged the periods and the velocities of comets in such a manner that, although occasionally crossing the planetary orbits, they should not pass these orbits at the time when the planets were in their immediate vicinity. And should such an event ever occur, we may rest assured that it is in perfect accordance with the plan and the will of Omnipotence, and that it is, on the whole, subservient to the happiness and order of the intelligent universe, and the ends intended by the Divine government.

> —Thomas Dick,
> *The Sidereal Heavens and
> Other Subjects Connected with Astronomy,
> As Illustrative of the Character of the Deity
> and of an Infinity of Worlds*, Philadelphia (1850)

I feel rather at a disadvantage in speaking of comets to you; comets nowadays are not what they used to be.

> —Arthur Stanley Eddington,
> "Some Recent Results of
> Astronomical Research,"
> Friday-evening discourse,
> The Royal Institution, London,
> March 26, 1909

STATED BALDLY, IT STILL SOUNDS LIKE SOMETHING from the early days of pulp science fiction: 65 million years ago, a comet came out of space, hit the Earth, and extinguished much of the life on this planet. The Alvarez discovery has created something like a scientific revolution, with consequences propagating through geology, astronomy, and evolutionary biology, and even making incursions into international politics and nuclear strategy. The connection between cosmic events and our own existence is stirring.

But the Cretaceous extinctions are not the only, or even the major, instance of mass extinctions in the history of life on Earth. And—as Laplace and others realized—there must have been many impacts of comets (and asteroids) with the Earth. Iridium concentrations have been found in another sedimentary layer, associated with another mass extinction (although Permian sediments—associated with a major mass extinction on everybody's list—shows so far not a hint of an iridium anomaly). The Alvarez discovery set many other scientists from a range of disciplines thinking about cometary impacts and mass extinctions, which has led to still more extraordinary claims. However, these recent developments are more speculative, and—since some of the ideas contradict each other, and for other reasons—we must here tread cautiously.

The new developments began at the University of Chicago, where the American paleontologist J. John Sepkoski, Jr., had been compiling a master list of all the families of marine life recorded in all the sedimentary rocks. By far the great preponderance of families of life identified in the history of the Earth are now extinct. Sepkoski tabulated the epoch in which each extinct family died out. He could see a background rate of extinction, more or less constant through geological time, due to many causes—mountain building, greenhouse effects, disease, Darwinian competition, and the like. Superimposed on this background were times when the extinctions all piled up, one on top of another, as at the end of the Cretaceous, and during the other Great Dyings (see table, Chapter 15, page 274). So far nothing much new, although the family extinction data were more encyclopaedic than anything compiled before.

But when Sepkoski, and David Raup, also of the University of Chicago, analyzed the extinction dates, they were surprised to find what looked like a periodicity. Every 26 million years or so, like clockwork it seemed, plants and animals drop dead all over the planet. At the family, genus and species levels, they concluded, a major fraction of all life on Earth becomes extinct at apparently regular intervals. Extinction. Scientists tend to talk about it with detachment, but there is something unnerving about these count-

The implications of periodicity for evolutionary biology are profound. The most obvious is that the evolutionary system is not "alone."...With kill rates for species estimated to have been as high as 77% and 96% for the largest extinctions, the biosphere is forced through narrow bottlenecks and the recovery from these events is usually accompanied by fundamental changes in biotic composition. Without these perturbations, the general course of macroevolution could have been very different.

—David M. Raup and
J. John Sepkoski, Jr.,
"Periodicity of Extinctions in the Geologic Past," *Proceedings of the National Academy of Sciences of the U.S.A.*,
Volume 81, p. 801, 1984

1980

less lives being snuffed out, their ancestral lines rendered meaningless, by some regular motion of the Grim Reaper's scythe. Naturally, we wonder if we may be next.

Paleontologists have for decades invoked a wild variety of explanations for the mass extinctions, but none of the proposed mechanisms had a period built in. The Earth does not seem to have any 26- or 28- or 30-million-year-long cycles—not in volcanos, not in plate tectonics, not in the weather. Periods this long fall in the domain of astronomy.

The iridium in the Cretaceous boundary has made the idea of periodic (or at least episodic) mass extinctions, visited on the Earth from space, more palatable; but for many scientists, and others, it remains a bitter pill to swallow. Even if for the moment we grant that small worlds fall out of the sky like clockwork, shouldn't this fact be evident from the ages of craters on Earth? The craters are there—little recent ones, not yet eroded away, and larger and older craters, the biggest being more than 100 kilometers across. Still larger craters must have been made, of course, but that was so long ago in Earth's history that the slowly crashing and diving plates that make up the Earth's crust would have destroyed them long ago. Thus newly motivated, several

A dinosaur contemplates its fate as the sky is ablaze with comets during the postulated late Cretaceous comet shower. The dinosaur depicted had something like hands and a larger brain size for its body weight than most of its contemporaries. If the dinosaurs had not been rendered extinct, perhaps the dominant form of intelligent life on Earth today would have descended from such a creature. Painting by Jon Lomberg.

Glass

Certain sedimentary layers of the Earth are strewn with glassy, geometrically smooth inclusions, ranging from a few centimeters in size to submicroscopic. They are called tektites. That tektites are produced when comets hit the Earth was first proposed by the American chemist Harold Urey in 1957, and—while not definitively decided—the hypothesis has stood the test of time. The comet strikes the Earth, digs out a large crater, and melts the underlying ground in the process; droplets of silicates are flung over huge distances, freezing in the process and producing streamlined, sometimes teardrop-shaped, forms. The tektites show signs that they themselves were cratered during these violent events (see opposite). Those shown here were produced a little over 35 million years ago. It has been claimed and disputed that layers of sediment containing the so-called microtektite horizon are associated with the Eocene extinctions. One of the leading experts on these tiny glass forms is named, appropriately enough, Billy Glass. He is at the University of Delaware.

...The evidence for [periodic mass extinctions] is strongly contingent on arbitrary decisions concerning the absolute dating of stratigraphical boundaries, the culling of the database and the definition of what is mass extinction as opposed to background extinction. This evidence becomes insufficient under other plausible geological timescales and other acceptable definitions of mass extinction. Analysis of the non-culled database shows that the reliability of identification of mass extinctions and their timing is at present extremely limited. It also suggests that the apparent periodicity of mass extinctions results from stochastic [random] processes.

—Antoni Hoffman,
"Patterns of Family Extinction Depend on Definition and Geological Timescale,"
Nature, Volume 315, p. 659, 1985

teams of scientists have examined the ages of the surviving craters, and have seemingly found—to simultaneous satisfaction and amazement—that craters tend to have been excavated, by the impacts of objects from space, around every 28 million years, very close to the inferred period of the mass extinctions. What's more, the events are reported to have been in phase: mass extinctions tend to happen when large craters are formed, the two events presumably caused by the same impacting bodies.

The correlation is not one-to-one. In sediments older than the Cretaceous, dating of these boundaries has an uncertainty of many million years or more; this is one reason—there are several—why a precise concordance between crater ages and extinction times could not exist. Some cosmic objects, as probably at the end of the Cretaceous, must have crashed into the oceans, leaving no visible crater. And, as we would expect, there are random craters that are not part of any 28-million-year cycle.

Since these heady announcements were made in the scientific literature, a number of scientists have gone back to the original data to check the reliability of the conclusions drawn. How complete is the list of craters? How accurately determined are the times of impact? How do you define a mass extinction? How reliable are the dates of cratering or mass extinctions? Suppose cratering and extinction events happened entirely randomly; how likely is it that you would by chance find spurious evidence of periodicity? The results of this reanalysis were not yet, in the mid-

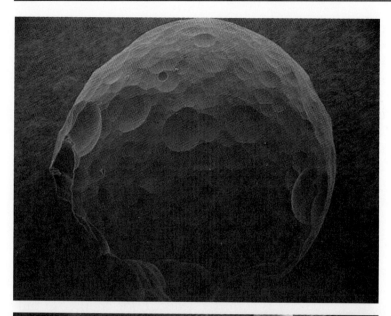

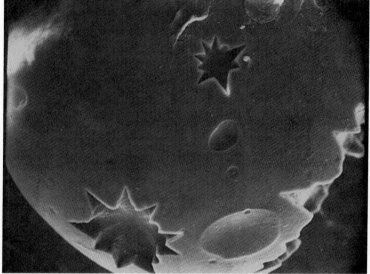

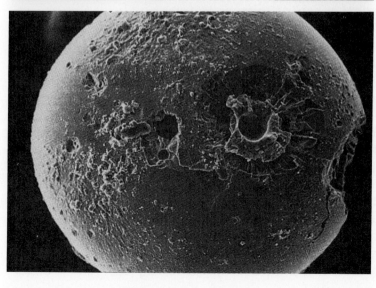

Three samples of microtektites. At top from Australasia, at a magnification of 50 times; at bottom from the Ivory Coast at magnification 190 times and at center; a microtektite about 240 microns in diameter with strange star-shaped craters. Courtesy B. P. Glass, University of Delaware.

dle 1980's, fully in, but in the heated debates that have ensued it is clear that the foundations of several sciences are being constructively reassessed. There are those who believe in an in-phase periodicity of cratering and mass extinctions. There are those who believe that cratering may be periodic but mass extinctions not, or the reverse. And there are many scientists who hold that, within the present uncertainties, there is no compelling evidence for periodicities either in cratering or in mass extinctions.

If this last group is right, then the remainder of this chapter is only a flight of fancy. But if there are such periodicities, a stunning connection between life down here on Earth and events up there in the sky has been uncovered. In that case, somewhere a cosmic doomsday clock is ticking away even now. Fortunately, we are today midway between mass extinctions. The next bombardment is not scheduled, if the periodicity is correctly deduced, for another 15 million years. Our job is to avoid becoming one of the background extinctions; in 15 million years we may be ready for the comets.

But how could any object in space know when its time had come to hit the Earth? What cosmic machinery could serve as a doomsday clock? Take, for example, the present population of short-period comets, or the hundreds of asteroids that cross Earth's orbit. They will strike the Earth from time to time, of course, and contribute to the background rate of cratering. Indeed much of the background iridium in rocks, unassociated with biological catastrophes, may also come from comets—the steady infall from meteor streams and the zodiacal dust. But this will be a constant pitter-patter; these worlds do not save themselves up for mass assaults on the Earth every 30 million years—or, for that matter, on any other time scale. Periodic impacts must be caused by other bodies, more distant.

By the mid-1980's two quite different proposals were in contention, each purporting to show how a cosmic metronome might beat out extinctions at 30-million-year intervals. Neither is fully satisfying; both have deficiencies. It is hard even to state the two hypotheses without seeming a little lurid, and the more sensational recountings in the popular press have in the minds of some scientists confirmed their initial misgivings. But together, they provide a good example of science in transition: A powerful conclusion is drawn from disputed data. But the conclusion poses an enigma. Two hypotheses attempt to explain the enigma in different ways. If science is to be served, the hypotheses should make different predictions about what you will find if you perform a particular sort of new experiment. Experimental confirmation of quantitative predictions is to sci-

Comet Showers

1981

Before a periodicity of extinctions and cratering was proposed, J.G. Hills of the Los Alamos National Laboratory suggested* that a passing star—not a companion star—in the inner Oort Cloud could stir up a storm of comets:

> The observed comet cloud may be only the outer halo of a much more massive comet cloud whose center of mass is well inside the observed inner boundary of the Oort Cloud.

Hills then goes on to deduce that a star passing only 3,000 Astronomical Units from the Sun would produce a cometary shower from the inner Oort Cloud that would generate one new comet per hour in the vicinity of the Earth. This would have many consequences, but the first one that Hills, an astronomer, mentions is this:

> Such a high comet flux would be a major nuisance to astronomical observers engaged in research on low-light objects!

This is another of the rare astronomical exclamation points. Hills then continues:

> The integrated comet flux from such a shower can be great enough that several comets will actually hit the Earth during the shower. This may show up in the geological record.

*J.G. Hills, "Comet Showers and the Steady State Infall of Comets from the Oort Cloud," *Astronomical Journal*, Volume 86, p. 1730, 1981.

ence what the fulfillment of prophecy is to religion.

In one of the two hypotheses, life on Earth is periodically extinguished because the Milky Way Galaxy has the shape first understood by Wright and, especially, Kant (Chapter 4). Our Galaxy is a thin disk that contains the spiral arms and that rotates about a bulging core of stars and dust (frontispiece, this chapter). It is at the center of the Milky Way that huge numbers of stars are concentrated; it is, in brilliance, mass, position, and explosive violence, the downtown of the Galaxy. Fortunately for us, we live nowhere near it, nor even in the suburbs. Our home is in an obscure galactic countryside, where stars like the Sun take 250 million years to go once around that distant hub.

But in addition to revolving around the galactic center, the Sun has another motion: it is bobbing up and down, each time passing through the imaginary plane of symmetry that cuts through the galactic center. When at its maximum excursion, about 230 light-years above the plane, the Sun is

The motion of the Sun in the Galaxy. The Sun orbits the massive, reddish, bright, bulging core of the Milky Way Galaxy once every quarter billion years. As the Sun slowly circumnavigates the center of the Galaxy it bobs up and down at a faster pace, making a bounce every 60 million years or so, and therefore crossing the plane of symmetry of the Galaxy roughly once every 30 million years. Does this 30-million-year period somehow trigger mass extinctions on the Earth? Diagram by Jon Lomberg/BPS.

gravitationally attracted by the gas and dust and stars beneath, slowly reverses direction, and falls back. But the galactic plane is an imaginary, not a real, surface, and when the Sun arrives there, it finds it has a sizable velocity and nothing to stop it. Accordingly, it plummets through and out the other side, slowing because of the gravity of the dust and stars it has left behind—until it is some 230 light-years on the other side of the galactic plane, stops, and then falls back again.

Since the Sun is traveling through a vacuum more nearly perfect than anything we know, there is no friction to retard its motion. Like a weight on a perfectly elastic spring, the Sun caroms up and down forever. The oscillating Sun is effectively a perpetual motion machine. As long as a great deal more matter doesn't move into or out of our neighborhood in space, this galactic bounce will continue forever. The period of oscillation depends only on how much mass there is in the space near the Sun, and this seems to be well-measured by astronomers. Also, the Sun's motion can be measured by examining the nearby stars with respect to which our system is moving. In both these ways it is found that the period from one crossing of the galactic plane to the next for the Sun and its entourage of planets and comets

is about 33 million years. In fact, all the nearby stars are bobbing up and down, crossing the galactic plane once each 30 million years or so. This period is fundamental for millions of suns. You look at the ages of fossils and craters on Earth, and you find a time scale of almost 30 million years. You look at the Sun bobbing in the Galaxy and you find another time scale of roughly 30 million years. It's hard not to think that the two periods might be related, that the oscillation causes the extinctions.

But how could the Sun's motion instruct comets to collide regularly with the Earth? There are giant molecular clouds spread out in dribs and drabs in our part of the Milky Way. They do not all move at the same speed as the Sun, and every now and then the Sun is bound to run into one of them. Extending over a much larger volume than the solar system, they are much more massive than the solar system. Passing by or through such a cloud would gravitationally produce a great flurry among the comets of the Oort Cloud, and pry loose a shower of comets to descend on the Earth and its neighboring worlds.

For this metronome to work, though, the clouds must be concentrated in one particular place—the galactic plane, probably. Then, every 30 million years, the Sun plunges

The Sun, at center, and its entourage of comets, shown in blue, encounter a massive interstellar cloud. The resulting gravitational perturbation sends a shower of comets into the inner solar system. Such showers must happen statistically from time to time. According to one view, they also happen periodically because of the Sun's motion above and below the galactic plane. Diagram by Jon Lomberg/BPS.

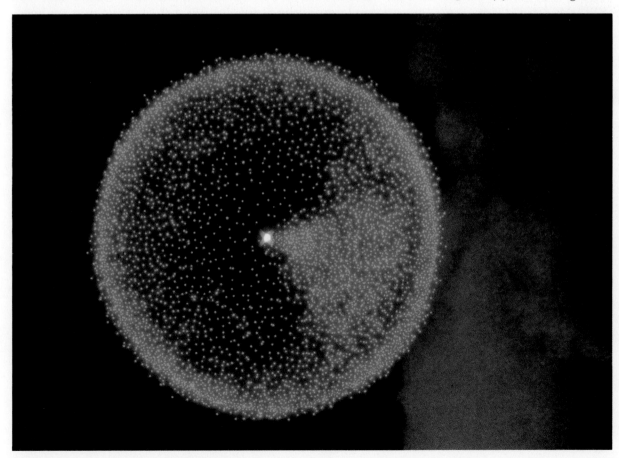

through the plane (alternately from above and from below), a shower of comets is launched toward the inner solar system, one or more of them strikes the Earth, the lights go out and the temperatures drop. If the interstellar clouds lived in the galactic plane, the oscillating Sun explanation of periodic mass extinctions would seem very promising.

However, for the whole 230 light-years of the Sun's maximum excursion from the galactic plane, the clouds are distributed essentially at random. There is no more reason to bump into one in the galactic plane than there is 230 light-years above or below. Indeed, in the present epoch the solar system is only about 25 light-years above (north of) the plane of the Galaxy, and there is no giant interstellar cloud at our doorstep. The extinctions should, therefore, occur at random. The roughly 30-million-year period of galactic plane-crossing by the Sun does not seem to translate into a roughly 30-million-year cycle of mass extinctions on Earth (even ignoring the differences among 26-, 28- and 30-million-year periods). Maybe there's some other way in which the Sun's periodic bobbing in and out of the galactic plane can stir the comets up and trigger global catastrophes on Earth. Maybe there's a flattened population of small primordial black holes in the Galaxy, through which the Oort Cloud runs as the Sun bobs. But if there is, no one has been able to find it so far,* and without something of the sort, this hypothesis, tantalizing as it is, remains unsatisfactory. What is the alternative?

Most of the stars in the sky are members of double or multiple star systems. In a typical binary system, two stars separated by several Astronomical Units are doing a stately gravitational fandango. Often the stars are more widely separated. In some instances we see two stars gravitationally bound to each other, but separated by 10,000 A.U. At least 15 percent of the stars in the sky seem to have a companion star at this distance. The nearest star system to the Sun— Alpha Centauri, 4.3 light-years away—is a double star with a third sun, a distant dim companion called Proxima Centauri, at 10,000 A.U. from the two bright stars. Often the companion star is very faint, suggesting that there may be many still undiscovered widely separated binaries. It is possible that most of the stars in the Galaxy are so dim that astronomers call them brown or black dwarfs. Most distant companions might be of this sort.

The solar system seems to be an exception. We do not know of any companion to the Sun. But if we were *not* an

---

*Although it is fair to note that almost half the mass in the Milky Way is "missing"—its gravitational influence is felt, but it corresponds to no known objects, neither stars nor gas nor dust.

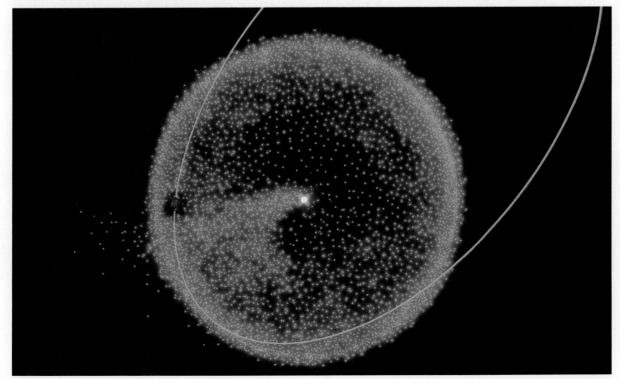

1983

exception, if the Sun *had* an invisible companion star in a very specific orbit, then the extinction clockwork again might be understood. Suppose there was a companion star on a very long elliptical orbit—so that, on average, it was 90,000 A.U. away, 1.4 light-years. But once each orbit it comes much closer to the Sun, maybe 10,000 A.U. or even a little closer. This would bring the star into the inner part of the Oort Cloud, where comets are not ordinarily jostled by passing stars. With such an orbit, once every 30 million years the companion would plow through the denser parts of the Oort Cloud, and shower the Earth and neighboring worlds with comets.

The year of this star would be 30 million of ours. Ten times around the Sun would carry it from the Permian, when there were dragons with sails on their backs, to now, when humans reconnoiter the planets and menace the Earth. At present, it would be near aphelion, the furthest point in its orbit from the Sun. Fifteen million years from now, it would run its next rampage among the comets. During its plunge through the inner Oort Cloud, the companion star would spray a billion comets into the inner solar system. But because they would be carried on slightly different trajectories, the comets would not all arrive at the same instant. Rather, they would be spread out over a million years or more. Thus, the companion star hypothesis predicts that the background cratering and extinction rates will be periodically interrupted by intervals lasting as long

The Sun, at center, and its entourage of comets encounters a small, dim star indicated by the red dot. The resulting gravitational perturbation sends a shower of comets into the inner solar system. Such encounters must happen statistically from time to time. In addition, one view holds that the Sun has a companion star on a highly eccentric orbit that enters the Oort Cloud around its perihelion, and therefore periodically sends comets cascading in toward the planets. Diagram by Jon Lomberg/BPS.

as millions of years, in which the Earth is hit by several, perhaps several dozens of comets. This would explain why mass extinctions do not occur instantaneously but sometimes over a period as great as millions of years.

The hypothetical companion star's period around the Sun cannot be exactly 30 million years; indeed, it cannot be exactly anything: gravitational perturbations by stars, interstellar clouds and the massive core of the Galaxy will tug it first one way and then another, altering its orbital period. Eventually, any such orbit will be changed greatly. A companion star in the present orbit could survive at most a billion years before it is stripped away by passing stars and the general gravitational pull of the Milky Way. This is long enough to account for periodic extinctions and impact craters back to the Permian—but, before that, the companion star would have had a very different orbit. If a companion star began much closer to the planets, it would have produced major and almost continuous cometary showers early in the history of the solar system, and might even be responsible for the "late heavy bombardment" of the Moon. However, such a greatly enhanced comet flux early in solar system history is expected anyway, as comets were cleared out from the region around Uranus and Neptune—being propelled both outward away from the Sun to populate the outer Oort Cloud, and inward to crater the terrestrial planets (Chapter 12).

But before we seriously consider such extravagant possibilities, the hypothesis must face one awkward fact: There is, to date, no evidence whatever of a companion star. It need not be very bright or very massive; a star much smaller and dimmer than the Sun would suffice, even brown or black dwarfs—giant planet-like bodies insufficiently massive to generate hydrogen fusion at their cores and burst forth with stellar light. It is conceivable that a companion already exists in one of the catalogues of dim stars without anyone noting something anomalous—an enormous apparent motion of the star each year against the background of more distant stars (parallax, Chapter 2). There is at least one major search for dark, cool stars now underway, designed specifically to catch a companion star, if one exists. If it *is* found, and in something like the right orbit, few will doubt that it is the principal cause of periodic mass extinctions on Earth.*   Otherwise, it will remain provocative but unproved.

---

*One of our scientific reviewers notes at this place in the margin, "If pigs had wings…" Indeed, not only has the companion star not been found, but there is no other evidence that requires its existence, beyond the debatable periodicities themselves.

1984

A light-year from the Sun, a still undetected and perhaps wholly mythical dwarf star companion to the Sun glowers redly. Nearby are a hypothetical planet and comets—of which there are trillions thereabout—all feebly illuminated. Painting by Anne Norcia.

But it is a notion of mythic power. If an anthropologist of a previous generation had heard this story from his informants, the resulting scholarly tome would doubtless use words like "primitive" or "prescientific" or "animistic." Consider:

The Sun has a Dark Sister. Long ago, before even great grandmother's time, the two suns danced together in the sky. But the Dark One was jealous that her sister was so much brighter, and in her rage she cursed us for not loving her, and loosed comets upon the world. A terrible winter came, and darkness fell and bitter cold, and almost every living thing perished. After many seasons, the Bright Sister returned to her children, and it was warm and light once more, and life was renewed. But the Dark Sister is not dead. She is only hiding. One day she will return.

This is why some scientists thought the theory of the death star a joke when they first heard of it. An invisible sun attacking the Earth with comets sounds like delusion or, at best, myth. But the theory is serious and respectable, if highly speculative, science, because the principal idea is testable: You find the star and examine its orbital properties. It is no wonder, though, that one of the groups of scientists that first proposed the theory reached to mythology to name the star, and agreed to call it Nemesis, after the Greek goddess who visited just punishment on the self-righteous. They then went on to add, "We worry that if the companion is not found, this paper will be our Nemesis."

If there is no Nemesis, some form of the galactic oscillation theory will become more probable, and we will have learned much about dim nearby stars, and perhaps something about the "missing mass" as well. Or it will strengthen the case of those who doubt there are periodic extinctions. There is no way that science loses. For the grandeur of its vision, and the new findings that will emerge even if there is no companion star, this theory is in the mainstream of science, a fact that is sometimes poorly appreciated in the press. The *New York Times*, on April 2, 1985, decided to take an editorial stand in this scientific debate, decrying the fact that discussion of extinctions

was suddenly snatched away by two brash Berkeley scientists and a crowd of astronomers…Terrestrial events, like volcanic activity or changes in climate or sea level, are the most immediate possible causes of mass extinctions. Astronomers should leave to astrologers the task of seeking the cause of earthly events in the stars.

In a letter dated April 14, 1985, Richard Muller and

A more disastrous situation than a comet shower would arise if Nemesis were perturbed into an orbit that takes it into the planetary system. In this case, it may strip some planets directly from the Solar System. It would leave the remaining planets in highly eccentric orbits which would produce a fast-moving, dangerous situation…A Nemesis with a semi-major axis of 90,000 A.U. is dangerous to life on Earth, but a Nemesis at the inner edge of the Oort Cloud would probably have produced a catastrophe of truly cosmogonical proportions.

—J.G. Hills,
"Dynamical Constraints on the Mass and Perihelion Distance of Nemesis and the Stability of Its Orbit," *Nature*, Volume 311, p. 636, 1984

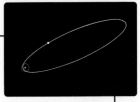

Walter Alvarez of the University of California's Berkeley campus replied to the *Times,* concluding:

> You say, "Complex events seldom have simple explanations." The entire history of physics contradicts you.
>
> You suggest "Astronomers should leave to astrologers the task of seeking the cause of earthly events in the stars." May we suggest it might be best if editors left to scientists the task of adjudicating scientific questions?

The debate is likely to continue.

As for the name for the death star, should it be found, the American paleontologist, Stephen Jay Gould, argues that Nemesis is not quite right:

> Nemesis is the personification of righteous anger. She attacks the vain or the powerful, and she works for definite cause...She represents everything that our new view of mass extinction is struggling to replace— predictable, deterministic causes afflicting those who deserve it...
>
> Mass extinctions are not unswervingly destructive in the history of life. They are a source of creation as well ...Mass extinction may be the primary and indispensable seed of major changes and shifts in life's history... Moreover, mass extinction is probably blind to the exquisite adaptations evolved for previous environments of normal times. It strikes at random or by rules that transcend the plans and purposes of any victim.

Gould proposes instead that the companion star be named after the Hindu deity Siva, the Destroyer, for whom destruction is prerequisite to creation, and vice versa:

> Siva does not attack specific targets for cause or for punishment. Instead, his placid face records the absolute tranquility and serenity of a neutral process, directed toward no one, but responsible for the maintenance and order of our world.

How characteristic of us, Gould is saying, that associating comets with "the wrath of heaven"—which Laplace denounced as superstition—should find its way into the most recent scientific views on the connection between comets and life on Earth.

These foreign planets, with their tails and their beards, have a terrible menacing countenance. It may be they are sent to affront us.

—Bernard de Fontenelle,
The Plurality of Worlds,
Paris, 1686

Waters flow briefly on Mars after a cometary impact. Painting by Don Davis.

Chapter XVII

THE ENCHANTED REGION

We have now advanced to the extreme boundary of the solid ground of our knowledge of comets. Before us lies the enchanted region of speculation.

—William Huggins,
On Comets,
Friday-evening discourse,
The Royal Institution
London, January 20, 1882

EVERYTHING YOU SEE AROUND YOU, ALIVE OR INANI-mate, has fallen from the sky—or, at least, the constituent atoms have. There was a time before the Earth was formed, when these atoms were parts of small worlds busily manufacturing planets as they collided. And there was a time before that, when all the atoms in and around you, every one of them, were in gas and microscopic grains floating in interstellar space. After the Earth was fully formed, more cosmic matter, chiefly from comets, glued itself to the Earth's surface. Everything on Earth is sky stuff—collected, for a while at least, on this lump of metal, rock, and water.

In the scientific literature, as in myth and folklore, the implications of comets for life tend to be on the dire side. Only occasionally is there a suggestion that comets may be benign. The earliest such statement seems to have been made by Isaac Newton in the *Principia*. He argues that cometary matter accumulates on Earth:

> The vapors which arise from…the tails of the comets may meet at last with, and fall into, the atmospheres of the planets by their gravity, and there be condensed and turned into water and humid spirits; and from thence, by a slow heat, pass gradually into the forms of salts, and sulphurs, and tinctures, and mud, and clay, and sand, and stones, and coral, and other terrestrial substances.

There is a clear suggestion here that comets are made chiefly of water, although no justification of this assertion is offered. Newton goes on to argue that whatever cometary "vapors" arrive on Earth just balance those that are being lost, presumably to space, and that this contribution is essential for the continuing fecundity of the Earth. He concludes surprisingly, in an almost mystic reverie:

> I suspect, moreover, that it is chiefly from the comets that spirit comes, which is indeed the smallest but the most subtle and useful part of our air, and so much required to sustain the life of all things with us.

Comets supply, Newton seems to be saying, something essential for life on Earth. But calling it "spirit" suggests it is not composed of matter. Demonstrating its presence in comets would then seem a considerable undertaking, and Newton does not attempt it.

Other hopeful views of the biological role of comets were still more dubious. E.-J.-F. Sales-Guyon de Montlivault proposed in 1821 that originally people came to Earth by falling from the sky. A comet was implicated. Montlivault noticed the absence of human remains in most rock strata rich in the fossils of animals. Writing almost forty years before *The Origin of Species*, it seemed plausible to him that humans had originally grown up on the Moon, that the gravitational pull of a passing comet had splashed some lunar ocean onto the Earth, and that in that ocean were our ancestors. Perhaps they sailed reed boats and spoke ancient Egyptian. Evolution freed us from the necessity of such extravagant hypotheses.

If the comets *were* to contain something special for life—something that starts it, say, or something that stops it—there would be ample opportunity for that something to take effect. All over the Earth, all the time, there is a fine rain of cometary debris, particles ultimately derived from the interstellar medium. They fall into the atmosphere, are carried out by rain and snow, and become incorporated into plants and animals. We are surrounded by cosmic particles in our everyday life. We eat them and breathe them. And we are made of them. Some of these extraterrestrial atoms are incorporated into sperm and egg cells and carried down to future human generations.

In addition to the fluffy, organic-rich particles found in the stratosphere (Chapter 13), there are other, slightly larger, tokens of the comets. For example, if you dredge the ocean bottoms and then, after sieving and magnetic separation, examine all fine particles smaller than a grain of sand, you will find tiny, metal-rich glassy spheres. They are present on land also, but because the land is now so polluted by fine particles of human manufacture, the spherules are harder to find there. Since their appearance is so similar all over the Earth, you might think they were the product of some widely distributed small animal—eggs, perhaps, or droppings. But they are made uniformly of a kind of glass and metal, and are certainly not biological. Their nickel/iron ratio shows them to be extraterrestrial in origin.

These spherules are the remnants of small, originally irregular objects that entered the Earth's atmosphere at high speed, were melted by friction with the air, and slowly fell through the atmosphere. They are called meteor ablation spheres—droplets that cooled and resolidified, teardrops falling on the Earth, a little like, but much more numerous than, the microtektites (Chapter 16). If you examine them, you find those chemical elements that are easily volatilized to be underrepresented. Carbon in organic matter is such an element, and sulfur. They have evidently been boiled off during entry. But other, more refractory elements are present in the usual cosmic proportions. The ablation spheres look like small pieces of cometary dust that melted, but otherwise survived high-speed entry into the Earth's atmosphere. When interplanetary particles collected at stratospheric altitudes are melted experimentally in the laboratory, they take on the physical and chemical properties of the little spheres found on the ocean bottoms, and, more rarely, in the stratosphere itself. And about 10 percent of the spheres collected from oceanic sediments contain grains of the parent material that have by accident escaped melting. To clinch the argument, the spherules contain radioactive isotopes such as manganese 53 that plausibly could be produced only by cosmic rays in space. Between the droplets and the flakes, we are awash in comet stuff.*

Through the effects of wind and water, temperature and the activities of microorganisms, all this cosmic debris becomes eroded, abraded, broken down, systematically converted into finer and finer grains, and ultimately reduced to individual molecules and atoms. The ponderous, multi-million-year circulation of the Earth's crust and upper mantle carries the remains of cosmic particles deep below our feet; hundreds of kilometers down, material that once resided in the comets is part of some great convection cell which, tens of millions of years into the future, will surface and create some new continent or mountain range. The comets may seem remote to the point of myth or fable; but, in fact, they have a most lucid reality—they are a part of our planet and ourselves.

* * *

Of all the materials imagined to fall from comets, the most enduring—both in folklore and in science—have been, oddly, germs, the causative agents of disease. The worldwide association of comets with pestilence is striking, transcending cultural differences, and it is tempting to consider whether comets might in fact and not just in fancy be the carriers of epidemics. We recall, of course, that through most of human history, disease has been rife, so that purely by chance the apparition of a comet must on occasion coincide with the outbreak of pestilence somewhere. For example, the belief that the Great Plague of London, and the Great Fire, were caused by comets was sufficiently widespread to be discussed by Daniel Defoe in his *Journal of the Plague Year.* In 1829 a book by T. Forster was published called *Atmospherical Causes of Epidemic Diseases*, which argued for the same connection.

---

*The ablation spheres are also found in dust samples collected at stratospheric altitudes, but they are dominated by the smaller, more fragile particles which were able to cool rapidly on entry and avoid melting. Fewer of the spheres fall on the surface; but they are comparatively hardy. Down on the ground, it's much easier to find ablation spheres than organic flakes.

The Comets of Plague and Fire

The comet before the Pestilence was of a faint, dull, languid colour, and its motion was very heavy, solemn, and slow; but... the comet before the Fire was bright and sparkling, or, as others said, flaming, and its motion swift and furious: and that accordingly one foretold a heavy judgment, slow but severe, terrible and frightful, as was the Plague; but the other foretold a stroke, sudden, swift, and fiery, as was the Conflagration. Nay, so particular some people were, that, as they looked upon the comet preceding the Fire, they fancied that they not only saw it pass swiftly and fiercely, and could perceive the motion with their eye, but even that they heard it; that it made a mighty rushing noise, fierce and terrible, though at a distance and but just perceivable. I saw both these stars, and must confess had I had so much the common notion of such things in my head...I could not but say, God had not yet sufficiently scourged the city.

—Daniel Defoe,
A Journal of the Plague Year,
London, 1722

1985

The idea is remarkably hardy, and has found a curious modern reincarnation. Fred Hoyle is a leading British astrophysicist who has made important contributions to cosmology, and to our understanding of the synthesis of chemical elements in the stars. He has also written popular science, fiction, and the libretto of an opera. Hoyle and the Sri Lankan astrophysicist N.C. Wickramasinghe have examined the timing and geographical distribution of some cases of epidemic disease, and conclude that the pattern is not that expected if the disease propagates by contagion on the Earth's surface, but rather is consistent with the pathogens falling from the sky. Most epidemiologists vigorously disagree.

Moreover, Hoyle and Wickramasinghe hold that bacteria and viruses are sprinkled throughout interstellar space; indeed, they boldly propose that interstellar grains—which have about the same size and atomic composition as bacteria—are in fact bacteria. If true, bacteria would have been incorporated into comets as they condensed out of the solar nebula, which would itself have been made partly of microorganisms.

While the sweep of this proposal is very grand, its supporting evidence is sparse. The elaborate molecular machinery employed by both sides in the continuing warfare between infection and immunity speaks of a long evolutionary relationship between humans and disease microorganisms; this would be hard to arrange if the microbes had

Next page: The early Earth at the time of the origin of life. The Moon, much closer then, hangs hugely over the horizon. Organic molecules are carried to the primitive oceans by the infall of cometary matter and are generated locally in the hydrogen-rich atmosphere by lightning and ultraviolet light. Painting by Kazuaki Iwasaki.

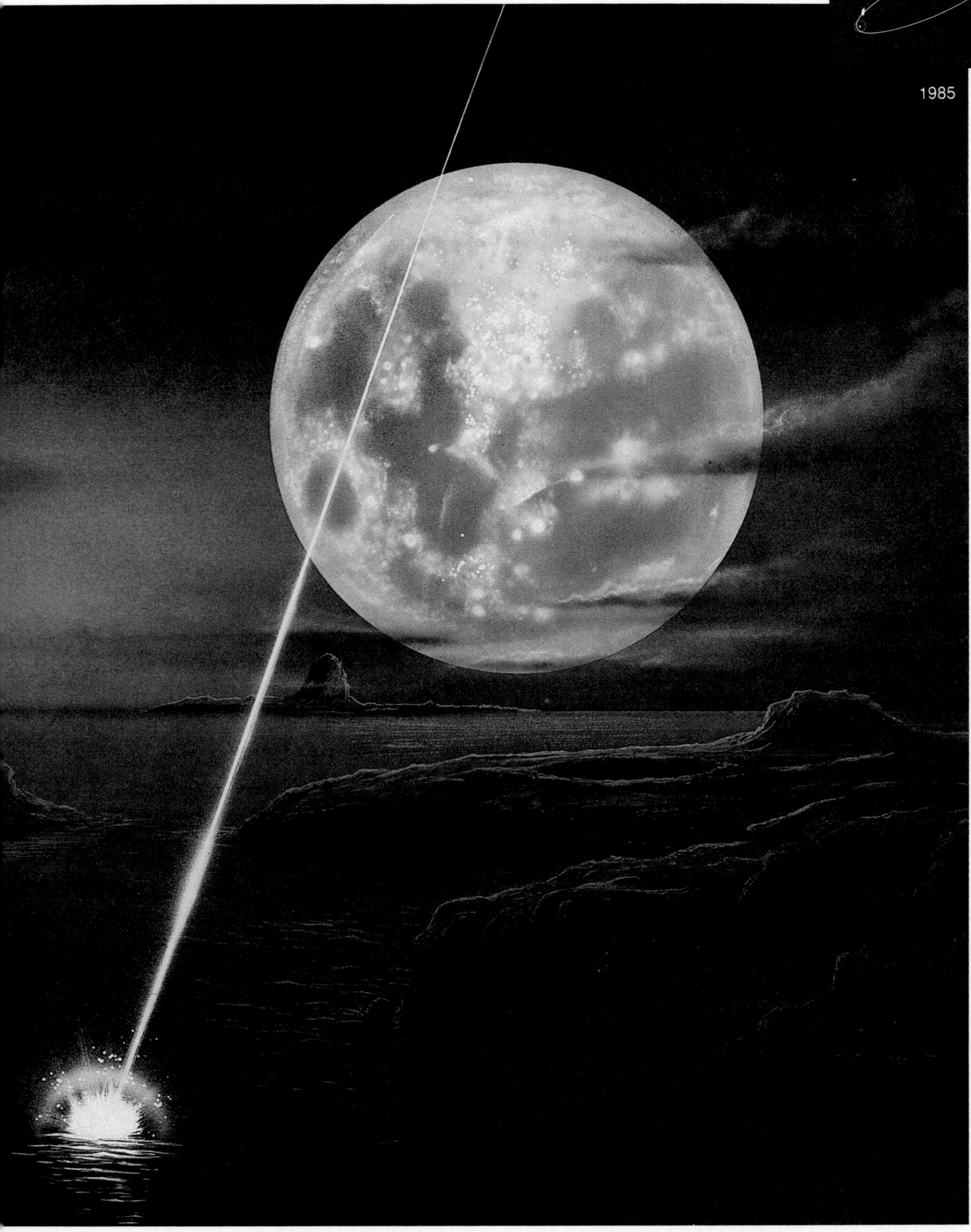

1985

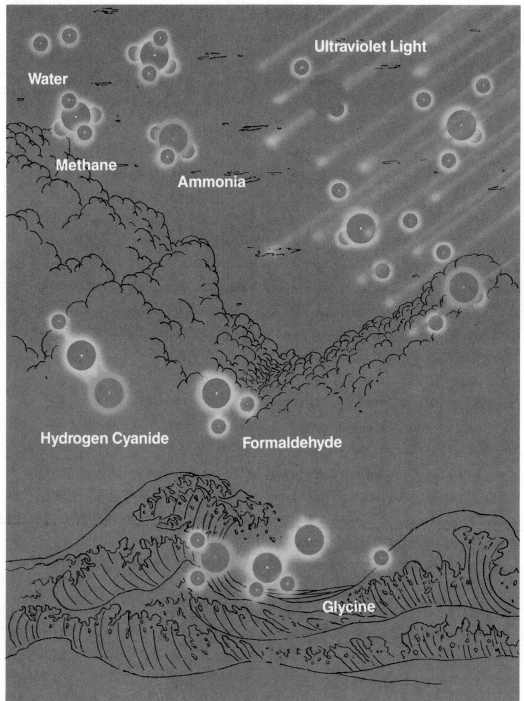

Steps leading to the origin of life on the primitive Earth, according to modern laboratory experiments. Some of the gases presumed to exist in the primitive atmosphere of the Earth are shown at top left. The hydrogen atoms are the small circles with yellow halos. Carbon is white, oxygen is orange, nitrogen is blue. Molecules shown at upper left are water vapor (H_2O), methane (CH_4), and ammonia (NH_3). These molecules are broken into pieces (*top right*) by ultraviolet light from the Sun or electrical discharges. The fragments recombine to form, among other molecules, hydrogen cyanide (HCN) and formaldehyde (HCHO). The recombination in turn of these molecules in the ammoniated waters of the primitive Earth produces the simplest amino acids (*lower right*), the building blocks of protein. Similar steps lead to the building blocks of the nucleic acids. Molecules much more complex than those shown in this diagram have been generated in experiments simulating the origin of life on the early Earth. Painting by Jon Lomberg.

recently fallen from a comet that had last seen the inner solar system four and a half billion years ago. Generating all those interstellar microbes seems very difficult; further, hardly all properties of interstellar grains are shared by microbes, and vice versa. There is no report of major epidemics associated—even with a one-year time delay to allow for the bugs drifting down from the stratosphere—with spectacular meteor showers. Moreover, the cometary debris collected by stratospheric aircraft (Chapter 13) is not reported to contain the charred and mangled bodies of, say, the diphtheria, polio, or cholera pathogens. Scientists examining recovered cometary particles do not seem to be dying of mysterious illnesses. Such facts must be considered among the deficiencies of the hypothesis.

In addition, Hoyle and Wickramasinghe explicitly propose that life has arisen inside the comet (rather than surviving from precometary eons), and is falling on the Earth to generate disease. Here again, the apparent absence of pathogens in recovered cometary grains poses a serious problem. But the idea that conditions inside a comet might be conducive to the origin of life is worth further consideration.

The chief problem is the cold. Comets in the Oort Cloud are so far away that the feeble sunlight is hardly able to warm them at all. They stay far below the freezing point of water or other abundant liquids. But with no liquids (and no atmospheres), it is hard for life to get going: the molecules are stolid, virtually immobile. The elaborate chemical processing that must have preceded the origin of life on Earth is unlikely to occur at 10° above absolute zero. Even if all the required organic building blocks were present, at these temperatures they simply could not interact sufficiently. But sunlight is not the only source of heat.

The interior of the Earth is warm, in part because of a smattering of radioactive elements—uranium, thorium, potassium. Every time one of these elements decays, a gamma ray or a charged particle is ejected which then strikes neighboring molecules, mainly silicates, and heats them up. This is, in part, why it is hotter at the bottom of a deep mine shaft than it is on the surface.

Now, as radioactive atoms decay, they fall to pieces with a statistical regularity. There is a characteristic time, called the half-life, in which half the atoms of a given radioactive species have decayed. The half-lives of uranium, thorium, and potassium are hundreds of millions to billions of years—comparable to the age of the Earth. This is no coincidence: All the radioactive atoms with shorter half-lives have already decayed. They may well have been here once, but they are gone now, decayed into other, more stable kinds of

Figure from a German horticultural publication shows an imagined connection between comets and life on Earth: Flowering planets and mushrooms are bursting through glass greenhouses to greet Halley's Comet. *Florists' Exchange*, April 30, 1910. Courtesy Ruth S. Freitag, Library of Congress.

atoms. So in the early history of the solar system, there must have been considerably more radioactive heating, and it is more likely that comets were internally warmed as the solar system started out.

The bigger the cometary nucleus, the longer it takes for heat to be conducted to the surface and for the interior to cool. Thus, if early heating had been provided by a now extinct radioactive atom called aluminum 26, the core of a comet twenty kilometers across could have been kept liquid for ten million years; and a two-hundred-kilometer comet for a billion years. So something like the Great Comet of 1729 might just conceivably have been carrying an ocean inside. A comet with a hundred-kilometer radius, made mostly of water and other liquids, would be enough to cover the entire surface of the Earth to a depth of ten meters. An interior ocean of such dimensions, rich in organic matter and given four billion years for subsequent chemical interaction, is an interesting sort of place. Is it likely that life could develop there?

The trouble is that everything is happening in the dark. Almost all forms of life on Earth work, one way or another, off sunlight. Ordinary plant photosynthesis is the best-known example. The plants absorb sunlight, and use its energy to convert water and carbon dioxide into carbohydrates and other organic matter. The animals live off the plants. Turn off the lights, and the entire ecosystem collapses. There are a few exceptions to this profound dependence of life on sunlight, but they do not seem relevant to the question of life in a cometary interior. Sunlight is a renewable resource. Not for billions of years will it be all used up. But if life were somehow to arise in such an underground sea, it would gobble up the nonrenewable energy sources until they were all gone. Then there would be nothing left to eat; and the life would become extinct. Life on the surface of such a comet is improbable because of the almost complete absence of an atmosphere or ocean. Just possibly, at some time in the future we will find the right sort of comet, and drill down to the still liquid ocean. It does not seem likely, though, that anything would be alive down there.

* * *

It is much easier to imagine the origin of life on the surface of the early Earth, where there was a planet-sized ocean and abundant sunlight. Let us, therefore, put aside the idea that life now or in the beginning fell to Earth with the comets, and ask about the origin of life from organic molecules on Earth. Life on our planet is built entirely around a hand-

1986

ful of molecular types, the most important of which are the nucleic acids and the proteins. If we could understand the large-scale production of these molecules on the early Earth, we would have made significant progress toward understanding the origin of life. The conventional scientific wisdom today—once wildly controversial—is that the key molecular building blocks were spontaneously formed on the early Earth in obedience to well-established laws of physics and chemistry. Molecules in the hydrogen-rich primitive atmosphere were broken apart by ultraviolet light from the Sun, lightning or even the shock waves produced when meteors enter the atmosphere at high speeds. The fragments are known spontaneously to recombine (see the diagram on page 314) to form the stuff of life.* It happens in the laboratory, and it must have happened on the early Earth. Ironically, cyanides, the deadly poisons that produced the great comet scare of 1910 (Chapter 6), seem to be an essential intermediary in the origin of life (see page 314). So could comets have contributed to the origin of life—not by carrying life to the Earth, but by conveying the building blocks out of which life arose?

We have seen that in the early history of the Earth—before the inner solar system was swept clean by collision and ejection—comets were much more common; an unmistakable record of these times is preserved in the impact scars on the moons and planets. In the first few hundred million years after the formation of the Earth, the flux of comets was much greater than in the more sedate present. It appears that there were at least thousands of times more comets then, and perhaps still more. These comets, made mainly of frozen water, would, on impacting the primitive Earth, of course have carried water with them. Over the first few hundred million years of Earth history, they would have deposited some 3×10^{15} (or 3,000,000,000,000,000) tons of water. But this is enough to cover the whole surface of the planet to a depth of almost six meters (20 feet). If the number of comets in the inner solar system was larger, then the Earth would have accumulated still more water. Just possibly, most of the water in the oceans arrived via comet special delivery after the Earth was fully formed.

Certainly, comets were not the only source of water. The clays in carbonaceous asteroids contain up to 20 percent water. Molten lava arriving at the Earth's surface from the

Comets cannot be homes of life; they are not sufficiently condensed; indeed, they are probably but loose congeries of small stones. But even if comets were of planetary size it is clear that life could not be supported on them; water could not remain in the liquid state on a world that rushed from one such extreme of temperature to another.

—E. Walter Maunder,
Are the Planets Inhabited?
London, 1913

---

*Simple nucleic acids in the right environment can make other nucleic acids out of smaller molecules. Simple proteins can regulate the chemical reactions in their neighborhood. Thus, while forming nucleic acids and proteins falls short of making life from scratch, it is a major step in that direction.

Layered terrain at the edge of the north polar caps of Mars. The martian polar snows are made of both frozen water and frozen carbon dioxide. The picture is about 100 kilometers across. *Viking Orbiter* photograph. Courtesy National Aeronautics and Space Administration.

interior is known to carry with it a few percent of water, and volcanic events were frequent billions of years ago. Even so, it is striking that comets could have done so much to fill the early ocean basins with water.

The waters that the comets brought were flavored with several percent of complex organic matter. So did comets bring the ingredients for the origin of life? Everything depends upon the timing. If the early comet flux fell off while the Earth's surface was still molten, then no matter how complex the organic matter in the comets may be, it would all have been fried in the magma; the origin of life would then have had to occur later, and without a significant role for the comets.

But if the Earth's surface cooled before the initial pulse of cometary impacts ended, everything would have been different. Then huge cometary nuclei would have fallen on the Earth, their contents of organic matter virtually undisturbed during entry. The comets would have struck the lower atmosphere and surface of the Earth, exploded, vaporized, and distributed fragments worldwide. Gradually, a warm, dilute soup of organic molecules would have accumulated, both the water and the stock for the broth supplied courtesy of the comets. This is not so different from the view of the Jukun people of Nigeria, who hold that meteors represent a gift of food from one star to another.

A few percent solution of organic matter is an excellent medium for life to start. Indeed, just such a dilution is imagined from entirely indigenous processes—the outgassing of water from our planet's interior, forming the oceans; irradiation of hydrogen-rich gases in the atmosphere; the slow

A comet over Vallis Marineris, Mars. Painting by Kim Poor.

1986

build-up of organic molecules in the waters of the primitive Earth; and their subsequent interaction, resulting in the origin of life. Thus, we have two competing processes—one from the inside, the other from the outside—both of which seem to lead, at roughly comparable rates, to the complex organic chemistry necessary for the origin of life.

You make a planet like the Earth and it will heat up all by itself and outgas. The water will make ponds and pools and lakes. If the rest of the atmosphere contains hydrogen-rich gases and the Sun is shining, huge quantities of organic matter will eventually be mixed in with the water. If a comet never fell to Earth in all the history of the planet, it would still be easy to understand where the original organic matter came from. And if the Earth never outgassed at all, comets may still have brought an atmosphere, an ocean and huge quantities of organic matter. Thus, in seeking the source of the organic molecules from which we come, we are in the embarrassing position of having two different, and apparently equally successful, hypotheses. It is a challenging task to devise an experiment that can distinguish between them, and both may have played an important role.

If the Earth's primitive atmosphere contained significant quantities of hydrogen-rich gases—CH_4, NH_3, H_2O, and the like—then an indigenous origin of life can fairly readily be understood. But if (and there is a school of thought that holds to this view) the early atmosphere was not hydrogen-rich—but contained say, N_2, CO_2, and H_2O—then it is difficult to understand the synthesis of organic molecules needed on the primitive Earth for the origin of life, and comets suddenly offer an attractive alternative. Even if indigenous syntheses were entirely adequate, a significant contribution may nevertheless have been provided by the comets. One recipe for life may, therefore, be: Take a million comets, warm gently, irradiate, and wait a billion years.

There is nothing in this argument unique to the Earth. Other nearby worlds, in *their* early histories, must also have been pummeled by the comets. Oceans must have accumulated on Venus and Mars too, even in the unlikely circumstance that no water from their interiors reached the surfaces of these planets. Indeed, for both Venus and Mars there is some evidence of ancient oceans.

On Venus, a world now depleted in water, the little hydrogen still left is swiftly escaping to space. Lightweight molecules escape more readily than heavier ones, because they are more likely to be induced, by random collision, to move in excess of the escape velocity. The heavy form of hydrogen, deuterium, escapes more slowly than the more common, lighter variety. As time goes on, the amount of deuterium relative to hydrogen on Venus should increase.

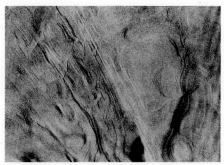

Layered deposits near the South Pole of Mars, signs of a complex and episodic (perhaps periodic) geology. The picture is almost 200 kilometers across. One day there will be machines and people examining these sediments to better understand the past history of Mars. *Viking Orbiter* photograph. Courtesy National Aeronautics and Space Administration.

An ancient river valley snakes across the battered and cratered martian landscape. *Viking Orbiter* photograph. Courtesy National Aeronautics and Space Administration.

Thus, from the present proportions of hydrogen to deuterium, it is possible to estimate how much water was once present. In this way Thomas Donahue of the University of Michigan and his colleagues have deduced an ancient ocean on Venus, now lost.

Today Venus is a desolate world, its surface temperature raised to a broiling 480° C (roughly 900° F) by a huge atmospheric greenhouse effect dominated by carbon dioxide. But this CO_2 was not outgassed into the atmosphere overnight, and it is possible to imagine a much more clement early environment, with oceans covering the surface, and dissolved cometary organic molecules colliding, interacting, growing still more complex. The turning off of the high early flux of comets and the outgassing of carbon dioxide from the interior may have changed Venus from a tropical paradise to an inferno. An important question for future exploration remains: Is there any trace today of the ancient oceans? Is it possible that life arose on Venus billions of years ago, with some durable remnant or fossil awaiting future explorers?

These questions can be asked with still greater seriousness for Mars, because there a range of evidence exists for abundant water today—frozen in the polar caps, buried subsurface, and chemically combined with the soil. Although liquid water is absent from Mars now, the data suggest that water ran in rivers and floodplains a billion years ago (see page 319). It is even possible that evidence of shorelines and other signatures of ancient martian oceans exists in images from the Viking orbiters. Mars is a smaller world than Venus or Earth, so a major fraction of its early atmosphere could have since escaped to space. It is also a colder world, so much of the surviving cometary water may now be locked away as ice. As for Venus, the question arises: If there once were extensive oceans, did life arise? Perhaps traces will be found by some twenty-first century expedition, or an exploratory party from the first permanent human outpost on Mars.

The worlds in this part of the solar system show the tokens and remnants of the great flux of comets that filled the inner solar system in those long-gone first times. We can think of the present modest infall of cometary debris as a feeble reminder of an epoch that lasted for hundreds of millions of years, and changed the face of the solar system.

If something close to the present oceans were carried as comets to the primitive Earth, simple calculations show that they would also have brought an amount of carbon comparable to that in the entire sedimentary column of the Earth—in all the carbon in the rocks for kilometers down, all the living things, all the humus in the soils, all the petro-

Viking 2 Lander photograph of the terrain in a region of Mars called Utopia. One of the footpads of this spacecraft that landed in 1976 can be seen in lower right. The metal cylinder above it covered the spacecraft's sampling arm (not in picture) until after landfall was achieved. Note the layer of thin frost that covers much of the picture, some of which is ultimately derived from cometary impacts. Courtesy National Aeronautics and Space Administration.

leum, coal, peat, graphite, and diamonds, all over the Earth. If anything like this is true, then it is fair to say that the surfaces of the Earth and the other terrestrial planets, after they were almost fully formed, were dusted with a layer of cometary stuff a few kilometers thick. This cometary coating is, relatively speaking, thinner than the confectioner's sugar on a jelly doughnut. But it means the world to us: On this planet, at least, the comet dust has come to life.

The early Earth was covered by impact craters, large and small; and still greater excavations called basins. Most of the water carried by the comets was vaporized on impact, and fell out as rain. So comets dig big holes, in effect fill them with water, and salt them with complex organic matter—performing these duties mainly in the earliest history of the planets when the maximum time is available for life to arise. Again, nothing in this recounting is restricted to the Earth; the same thing may have happened on countless other worlds in the Milky Way. It is hard not to think of the comets as cosmic elves, dashing through space and conferring the potential for life on the worlds of this solar system and countless others.

So take a look around you. If that comet flux after the surface of our planet cooled was many orders of magnitude more than it is now, what comes from Earth, and what from comets? Everything alive would derive ultimately from comets—all the plants and animals and microbes and men and women. All the buildings, railroads, highways, sculpted farmland, songs, submarines, space vehicles—these are all made by humans, and for this reason alone are derived from comets. Even the daytime sky comes from the comets, because O_2 and N_2 are made by life. Well, at least the rocks are from the Earth, we might argue, at least the mountains. Yes, but the rocks are oxidized, chemically altered and broken down by water and by life. The mountains are sculpted and eroded by water. When a lump of the underlying planet tries with exquisite slowness to poke up through the comet stuff, it is ruthlessly worn down. Even in the wasteland of Antarctica, the landscape appears to be, one way or another, comet-derived. It seems possible that the only things we see that exist entirely apart from comets are the Sun and the stars. A layer of comet stuff once powdered this world, and during the subsequent four and a half billion years the dust has made a little progress. It has developed complex beings, aspirations, a first try at intelligence. At last it is turning its attention to the comets from which it came.

PART III

COMETS AND THE FUTURE

A Soviet *Vega* spacecraft about 10,000 kilometers from the nucleus of Halley's Comet in early March 1986. Far away, only a few hundred kilometers from the comet, is the tiny form of *Giotto*, the Halley's Comet explorer of the European Space Agency. Painting by Rick Sternbach.

Chapter XVIII

A FLOTILLA RISING

By the middle of the Twentieth Century, it will have rounded its remotest stake...and will once more have begun the long journey sunward, which will not end 'til 1986. Then again will telescope and camera, spectroscope and photometer, be pointed Halley-ward, just as eagerly as today.

> —David Todd,
> *Halley's Comet,*
> American Book Company,
> Cincinnati, 1910

SHORTLY BEFORE DAWN, A RIFLE SHOT RANG OUT and a moment later the whiz of the bullet could be heard, narrowly missing the three occupants of the gondola. Despite the fact that it was May, they were heavily bundled up. In the early hours of the morning it gets very cold a kilometer or two above the Earth. But they were only 500 meters high when they were fired at. It must have been an unusual sight—the great balloon and its suspended gondola illuminated by the rising Sun when it was still dark in the Connecticut countryside below. Perhaps the balloon had startled some farmer who felt his only recourse was to shoot it down, whatever it was. Or perhaps the bullet was merely a safe protest against the idle rich and their expensive diversions.

But this was not a pleasure cruise. Aboard was Dr. David Todd, professor of astronomy and navigation and director of the observatory at Amherst College. Todd's specialty was the astronomical excursion. During his professional career, he led expeditions to the Dutch East Indies, South Amer-

"An astronomical observatory in the upper air: taking observations of Halley's Comet from a balloon." Drawing by Henri Lanos, published in the magazine *Graphic*, May 28, 1910. Courtesy Ruth S. Freitag, Library of Congress.

1988

ica, the Barbary Coast, Russia, Japan, and West Africa—in each case to observe an eclipse of the Sun. He once dismantled the college's observatory and crated the telescope to the Andes, in part to look for signs of life on Mars. Now, in May 1910, Comet Halley was in the skies, and David Todd had risen to the occasion—not very far, it is true, but high enough to get above some of the distortion caused by the Earth's atmosphere. He had taken aboard a small two-and-a-half-inch telescope, magnification 30 times, with which he hoped to observe the comet. The night was clear and the vibrations from the balloon minor.

Accompanying Todd was Charles Glidden, the pilot, and Mabel Loomis Todd, the astronomer's extraordinary wife. A recent historian describes her as follows:

> Cheerful, talented, sociable, popular enough to arouse jealous gossip, she was capable of sustaining affectionate and amorous ties; effective as a writer, lecturer, and editor, as wife, hostess, and lover, she constructed for herself, more than most, an enviably robust and resilient character. *

It was Mabel Todd, the daughter as well as the wife of an astronomer, who saw the comet first. The view, her husband said, was much better "than through the big 18-inch telescope at Amherst Observatory," at the observatory he had himself designed and built. The head of the comet was near the horizon, and Todd made four sketches of the tail. This seems to have been the first successful astronomical observation from an artificial platform above the Earth's surface.

Fifteen years later, Todd went on to obtain the first photograph of the solar corona from an airplane. But afterwards, his behavior became increasingly erratic, and Todd was intermittently confined to mental institutions until his death at age eighty-four, in 1939. Mabel Todd predeceased him, mourned by their only child, who remembered especially "her indefatigable energy and her infectious gaiety."

Today, new balloons are rising. 1985/1986 represents a historic moment in the study of comets. Previously, the comets came to us. Now, for the first time, we are going to them. The occasion is the apparition of Comet Halley between fall 1985 and spring 1986. Apart from its historic significance (Chapters 3 and 4), Comet Halley is the only vigor-

Space travel to investigate Halley's Comet, as conceived in a Chicago newspaper, April 30, 1910. Courtesy Ruth S. Freitag, Library of Congress.

---

*Peter Gay, *Education of the Senses,* New York, Oxford University Press, 1984. Gay, professor of history at Yale University, does not mention this balloon ascent; Mabel Loomis Todd interests him chiefly because of the light her candid diaries shed on Victorian sexual behavior.

The *Astro 1* mission of the Space Shuttle: Specially designed instruments are carried to Earth orbit to survey Halley's Comet in early 1986. Courtesy National Aeronautics and Space Administration. Painting by William K. Hartmann.

ously active comet with an orbit well enough known to permit detailed scientific planning years in advance. Following David Todd's lead, aircraft will fly above most of the Earth's atmosphere and rockets will briefly enter space to glimpse the visitor. The unmanned Solar Maximum Observatory, repaired by Shuttle astronauts in April 1984, will examine Comet Halley; special instruments will be flown on the Space Shuttle to scrutinize the comet; and the attention of the U.S. *Pioneer-Venus Orbiter* will be redirected from Venus to the comet as it passes by. Compared to the usual levels of effort, this would already represent an extraordinarily concerted study. But, in fact, all this represents something of a sideshow—because a flotilla of five spacecraft is scheduled to fly by (and one of them perhaps into) Halley's Comet in March 1986.

The close-up examination of comets from space is the fulfillment of an astronomer's dream, because we are still painfully ignorant about the most fundamental aspects of comets. We live in the inner solar system, so that cometary nuclei approaching the Earth tend to outgas and cover themselves with coma. But if we could fly close to a comet, we might for the first time clearly view a cometary nucleus. What does it look like? What is its shape, appearance, color? Are there patches of ice, dark organics, rock outcroppings? Is it surrounded by a swarm of small boulders? Is there a lag deposit? Craters? What about signs of past surface melting? Hills? Any hint of layering as in the Earth's sedimentary column? You might be able to tell a great deal about the nature and evolution of comets if you could photograph a nucleus close up.

Then there is spectroscopy. If we are restricted to the surface of the Earth, we can examine comets only in the visible spectrum, as Huggins did, and at a few "windows" at infrared and radio frequencies. If we wish to examine comets at other frequencies, we must get above the Earth's absorbing blanket of air. But what can be lifted even into near orbit must be much smaller than major astronomical facilities on the surface of the Earth; the intrinsic capability of instruments in Earth orbit tends, therefore, to be inferior to their groundbased counterparts. Nevertheless, instrumentation in Earth orbit can make important new findings; the discovery of the hydrogen coma around comets (Chapter 7) is an example. But very few bright comets have come close to the Earth since the advent of astronomy from orbit. What you really need is to take the spectrometers close to the comet. Then you might find yourself discovering how various molecules are distributed through the nucleus, coma, and tail.

Generally, spectroscopy performed from Earth reveals the presence of molecular fragments, not the parent molecules

1989

A comet probe would certainly solve most of the cometary problems without ambiguity.

—Pol Swings,
University of Liège,
Belgium,
August 1962

of the cometary nucleus from which they come. For example, the nature of the organic molecule from which C_3 derives is entirely unknown. There is an elaborate machinery of outgassing, dissociation by ultraviolet light and further chemistry that occurs before we detect the daughter fragment. Many of these chemical mysteries could be resolved if we could fly close to the nucleus, into the cloud of parent molecules and there measure directly, before they are dissociated by sunlight, the molecules, organic and inorganic, that have just been evaporated off the comet.

To detect some sign of the material in the tail of Halley's Comet in 1910, many studies were made at ground level. In France, a large volume of air was brought to very low temperatures, the oxygen and nitrogen liquified, and the residuum examined for exotic constituents. Nothing was found. Metal plates coated with glycerin were attached to the struts of a tower at the Mount Wilson Observatory in California, but no cometary dust motes were reported. The experiment is the forerunner of modern studies in which similar glycerin-coated plates are attached to aircraft and flown to stratospheric altitudes—where cometary debris is much easier to come by (Chapter 13). But now it is possible to fly spacecraft with mass spectrometers aboard that can measure directly the parent molecules of the comet.

We think the magnetic field in the solar wind drapes itself over the cometary nucleus, while solar flares produce gusts in the solar wind that generate elegant patterns in the ion tails. But our understanding of comets would be far superior if we flew instruments very near the comet, and measured charged particles and magnetic fields directly.

Visual observers, straining at the limit of the resolution of their telescopes, have detected great fountains of gas and dust shooting into space from the cometary nucleus. We think much of the dust released by comets to space arises in this way. We need to sidle up to a comet, photograph the jets—obtaining a motion picture if possible—and fly through the dust cloud, counting and measuring the fine particles.

All this and more is now underway. At the Japanese Space Center at Uchinoura in Kogoshima Prefecture, Kyushu, two new launch vehicles of the Mu class lift off. The first is called *Sakigate* (Japanese for "pioneer"); the second bears the simple designation, *Suisei* (Japanese for "comet"). They are the first interplanetary vehicles ever launched by Japan.

On the equatorial island of Kourou, off French Guiana in South America, an *Ariane* rocket is launched. In its nose cone is the first interplanetary spacecraft of the European Space Agency, a consortium of Belgium, Denmark, France,

Halley's Comet as seen in 1301, depicted as the Star of Bethlehem. From Giotto di Bondone's *Adoration of the Magi*, a fresco completed in 1304.

the Federal Republic of Germany, Ireland, Italy, the Netherlands, Spain, Sweden, Switzerland, and the United Kingdom. It is named *Giotto*—after the Florentine painter who incorporated his own observation of Comet Halley, during its apparition in the year 1301, into his *Adoration*, a celebrated fresco on the Arena Chapel in Padua (above).

In Tyuratam in the Kazakh Soviet Socialist Republic, two *Proton* launchers rise to the heavens. Their ambitious mission: to fly to Venus—depositing two spacecraft to perform a night landing there, and dropping off two balloon stations to examine the meteorology of the middle atmosphere*— and then continue on to encounter Halley's Comet eight months later. The spacecraft are called *Vega*—*Ve* from Venera, the Soviet word for Venus, and *Ga* from Galley, the Russian language having an aspirated G, but no H. There are twelve separate scientific instruments aboard *Vega's* cometary payload. In addition to Soviet instrumentation, the *Vega* missions carry equipment from Austria, Bulgaria, Czechoslovakia, the German Democratic Republic, the German Federal Republic, Hungary, Poland...and the United States.

The United States of America has played a key role in

---

*Tasks successfully carried out in June, 1985.

International Cometary Explorer

In 1978, a satellite called *International Sun-Earth Explorer 3 [ISEE 3]*, designed to study how the solar wind interacts with the magnetic field of the Earth, was placed in an orbit between the Earth and the Sun. After its prime mission was over, *ISEE 3* was redeployed to a quite different task. Robert Farquhar of NASA's Goddard Spaceflight Center designed an ingenious orbital maneuver by which the spacecraft would encounter an extremely interesting object called Giacobini-Zinner, an active short-period comet which some astronomers suggest is shaped like a rapidly rotating pancake—with its equatorial radius eight times its polar radius. The comet is also the source of the fluffy particles that make up the Draconid or Giacobinid meteor stream. It would be a joy to see close-up pictures of Giacobini-Zinner, but that is not in the cards. *ISEE 3* does not have a camera.

To get to Giacobini-Zinner, the spacecraft's thrusters altered its orbit to take it on two leisurely transsections of that part of the Earth's magnetic field that points, like a comet's tail, away from the Sun; and then on five consecutive close approaches to the Moon, the last of which, in late 1983, took it to 120 kilometers from the lunar surface. A minor malfunction in the spacecraft's small rocket engine could have crashed it into the Moon. But instead, the cumulative effect of successive passes by the Moon's gravity flung the spacecraft (like a comet from the Oort Cloud passing close by Jupiter) into a very different trajectory which successfully passed through the tail of Comet Giacobini-Zinner on September 11, 1985.

This was almost six months before the spacecraft flotilla reaches the vicinity of Halley's Comet. So you might think that the exercise was mainly political—like the unsuccessful attempt, in July 1969, by the unmanned Soviet *Luna 15* spacecraft to return a sample from the Moon a few hours before the American manned *Apollo 11* could do so. And you'd be right. But *ISEE 3*—now renamed *International Cometary Explorer*, with the agreeable acronym *ICE*—obtained much more useful information than if it had merely continued to languish in the vicinity of the Earth. It discovered the interplanetary magnetic field draped over the cometary nucleus, and unexpected energetic particle and field events in the tail. *ICE* is a testament to the remarkable abilities to tool around the inner solar system that are now at hand—if you know how to use a small rocket motor, the masses of the Moon and planets, and Newton's laws of motion.

1991

The *International Cometary Explorer* approaches Comet Giacobini-Zinner in September 1985. Painting courtesy National Aeronautics and Space Administration.

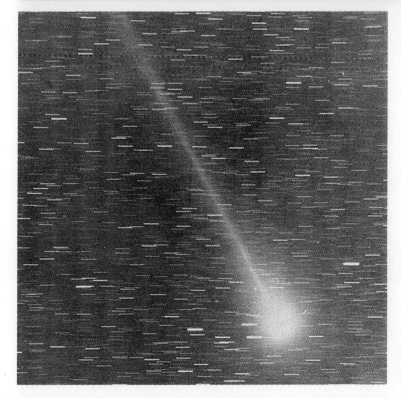

Comet Giacobini-Zinner, photographed by Elizabeth Roemer, U.S. Naval Observatory, Flagstaff, Arizona. Official U.S. Navy photograph.

The trajectories of Venus, Earth, Halley's Comet, and the *Vega* spacecraft. The inner blue circle is the orbit of Venus. The outer blue circle is the orbit of the Earth. The multicolored arrow is the orbit of Halley's Comet, and the dashed black arrows represent the orbit of the *Vega* spacecraft: They are launched from Earth (in the ten o'clock position), encounter Venus (five o'clock), then cross the Earth's orbit twice more until intercepting Halley's Comet (three o'clock). The *Giotto* spacecraft, which does not have a planetary component to its mission, employs a slightly simpler trajectory to get to Halley's Comet. Diagram by Jon Lomberg/BPS.

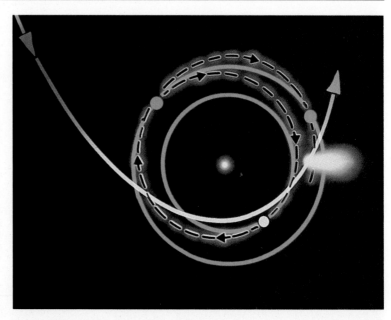

The *Suisei* spacecraft remains at a discreet distance from Halley's Comet while it photographs its hydrogen coma in the ultraviolet. The halo around the Sun is the solar corona. Painting by Kazuaki Iwasaki.

spacecraft exploration of the solar system—examining for the first time every planet from Mercury to Uranus, and dozens of moons. But the United States does not have a spacecraft to Comet Halley. Several innovative mission

1991

A *Vega* spacecraft on its way to Venus and Halley's Comet. The large sphere at top contains the Venus atmosphere descent package, including the balloon stations. The blue panels are solar cells for the conversion of sunlight into electricity. Courtesy Institute for Cosmic Research, Soviet Academy of Sciences, Moscow.

designs were proposed by American scientists and engineers; they would have procured a range of essential data that *Giotto*, *Vega*, and the Japanese missions will not. But the proposals were turned down by Democratic and Republican administrations both. There was not enough money. The United States had more important things to do. The cost of a major mission to Comet Halley is roughly that of a single B-1 bomber. The United States is committed to 100 B-1 bombers. We couldn't have gotten by with 99. Perhaps in the year 2061, when Halley's Comet next returns, there will be a different American response. But while the United States has in this respect opted out of cometary exploration, the flotilla of five spacecraft from twenty nations represents a stirring response by the human species to this emissary from the depths of space and from the early history of the solar system.

Through a bit of orbital legerdermain, the United States was nevertheless the first nation to examine a comet close up by spacecraft. It is in its instrumentation a rudimentary spacecraft, unable to obtain either pictures, spectra, or much compositional information, but mainly data on charged particles and magnetic fields. (See the box on page 332.) However, without a doubt, it was first.

Logo of the *Vega* missions. As portrayed, this spacecraft has dropped off the Venus descent module and is on its way to Halley's Comet. National flags of participating nations are shown. Courtesy Institute for Cosmic Research, Soviet Academy of Sciences, Moscow.

The *Suisei* spacecraft on a test stand in Japan, before launch. Courtesy Japanese Space Agency and Astronomical Society of the Pacific.

There is also an American experiment aboard *Vega*—there entirely through the individual initiative of John Simpson, a professor of physics at the University of Chicago and a veteran of dozens of U.S. unmanned space missions. In a time when the United States had allowed its agreement on cooperation in space science with the Soviet Union to lapse out of displeasure with Soviet foreign policy, Simpson designed a novel analyzer of cometary dust. He discussed his device at an ESA meeting in the Netherlands, hoping to have it included in *Giotto*; less than a month later, however, Simpson was informed by Roald Sagdeev, director of the Institute for Cosmic Research of the Soviet Academy of Sciences, that the instrument had been accepted for the *Vega* mission. Simpson had never even proposed his instrument for the *Vega* spacecraft. After obtaining permission from appropriate American authorities, Simpson put together a device that employed technology at least a decade old; he did not wish to violate U.S. strictures on "technology transfer." When the time came to integrate payloads, Soviet engineers asked Simpson why his instrument did not have a computer microprocessor as all of theirs did. Simpson smiled.

All spacefaring organizations involved in the Halley encounter have pledged to make the results available to scientists worldwide. The missions have been jointly organized —not only so that the various scientific instruments complement each other and the data are exchanged rapidly, but also so that the results of one mission will help assure the success of the others.

The most modest are the Japanese missions. *Sakigake* is considered mainly a test of the machinery for getting there, and will not pass closer than a million kilometers to Comet Halley. Even so, it will measure the distant solar wind for comparison with measurements made cloaser in of the interaction of the comet with the solar wind. *Suisei,* on the other hand, will approach to within about 200,000 kilometers, and includes an ultraviolet television camera to photograph the hydrogen coma for a month or more before closest approach. Since the hydrogen is produced by dissociated cometary water ice, we will have a record of the outgassing history of the main cometary volatile.

The orbit of Halley's Comet is inclined 162 degrees to the plane of the zodiac or ecliptic, the plane that includes the Earth's orbit. Thus because of limitations in the propulsion systems of modern interplanetary spacecraft, the comet can be intercepted only when its orbit and the plane of the Earth's orbit intersect. For this reason, the trajectories of the *Vega* spacecraft take them by Comet Halley on March 6 and March 9, 1986. The closest approach of *Vega 1* is designed to

1992

be about 10,000 kilometers, and perhaps a little less for *Vega 2. Giotto* is designed to encounter the nucleus of Halley's Comet on March 13, 1986, and pass a few hundred kilometers above the sunlit side. Because of the enormous relative velocity, the encounter will last only a few hours; and some of the key measurements only a few minutes.

But to pass a few hundred kilometers from the comet, you have to know where this rapidly moving iceberg is to within a few hundred kilometers, and nobody yet knows its orbit to that accuracy—you cannot see the nucleus, enveloped in all that coma, well enough from Earth to measure its precise position. Accordingly, a cooperative navigation scheme between the Soviet Union, the United States, and the European Space Agency has been organized. It is called Pathfinder. Radio telescopes of the U.S. National Aeronautics and Space Administration will listen, with Soviet cooperation, to the radio transmission from the *Vega* vehicles, against the background of much more distant quasars lying far beyond our Milky Way Galaxy.* The position and motion of the spacecraft relative to the Earth can thus be determined with high accuracy. From the first *Vega* pictures, Soviet space scientists will know to high accuracy the direction of the nucleus of Halley's Comet. But if we know where *Vega* is relative to the Earth, and where Halley's Comet is relative to *Vega*, we know where Halley's Comet is relative to the Earth. This information must be extracted from the observations very rapidly so *Giotto* will be able to effect a most delicate maneuver: To optimize the fine detail that the *Giotto* cameras will see in the cometary nucleus, the spacecraft must pass very close to the comet. But if it passes on the night side, there will be almost nothing to see. Without the cooperation by American radiotelescopes and Soviet spacecraft, it would be impossible for the *Giotto* scientists to plan their close fly-by, and the miss distance might be 1,000 kilometers or more. But there is only a couple of days for the *Vega* pointing data to be received and used to calculate a revised trajectory for *Giotto*. This will be a race against time and the comet.

At a 500-kilometer range, the smallest detail visible will be about 30 meters; if pictures are taken at closer ranges, the resolution will be better. Since *Giotto* will fly as close as possible to the day-side of the nucleus of Comet Halley, there is then a possibility that the spacecraft will hit the comet, although there is no plan to do so. If the impact occurs,

---

*The French/Soviet balloons released into the atmosphere of Venus were also tracked by American radio telescopes. They lasted for about two days before they were destroyed by the fierce Venus environment.

The *Giotto* spacecraft gingerly approaches the nucleus of Halley's Comet. Courtesy of European Space Agency.

Giotto will be splattered all over the cometary landscape; the relative velocity of spacecraft and comet is about 68 kilometers per second (153,000 miles per hour). If there is such a collision, a few scattered fragments of a machine from Earth will be embedded in the cometary snows and carried out past the orbit of Neptune—like Ahab's corpse, pinioned to the great white whale. It would return in another 76 years, perhaps to be recovered by future historians of technology.

Even if *Giotto* does not strike the nucleus, there are other dangers. Ordinarily, the outer skin of a spacecraft is only a few millimeters thick; with no other provisions, such a skin would doubtless be punctured near a cometary nucleus. Radar evidence suggests that comets may be surrounded by a cloud of debris with some individual particles the size of your fist or even larger; and there will be many more small particles than large ones. For safety, the spacecraft has a meteor bumper, of the sort first proposed in the 1950's by Fred Whipple, the father of the dirty-ice model of the cometary nucleus. Whipple advocated a thin outer skin surrounding a thicker shield, the two layers in tandem serving to protect the spacecraft from all but the largest particles.

1993

The outer surface of the *Giotto* meteor bumper is made primarily of aluminum; the inner surface is a mainly aluminum honeycomb with layers of polyurethane foam, epoxy, and mylar—similar to a bulletproof vest. The two *Vega* spacecraft have something similar.

International cooperation is involved in the Halley encounter in yet another way. Suppose you are interested in, say, an hour-by-hour record of the jetting fountains in a nucleus or the structure of the tail for months around perihelion. This is simply too much data for the spacecraft to handle; you will have to do the observations from the Earth's surface. But the Earth turns, the comet rises and sets, and you must have observers distributed all over the planet. More than 900 professional astronomers from 47 countries, and many amateurs, have formed an organization called the International Halley Watch, designed to coordinate observations. Originally conceived by Louis Friedman, executive director of The Planetary Society—the largest space interest group on Earth—the International Halley Watch represents the first major organized worldwide effort to observe a comet. However, international cooperation has been central to the understanding of comets for centuries: Tycho's determination that the Comet of 1577 was far above the Earth's atmosphere (Chapter 2) and Newton's calculation (Chapter 3) of the trajectory of the Comet of 1680 both relied fundamentally upon observations made in several nations.

Both the *Vega* and the *Giotto* television cameras will produce color photographs. In addition, they have filters that are centered on the frequencies of specific cometary emissions—C_2, for example, and CN (Chapter 8). We will have pictures of the comet close up, painted by the light emitted by these molecular fragments; William Huggins would have been pleased.

Scientists hope, from photographs obtained far from encounter, to map the positions of the spouting nuclear jets and position the cameras so they can photograph between the jets. The situation is like seeing the house down the road through a momentary clearing in a snowstorm—but is probably not as difficult. The nucleus, coma and tail will be examined at many frequencies through the ultraviolet, visible, and infrared. Dust particles will be counted and measured. Mass spectrometers will determine something of the nature of the parent molecules. Charged particles and magnetic fields will be monitored. It seems likely that we are on the verge of a revolution in our understanding of the nature of the comets. But if the past history of space exploration is any guide, the most interesting findings will provide answers to questions we have not yet asked.

There was a time—and very recently—when the idea of learning the composition of the celestial bodies was considered senseless even by prominent scientists and thinkers. That time has now passed. The idea of a closer, direct study of the universe will today, I believe, appear still wilder. To step out onto the soil of asteroids, to lift with your hand a stone on the moon, to set up moving stations in ethereal space, and establish living rings around the Earth, the moon, the Sun, to observe Mars from a distance of several tens of versts, to land on its satellites and even on the surface of Mars—what could be more extravagant! However, it is only with the advent of reaction vehicles [rockets] that a new and great era in astronomy will begin, the epoch of a careful study of the sky.

—Konstantin Tsiolkovsky,
*Investigation of World
Spaces by Reaction Vehicles,*
Moscow, 1911

Astronauts of the twenty-first century examine the surface of a cometary nucleus. A jet of volatilized ices shoots off into space from beyond the horizon as the visitors' spacecraft keep pace with the comet's motion. Painting by Pamela Lee.

Chapter XIX

STARS OF THE GREAT CAPTAINS

Whatever opinions we may adopt as to the physical constitution of comets, we must admit that they serve some grand and important purpose in the economy of the universe; for we cannot suppose that the Almighty has created such an immense number of bodies, and set them in rapid motion according to established laws, without an end worthy of his perfections, and, on the whole, beneficial to the inhabitants of the system through which they move.

—Thomas Dick,
*The Sidereal Heavens and
Other Subjects Connected
with Astronomy, As Illustrative
of the Character of the Deity
and of an Infinity of Worlds,*
Philadelphia (1850)

WHEN HUMANS DISCOVER SOMETHING NEW, THERE is a natural inclination—probably hard-wired into our brains—to put it to use. This odd proclivity, shared in a systematic way by none of the other beasts and vegetables on Earth, is both a principal cause of human success and a major reason that so much of the surface of the Earth is being despoiled. Birds know, better than humans, not to foul the nest.

If the trends continue, and if we do not destroy ourselves first, it seems likely that sometime in the twenty-first century we will begin utilizing the comets. The environment of empty space, the surfaces of large worlds, and the entire volumes of small ones will be visited and made to serve the purposes of future explorers and settlers, robot and human. The resources, the isolation, and the perspective provided by this new arena may have a benign influence on the quarrelsome human species, but it is at least equally probable that the most irresponsible environmental ethics will be extended to the cosmos. There are bags of human excrement lying on and splattered over the Moon, vivid symbols of the misuse of space technology. The proposed Star Wars program, the so-called Strategic Defense Initiative, is another.

One of the attractions of the comets is that there are so many of them. Even if the human tendency to litter, contaminate, and obliterate were to continue unchecked down the centuries, it would not readily be possible to despoil trillions of small worlds, or even the mere millions that exist in the planetary part of the solar system. Perhaps we can exercise our ethical muscles more effectively on the comets than on the Earth. However, it would be wholly foolish to begin "utilizing" the comets—whatever that might entail—without first understanding them thoroughly. After that, an argument can be made even for grinding up and reprocessing whole comets—because in tens of thousands of years most of the short-period comets will be ejected from the solar system anyway, if they do not first collide with the planets, or disintegrate, or transform themselves into asteroids. So there is a certain justification here—if not in the much more outre subject of battlefield nuclear tactics—for the admonition "Use 'em or lose 'em."

How could you "use" a comet? You could destroy it or change its orbit if it posed a threat to the Earth, you could exploit it for its resources, or you could colonize it. In this chapter, we discuss each of these possibilities. But one of the earliest practical connections between comets and spaceflight was quite different. Konstantin Tsiolkovsky was a gifted Russian schoolmaster who lived when balloons went higher than rockets. In a series of writings around the turn of the century, he outlined on a grand scale how the rocket engine would eventually carry the human species

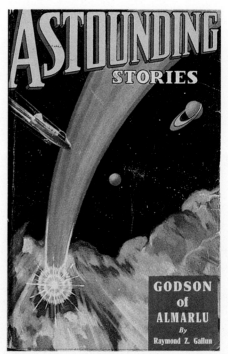

1994

into space. His influence on the Soviet space program is still palpable, and a heroic bust of Tsiolkovsky adorns the lobby of the Institute for Cosmic Research of the Soviet Academy of Sciences in Moscow. In a 1911 publication called "Investigation of Cosmic Space by Reaction Vehicles," Tsiolkovsky wrote:

> Comets have long been expected to bring an end to the Earth, and not without reason, though the probability of this end is extraordinarily small; still and all, it might happen tomorrow or in trillions of years. [But] it will be rather difficult for a comet and other accidental, highly improbable but terrible and unexpected enemies of living beings to strike down in one blow all creatures that have formed [extraterrestrial] habitations.

A comet might conceivably wipe out life on one planet, Tsiolkovsky is saying, but not on many simultaneously; therefore, to ensure the long-term survival of the human species, we should colonize the solar system. The argument loses some of its force, however, if it turns out to be true that cometary catastrophes arise from comet showers—whether induced by sporadic passages of stars or interstellar clouds, or by Nemesis, should it exist. (Imagine the astronomers on a dozen adjacent worlds tracking some red dwarf star which raises havoc in the inner Oort Cloud, and a swarm of comets—more than enough to impact each of the inhabited worlds—is detected approaching the planets.)

However, the general point that Tsiolkovsky makes is certainly valid: the more worlds on which there is a self-sustaining human presence, the less likely it is that random catastrophe will render our species extinct. Tsiolkovsky had not envisioned that humans might, two generations after his time, be able to use the rocket motors about which he was so optimistic to destroy civilization. The history of rocketry has many such ironies. Tsiolkovsky's argument for colonies in space will doubtless be employed to argue that global nuclear war need not be as dreadful as some alarmists say.

* * *

The idea that humans should be able to do something about a comet on collision trajectory with the Earth has many antecedents in science fiction, as does the notion of capturing an asteroid or a comet and employing it for human benefit. In *The Comet Hunters*, an early-twentieth-century novel by Jean Kerouan, a Professor Granger invents a ray gun to deflect a fictitious Comet Swanley into Earth orbit, whence a range of benign climatic and agricultural conse-

Science fiction depiction of a comet in the Saturn system. From a 1936 issue of one of the leading science fiction magazines of the time. Courtesy Science Fiction Library, Massachusetts Institute of Technology.

A small cometary nucleus over the New York metropolitan area a moment before impact. Painting by Michael Carroll.

quences will ensue. But his treacherous assistant steals the plans as well as a small bundle of cash from an American billionaire, adopts the name Khan Zagan, and assembles the ray gun on the Tibetan border of his native Mongolia. One thing leads to another, and Khan Zagan holds the Earth for ransom, threatening to deflect the comet so it will impact the Earth. But Granger and a handy Tibetan Lama, adept in soul transfer, foil the evil plot, and reroute the comet to the Moon—which it hits with some force, rousing our natural satellite from a long geological lethargy. Everything works out satisfactorily, except, of course, for Khan Zagan. The story is more or less typical of pre-1940 pulp science fiction.

Something similar was described in 1917 in a novel called *The Second Moon*, by R.W. Wood and A. Train (Wood was professor of optics at Johns Hopkins University, and a leading physicist of the day). They imagined an asteroid called Medusa whose orbit is by chance perturbed by a near encounter with a comet into a trajectory which would impact the United States. A spacecraft is promptly dispatched with a newly-invented disintegrating ray gun, which converts Medusa into a ball of fire, no less. Its trajectory obligingly altered, what remained of the asteroid retreats demurely into a safe orbit about the Earth, where it can be gainfully employed.*

---

*Ronald Reagan was an impressionable seven-year-old when this story was published, although there is no evidence that it influenced the future President's strategic thinking.

1995

Every ten thousand years or so, a comet or asteroid the size of a football field strikes the Earth. The impact energy is about ten megatons, the equivalent of a single large nuclear weapon, and about 1,000 times more powerful than the bomb that destroyed Hiroshima. Because of the absence of significant radioactivity, the explosion would be somewhat less deadly than a single groundburst by a nuclear weapon of comparable yield. Impacting randomly on the Earth, it would also be much less likely to fall on a city. The resulting crater, if excavated on land, might be as much as a kilometer across. Beyond the crater itself, the damage would be slight. Because the chance of such an object falling on a populated area is small, it is unclear that extraordinary efforts to prepare for such a contingency, or to counter it, are justified.

But once every million years or so, a comet or asteroid a

A small cometary nucleus improbably makes a direct hit on Manhattan Island. Painting by Michael Carroll.

kilometer across will by chance run into the Earth. The impact energy would now be about 10,000 megatons, the equivalent of the instantaneous detonation of ten billion tons of TNT. (If you collected ten billion tons of a chemical explosive into a cube, the aggregate would be more than a kilometer on a side—a small city solidly filled, to a height of three hundred stories, with TNT.) This is only a little less than the cumulative yield of all the nuclear weapons that humans seem generally content to have sharing the planet with them. (The world arsenals now exceed 13,000 megatons). An explosion of this magnitude would put so many fine particles into the atmosphere that serious climatic consequences might follow, like the impact events that seem to have been responsible (Chapter 16) for the extinction of the dinosaurs and most species on Earth 65 million years ago.*

If such an event happens on average once every million years, there is one chance in a million that it will happen in the coming year. But one chance in a million is roughly the risk you take on a commercial airline flight. Many people take such risks seriously. Thus it makes sense first to refine our census of nearby comets and Earth-approaching asteroids that might one day impact the Earth, and second, to see whether, with existing technology, we might contrive a means to deflect or destroy any object that we discover to have an uncomfortably high probability of impacting the Earth.

Such objects are searched for with groundbased wide-angle telescopes, such as the Schmidt telescopes at Palomar Observatory in California, where the American astronomer Eleanor Helin is now discovering them at the rate of several a year. But such research is woefully underfunded. There are probably two thousand nearby worlds around a kilometer in diameter, of which only a few dozen are known; and a hundred thousand around 100 meters in diameter, of which only a few are known—and all of them were discovered recently. It is not only good science, but also, as Eugene Shoemaker has emphasized, simple prudence to step up the discovery rate of bodies in the Earth's neighborhood—with special emphasis on the largest and darkest of them. (Most of the big bright ones have already been found.)

As for "doing" something about an errant comet or asteroid, some provocative studies have been performed. The

Intrepid but involuntary visitors to a comet. From the Jules Verne novel *Hector Servadac*.

---

*Over longer periods, still more catastrophic collisions must occur. Impacts of objects ten kilometers across, the size of a large cometary nucleus, will happen roughly once every hundred million years, producing an explosion a thousand times larger than would result from the simultaneous detonation in a given place of all the nuclear weapons now on Earth.

1996

Verne's cometary explorers leave the comet as it returns close to the Earth and descend to Europe by balloon. Engraving by Frank R. Paul for *Off on a Comet*, as it appeared in *Amazing Stories*, No. 2, June 1929.

pioneering work emerged from a student project at the Massachusetts Institute of Technology in 1967, in a graduate problem course called "Advanced Space Systems Engineering." The problem was Icarus, an Earth-approaching asteroid that was scheduled, the following year, to pass within some six million kilometers of the Earth. Suppose, it was posited, that Icarus instead had been making a beeline for the Earth. What could we do about it? Subsequent studies have confirmed the general conclusions of Project Icarus. Atypically for graduate seminars, the joint student/faculty work in this course was published as a book, and led to a high-budget Hollywood motion picture.*

---

*Called, erroneously, "Meteor." The word "meteorite" probably sounded too small-scale.

Abandoning the ray guns of an earlier generation, the MIT engineers opted for high-yield nuclear weapons and *Saturn 5* rockets. The higher the yield of the explosion, the bigger the resulting crater. But once the crater depth is comparable to the size of the intruding object, you've reached a point of limiting returns; still higher-yield weapons will only break the object into many fragments. In this way, the MIT team estimated that Icarus might be "destroyed" by the explosion of as few as a hundred megatons, but more likely a few thousand megatons—a fair fraction of the present global arsenal. However, it is not clear that fragmenting a big comet, on a collision trajectory, into hundreds or thousands of pieces does much for the inhabitants of the target planet. It is a commonplace in science fiction movies for the intruding object to be "disintegrated" and the threat thereby removed; in reality, all the fragments would continue to move with the original center of mass, and many would collide with the Earth, spreading the destruction over a much larger area than if we had simply left it alone. It makes more sense merely to nudge the comet or asteroid into a new orbit that does not take it dangerously close to the Earth. This also, the MIT group showed, could be done with nuclear weapons.

Suppose it is only at the last moment that we become aware of the reality of an imminent impact, and are required to deflect the intruder on its fateful last approach to Earth. Then high yields may be required, perhaps thousands of megatons. You would emplace your nuclear weapons close together on the surface of the cometary nucleus, and detonate them all at once.* The vaporized ice and dust and organics would be shot off into space from one side of the nucleus, making a short-lived but effective rocket motor of unprecedented scale. The comet could be speeded up or slowed down in this way, its orbit changed, and the Earth saved. But it would be better if you had significant advance notice of the collision. Then, you could finesse the problem during a previous orbit, by setting much smaller explosions at aphelion or perihelion. A small change in velocity would be multiplied in successive orbits. Doing it this way, the required thrust can be so small that conventional rather than nuclear explosives might suffice, or even a real rocket engine, affixed to the offending asteroid or comet and using local material for fuel, oxidizer and reaction mass. This

A comet passed quite near...so they sprang upon it, together with their servants and their instruments.

—Voltaire,
Micromegas,
1752

---

*Whatever happens to the comet or asteroid, this response would have the desireable side effect of making deep cuts in the bloated nuclear arsenals of Earth—which pose just the same sort of threat as a middling asteroid or comet on a collision trajectory. The key question is whether we humans can put our house in order in the absence of a threat from without.

1997

more subtle solution to the problem was discovered after the pioneering MIT study. The lesson seems to be that a little knowledge is worth a great many megatons.

Once we have gone this far, and considered rerouting asteroids and comets to prevent them from impacting the Earth, we can also contemplate rearranging their orbits for other purposes. Before humans invented mining, the only ready source of iron for metallurgy was from meteorites. This fact is still commemorated in many languages, where the word for iron is connected with the word for the sky. (The Latin word *siderus* comes from a word for star or constellation, as in the English word sidereal; and the Greek word for iron is *sideros*.) Billions of years ago an asteroid melts, and forms a nickel-iron core under a rocky mantle; a violent collision with another asteroid strips off the mantle, leaving a bare metallic core, and a subsequent collision sends a shard of iron careening through space. Millions of years later, it falls on the Earth, startling some itinerant tribesperson and providing another inducement toward technology and civilization. We humans are already experienced in working with extraterrestrial resources.

Since the platinum group of metals (Chapter 17) is relatively more abundant in asteroids (and probably comets) than on the Earth, there may be real economic motivations for mining small nearby bodies. An asteroid 100 meters across has a mass of a million tons. There are some hundred thousand of them in the vicinity of the Earth. On a planet with dwindling supplies of accessible metals—among them, nickel and the platinum group—this may provide a significant incentive for commercial utilization of the asteroids. A rocket motor is towed to the asteroid, attached to it, ignited, and used to change its trajectory. Perhaps there will be a gravity swingby of the Moon or a nearby planet to bring it to a convenient orbit about the Earth; there it would be dismantled, and great cargo ships would carry the valuable metals to the Earth.

From a purely economic standpoint, the question is whether the money spent to capture, dismantle, and transport asteroidal material to Earth would be better spent on extracting ores already on Earth, but in less tractable deposits or at greater depths. Where the commercial benefit of space mining becomes most apparent is in building large structures in Earth orbit—were such constructions otherwise justified, for the manufacture of pharmaceuticals and exotic alloys, for example; for scientific investigations; as a waystation to the planets; or—if we are so foolish as to permit it to happen—for the introduction of weapons systems into space. It is very expensive to carry material up from the surface of the Earth against the pull of gravity, and far cheaper

Where there are such lands, there should be profitable things without number.

—Christopher Columbus,
Journal of the First Voyage,
November 27, 1492

The Trojan asteroids. Shown here are the orbit of the Earth (background) and the orbit of Jupiter (foreground), both planets circling the Sun. In Jupiter's orbit, 60 degrees ahead of it and 60 degrees behind it, are two catchment regions for interplanetary debris, part of a class of relatively stable positions called Lagrangian Points. It is possible that a sizable collection of both asteroidal and cometary bodies are sitting in these Lagrangian Points awaiting future explorers. Diagram by Jon Lomberg/BPS.

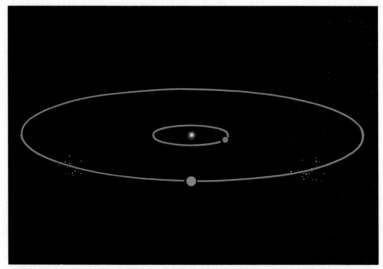

to utilize materials that are up there already. The Moon is already in orbit about the Earth, but it seems woefully depleted in a number of key substances, especially water and organics. If we humans begin large orbital constructions in the next century, it will make sense to use asteroids and comets as raw materials. In the process, we will gain considerable practice in living on small worlds, and moving them about the inner solar system at will. This capability then seems likely to open a startling vista:

Crossing the Earth's orbit, as we have said, are roughly a hundred thousand worlds, each around 100 meters across. Many of them are of cometary origin. The total area of all these worlds is ten thousand square kilometers, a sizable chunk of territory—although, individually, each is only as big as a square block in an urban center. If there are a hundred trillion cometary nuclei in the inner plus outer Oort Clouds, then the total surface area of the comets is equivalent to hundreds of millions of planets each the size of the Earth. A hundred million Earths parked conveniently in our backyard is a heady prospect.

The comets are traveling so slowly that it would be possible, even with current technology, to overtake one. A million years or more must pass for a comet to get here from the outer Oort Cloud; the *Voyager* spacecraft will traverse the same distance in ten thousand, and likely future technologies will lower the transit time to less than a human life span. If we are planning where humans might live in the far future, the comets offer by far the greatest range of possibilities. But since a few square kilometers will not accommodate many people, we must imagine a large number of small worlds, each sparsely populated. But in what sense can a comet be considered habitable?

Certainly all the molecular necessities of life are to be found on comets. Humans, like most other lifeforms on

1997

Earth, are made largely of water, and except possibly for a few of the moons in the outer solar system, there are no known worlds richer in water than the comets. Comets also contain large quantities of organic molecules, useful for agriculture and biological engineering, and rock and metal probably sufficient for practical purposes. The great quantities of water also mean that oxygen for breathing should be easily extractable, and rockets working as the *Centaur* booster does, off liquid hydrogen and oxygen, could be readily refueled on a cometary surface. In each of these respects comets are much more obliging bases and habitats than, say, the rocky and metallic asteroids.

But life on Earth, almost all of it, runs off the energy of sunlight. The plants harvest the sunlight and the animals harvest the plants. The inner solar system is flooded with light but, except for the Earth and Mars, is depleted in water. The outer solar system, by contrast, is rich in (frozen) water, but poor in sunlight. Equatorial noon on a cloudless world in the Saturn system is no brighter than twilight on Earth. The water is where the light isn't, and vice versa—a point emphasized many years ago by the American science writer Isaac Asimov.

With modern technology, we can now imagine doing something to redress this oversight in solar system design. Comets (and icy boulders from the rings of Saturn) can be driven or towed to the inner solar system, where the ice would be mined directly from the surface or, for nearby extinct comets, by drilling through the lag deposit to the icy core below. The water would be dissociated to make rocket fuel and oxidant, and to supply oxygen to human outposts in space and on the other terrestrial planets. There is so much water available from comets that it is even possible to imagine water provided to selected regions of parched worlds, permitting life to be transplanted to previously desolate environments. The organic matter in a dead comet or a carbonaceous asteroid, if finely pulverized, might also be used—as a growth medium for living things, and to moderate the hellish climate of Venus, by the same mechanism invoked (Chapter 16) in studies of the extinction of the dinosaurs and of Nuclear Winter. Since extinct comets with icy cores may be already at our doorstep, they may prove to be a critical factor in human utilization of space during the next century or two.

The biological essentials that are not supplied by the comet, at least not directly, is heat, warmth, energy, power. These are ordinarily supplied to comets only when they come close to the Sun. We can readily envision vast arrays of solar panels deployed on and around a comet fairly near the Sun—conceivably, even as far away as the orbit of Sat-

You have to bundle up when your comet is far from the Sun. Jupiter is seen with its four large moons in the sky, top left. From Jules Verne's *Hector Servadac*.

In the vicinity of Saturn, some centuries from now, cometary nuclei support the growth of immense, genetically engineered arboreal forms. Painting by Jon Lomberg.

1998

Forested Comets

I propose to you then an optimistic view of the Galaxy as an abode of life. Countless millions of comets are out there, amply supplied with water, carbon, and nitrogen, the basic constituents of living cells. We see when they fall close to the sun that they contain all the common elements necessary to our existence. They lack only two essential requirements for human settlement, namely warmth and air. And now biological engineering will come to our rescue. We shall learn to grow trees on comets…From a comet of 10-mile diameter, trees can grow out for hundreds of miles, collecting the energy of sunlight from an area thousands of times as large as the area of the comet itself. Seen from far away, the comet will look like a small potato sprouting an immense growth of stems and foliage. When man comes to live on the comets, he will find himself returning to the arboreal existence of his ancestors. We shall bring to the comets not only trees but a great variety of other flora and fauna to create for ourselves an environment as beautiful as ever existed on Earth. Perhaps we shall teach our plants to make seeds which will sail out across the ocean of space to propagate life upon comets still unvisited by man.

—Freeman Dyson,
"The World, the Flesh, and the Devil," Third J.D. Bernal Lecture, delivered at Birkbeck College, London. Reprinted in *Communication with Extraterrestrial Intelligence (CETI)*, C. Sagan, ed., MIT Press, Cambridge, Massachusetts, 1973

urn. Farther out, we can contemplate large nuclear reactors powering cometary bases. If fusion reactors—powered by water itself—become commercially feasible in the middle of the next century, as some experts predict, they would represent an ideal power source for cometary bases because of the abundance of ordinary water ice, as well as frozen heavy water, HDO and D_2O (where here D stands for deuterium, the form of heavy hydrogen that has a neutron as well as a proton in its nucleus).

A more romantic idea has been proposed by the British-born physicist Freeman Dyson, who suggests that through genetic engineering we will one day be able to design a special tree of unprecedented size to grow on comets far from the Sun. It would be planted in the organic snows, and grow enormously, so that its leaves could gather in enough of the sparse sunlight. Various requirements must be fulfilled, including heat insulation, no loss of gases to the adjacent vacuum, and the like. Because of the low gravity, growth of the trees is not restricted by their weight, and Dyson envisions forests larger than the comets from which they sprout (opposite). The oxygen produced in photosynthesis "will be

A forested comet leaves the Oort Cloud on its way to the stars. Painting by Jon Lomberg.

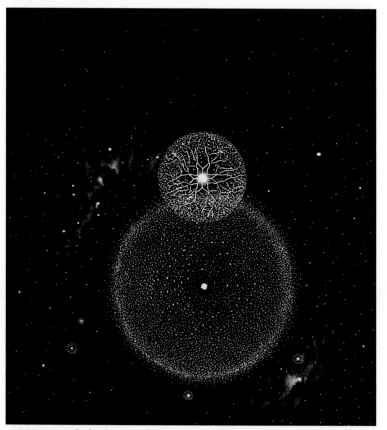

What I conceive to be one of the main designs of the Creator in the formation of such a vast number of splendid bodies is, that they may serve as habitations for myriads of intelligent beings...If this position be admitted, then we ought to contemplate the approach of a comet, not as an object of terror or a harbinger of evil, but as a splendid world, of a different construction from ours, conveying millions of happy beings to survey a new region of the Divine empire...

— Thomas Dick,
The Sidereal Heavens and Other Subjects Connected with Astronomy, As Illustrative of the Character of the Deity and of an Infinity of Worlds, Philadelphia, 1850

transported down to the roots and released into the region where men will live and take their ease among the tree trunks." "Will" sounds a little optimistic, but there seems to be nothing impossible about the proposal. But very far from the Sun—within the Oort Cloud, for example—even such heroic measures will be unavailing; the Sun is simply too dim there, and something like a fusion reactor will be needed to power the biological cycles and keep the place tolerably warm.

It is a law of biology as well as of social relations that isolation fosters diversity. Picture a time in the far future with millions of inhabited comets, each harboring no more than a few hundred individuals. Within the Oort Cloud, it would take a day or more for a radio message traveling at the speed of light to reach one colonized comet from another. That might be sufficient to maintain some cultural homogeneity among these many worlds, but the absence of frequent visits would permit a slow divergence of cultural and behavioral norms, and an enormous diversification of social, political, economic, religious, and other views—a development that might be of major benefit for the human species. It is hard to see, though, what advantage might thereby accrue to individual nation states, the only current entities wealthy enough to foot the bill; and the nation states have a history of preferring their own short-term advantage to the well-being of the species. Thus, the time when the majority of the human species will have been scattered among the comets is, for this and other reasons, far off. But in the long term, if space technology continues to develop, we will go to where the surface area, the water, and the organic matter are—to the comets.

* * *

If, in the remote future, we have populated not only all the nearby small worlds, but also the comets out to the Oort Cloud, we will have gone, through a succession of slow steps, halfway to the nearest star. A natural progression exists from there to the rest of the Galaxy. The colonization of the Galaxy would happen all by itself if the Oort Cloud becomes populated. Individual comets are so loosely bound that casual gravitational perturbations by passing stars release enormous numbers of them (Chapters 11 and 16) from bondage to the Sun; they then slowly pirouette through interstellar space on their own. In the far future, inhabited comets in the Oort Cloud would be shaken free from the shackles of the Sun's gravity, to begin seeding at least the nearer parts of the Galaxy with humans.

The comets of the Oort Cloud have a natural slow leakage to interstellar space, which in the far future may be accelerated by human technology. Diagram by Jon Lomberg/BPS.

1999

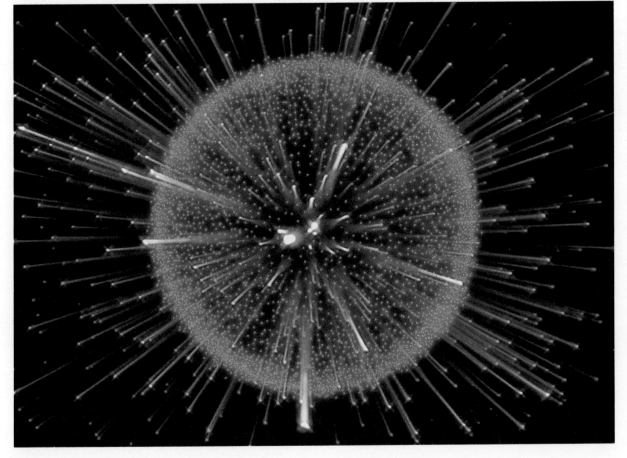

Thomas Wright's vision of innumerable suns surrounded by cometary orbital rosettes. From his *An Original Theory of the Universe* (1750). Courtesy Michael A. Hoskin.

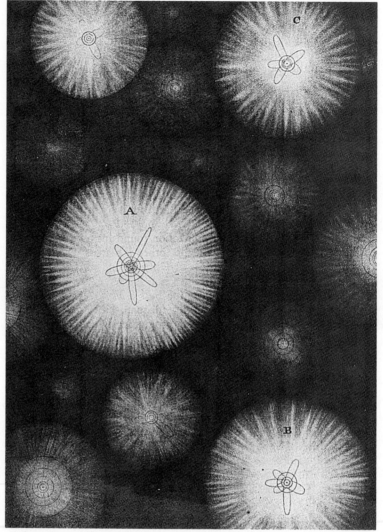

At present, it seems that as much as forty or fifty Earth masses of cometary material has been ejected from the Oort Cloud since its formation. Much larger masses must have been ejected from the planetary part of the solar system when the vicinity of Jupiter and Saturn was populated by enormous numbers of small icy worlds (Chapter 12). Estimates of these masses range from a hundred to as much as a thousand Earth masses. All this material, originally derived from our neck of the galactic woods, is now tumbling through space between the stars, randomly perturbed by passing cosmic objects over billions of years of time, and diffusing until they are distributed through a major sector of the Milky Way.* But up to now, all these comets have been, so far as we know, uninhabited.

---

*Occasionally, one may briefly enter the inner regions of some other planetary system and streak across alien skies, perhaps provoking some amalgam of fear and wonder in creatures very different from us.

No comet has ever been observed on a trajectory originating outside the gravitational influence of the Sun. And yet, sooner or later, such comets should be seen. In our own system, we conclude that many comets have been ejected into interstellar space after close passages by the Sun or the major planets. Especially with the discovery of debris rings around nearby stars (Chapter 12), it is reasonable to think that many—perhaps most—of the stars in the sky are similarly enveloped in clouds of comets, which are also being ejected into interstellar space.*

Newton seems to have been the first to imagine comets around other stars:

> This most beautiful system of the sun, planets, and comets, could only proceed from the counsel and dominion of an intelligent and powerful Being. And if the fixed stars are the centres of other like systems, these, being formed by the like wise counsel, must be all subject to the dominion of the One.

And Laplace envisioned comets "which, moving in

---

*Martin Harwit and E.E. Salpeter, astrophysicists at Cornell University, have proposed that celestial gamma ray bursts, observed by satellite observatories, are due to the impact of comets with neutron stars. If this is true, comets must surround many stars, because these still unexplained bursts come from all over the sky.

A bifurcated cometary tail is produced in the planetary system of a contact binary—two stars, of different properties, so close together that matter flows from one to another. Painting by Don Davis.

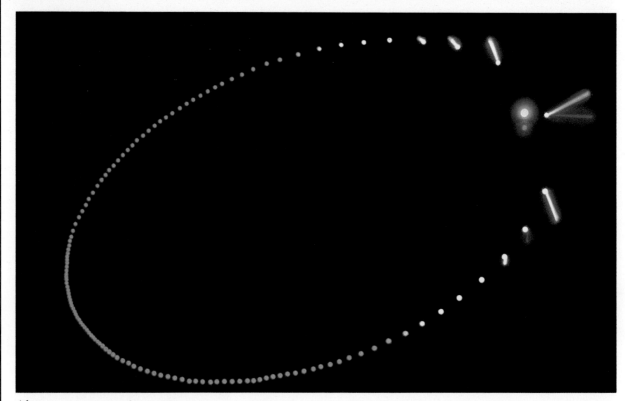

Above, a comet orbits a binary system, composed of red and blue component stars. The length, color, and multiplicity of the tails varies with position in the orbit. Opposite, a comet orbits a widely separated binary star in a figure-eight orbit. Again, the tails are shown for various orbital positions. Diagrams by Jon Lomberg/BPS.

hyperbolic orbits, can wander from system to system." But space is very empty and the stars are far apart. If every star in the Milky Way Galaxy had an Oort Cloud like ours, and a cometary ejection rate like ours, the average time between arrivals here of true interstellar comets would be hundreds of years. Astronomers are eagerly waiting.

As time goes on, the Milky Way acquires more and more interstellar comets. If every star in the Milky Way ejects a thousand Earth masses of comets into interstellar space every 4.5 billion years, as ours has, then there may be the equivalent of the mass of a hundred million suns floating as comets, undetected in the space between the stars. As great a mass as this represents, it is much less than one-tenth of a percent of the mass of the Milky Way itself.

There seems to be a community of comets that fills the Galaxy. Comets are likely to form in the accretion disk around every proto-star. If the case of the solar system is typical, every star will eject something like a trillion comets into interstellar space—mainly in the nebular stage, with a trickle of cometary ejections continuing for the life of the star. If there are a few hundred billion stars, each of which has ejected something like several trillion comets, the number of interstellar comets in the Galaxy is some 10^{24} [or 1,000,000,000,000,000,000,000,000], more than the number of stars in the universe. (The number of comets still bound to stars would be larger.) A multitude of interstellar

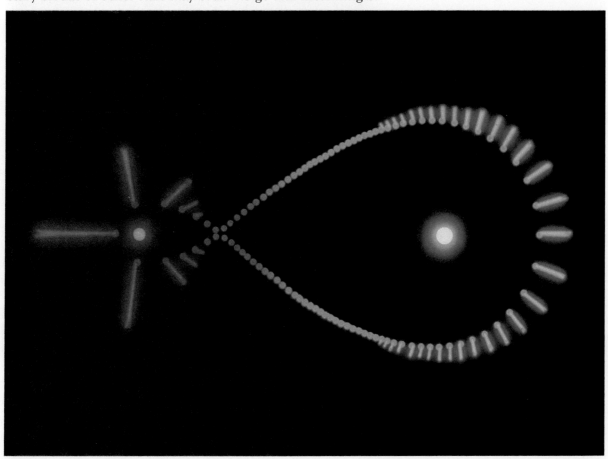

comets should by now be distributed entirely at random, in and between the galactic spiral arms. The average distance between them, even far from stars, would be some tens of Astronomical Units, the same sort of distance that separates the comets in the outer Oort Cloud.

The Galaxy can then be pictured as a vast flattened disk of comets, in which are embedded the more massive but less abundant interstellar clouds, stars and their planetary companions. In the vicinity of the stars, the concentration of comets is greater. Accordingly, nowhere in the Galaxy would there be a place more distant from a comet than the Earth is today from the planets we have already visited. Advanced civilizations may, for all we know, be able to cross the Galaxy in one fell swoop on some kind of interstellar express; but the comets provide an opportunity for backward civilizations like ours to construct a Toonerville Trolley that chugs off for a few years, stops at a comet, explores, refurbishes, and then squeaks off to the next comet. The principal problem is to find and catalogue the nearby interstellar comets. The explorers may have to take a large radar telescope with them.

The comets in the Oort Cloud of the Sun and the cometary clouds of other stars may even merge and intermingle.

2000

"We must not let our people," said he, "lose all interest in Earthly matters. They are still the children of the Earth. They are someday to return to the Earth, but even if they never return, it will be useful to attach them to our dear old mother world by at least memory and grateful recollection."

—Jules Verne
Off on a Comet,
Paris, 1878

There may even be comets on figure-eight orbits, bound alternately to two different stars, if the stars themselves are close enough together. The comets then, are moving steppingstones to anywhere—like the ice floes on which Liza was able to cross the raging river in *Uncle Tom's Cabin.*

Even if there is no colonization of the comets, we will one day set out to explore the trans-Plutonian spaces, and it will make sense to refuel on the comets. They would still be steppingstones to the stars. Perhaps eventually comets will themselves be converted into spaceships bound for other star systems, taking thousands of generations or more before they approach a new star and the sleeping cometary forest awakens to long-forgotten sunlight. The prospect recalls the image sketched by the German astronomer J.H. Lambert, who wrote in the eighteenth century:

> Thus we can conceive comets which, being attached to no particular system, are in common to all, and which, roaming from one world to another, make the tour of the universe...I love to figure to myself those traveling globes, peopled with astronomers, who are stationed there for the express purpose of contemplating nature on a large, as we contemplate it on a small scale ...We may suppose that their year is measured by the length of their route from one sun to another. Winter falls in the middle of their journey; each passage of a perihelion is the return of summer; each introduction to a new world is the revival of spring; and the period of quitting it is the beginning of their autumn.

But there is nothing peculiarly human about this story, not even the urge to colonize. We can imagine the comet clouds making every star a dandelion; periodically or episodically, some of the seeds are shaken loose, carrying the local life forms out into the Galaxy. Occasionally, a seed travels of its own volition. Sooner or later, it would seem, unless all technological civilizations have an unerring drive for self-destruction, the expanding inhabited cometary clouds will encounter one another somewhere in interstellar space. In our explorations of the Oort Cloud, will we find a stray object on which some alien technology is evident?

There are a great many cometary distractions between here and the nearest star. Some of them may be substantial worlds, hundreds or even thousands of kilometers across. If there are a hundred trillion comets filling the spaces between adjacent stars, any advanced civilization will expand outward very slowly. This may be why we have not found evidence of alien visitors in our solar system: The Galaxy is too interesting, and they have not discovered us yet.

* * *

1985/86 is the time of the historic maiden voyages by the human species to the comets. There will be more elaborate missions to a wider variety of comets, eventually extending to great distances from the Sun. One day—probably in the next century—these spacecraft will carry human crews. We will be living on the comets, and with the aid of rocket engines and Newton's laws, piloting the comets. If such a day comes, we will have fully justified the faith implied by the !Kung people—that perhaps unique surviving culture that considers cometary auspices benign. In the !Kung language, comets are "The Stars of the Great Captains."

From an outpost on the planet Mars, two humans view the return of Halley's Comet in July 2061. The Earth is the bright blue point of light in the sky. Its representation on the construction indicates the settlers' planet of origin. Painting by Don Davis.

Chapter XX

A MOTE OF DUST

I heard much of the comet of that year, 1577, and was taken by my mother to a high place to behold it.

> —Johannes Kepler,
> remembering an event that
> took place when he was
> six years old

Contemplated as one grand whole, astronomy is the most beautiful monument of the human mind; the noblest record of its intelligence. Seduced by the illusions of the senses, and of self-love, man considered himself, for a long time, as the centre of the motion of the celestial bodies, and his pride was justly punished by the vain terrors they inspired. The labour of many ages has at length withdrawn the veil which covered the system. Man appears, upon a small planet, almost imperceptible in the vast extent of the solar system, itself only an insensible point in the immensity of space. The sublime results to which this discovery has led may console him for the limited place assigned him in the universe.

> —Pierre Simon, Marquis de Laplace,
> *System of the World,*
> Part 1, Chapter 6 (1796)

Thirty-Two Perihelion Passages of Halley's Comet

239 B.C.	MARCH 30
163	OCTOBER 5
86	AUGUST 2
11 B.C.	OCTOBER 5
A.D. 66	JANUARY 26
141	MARCH 20
218	MAY 17
295	APRIL 20
374	FEBRUARY 16
451	JUNE 24
530	SEPTEMBER 25
607	MARCH 13
684	SEPTEMBER 28
760	MAY 22
837	FEBRUARY 27
912	JULY 9
989	SEPTEMBER 9
1066	MARCH 23
1145	APRIL 22
1222	OCTOBER 1
1301	OCTOBER 23
1378	NOVEMBER 9
1456	JUNE 9
1531	AUGUST 25
1607	OCTOBER 27
1682	SEPTEMBER 15
1759	MARCH 13
1835	NOVEMBER 16
1910	APRIL 20
1986	FEBRUARY 9
2061	JULY 28
2134	MARCH 27

Every apparition listed (except, of course, the last two) was recorded by the astronomers of Earth.

COMETS MAY ACT AS THE CREATORS, THE PRE-servers, and the destroyers of life on Earth. A surviving dinosaur might have reason to mistrust them, but humans might more appropriately consider the comets in a favorable light—as bringers of the stuff of life to Earth, as ocean-builders, as the agency that removed the competition and made possible the success of our mammalian ancestors, as possible future outposts of our species, and as providers of a timely reminder about large explosions and the climate of the Earth.

A comet is also a visitor from the frigid interstellar night that constitutes by far the greatest part of the known universe. And a comet is, further, a great clock, ticking out decades or geological ages once each perihelion passage, reminding us of the beauty and harmony of the Newtonian universe, and of the daunting insignificance of our place in space and time. If, by chance, the period of a bright comet happens to be the same as a human lifetime, we invest it with a more personal significance. It reminds us of our mortality:

Like hundreds of other little boys of the new century, I was held up in my father's arms under the cottonwoods of a cold and leafless spring to see the hurtling emissary of the void. My father told me something then that is one of my earliest and most cherished memories. "If you live to be an old man," he said carefully, fixing my eyes on the midnight spectacle, "you will see it again. It will come back in 75 years. Remember," he whispered in my ear, "I will be gone, but you will see it. All that time it will be traveling in the dark, but somewhere, far out there"—he swept a hand toward the blue horizon of the plains—"it will turn back. It is running glittering through millions of miles."

I tightened my hold on my father's neck and stared uncomprehendingly at the heavens. Once more he spoke against my ear and for us two alone. "Remember, all you have to do is be careful and wait. You will be seventy-eight or seventy-nine years old. I think you will live to see it—for me," he whispered a little sadly with the foreknowledge that was part of his nature.*

Comet Halley is unique in our epoch—a bright, periodic, sometimes spectacular comet that sews the generations together, stitching back through history and forward into epochs to come, awakening us from the delusion that we exist separate from our past and our future, time-binding the

*Loren Eiseley, The Invisible Pyramid, New York, 1970. We wish that Loren Eiseley had lived to see it.

Late Babylonian tablet, with an account in cuneiform writing of the apparition of Halley's Comet in 164 B.C. This is a portion of a systematic Babylonian astronomical and astrological textbook, and reads in part, "The comet which previously had appeared in the east in the path of Anu in the area of the Pleiades and Taurus, to the west…passed along in the path of Ea." Courtesy The British Museum.

human species. You look up and see Halley's Comet—through a pair of binoculars, perhaps—and you are looking at the same comet* that Edmond Halley saw the summer after he and Mary Tooke were first wed. This is the comet all but one of whose apparitions back to the year 239 B.C. were carefully noted by Chinese astronomers. It has been described on tablets, silk, bamboo paper, parchment, newsprint and computer disks. This is the comet that cheered the !Kung hunter-gatherers and frightened nearly everyone else in all those cultures far-flung over the surface of the Earth and back at least millennia in time. We share this comet with many others.

A million years ago or so, as our hominid ancestors were hunting game and figuring out how to build a house, a passing star sent a gravitational ripple through the cloud of comets that envelops the Sun, and a pristine small world of ice went careening in toward the Sun. Ten or twenty thousand years ago, as our ancestors were dealing with the Wisconsin Ice Age, the comet finally arrived in the planetary part of the solar system, made a close encounter with one of the major planets, and began the conversion into its present orbit—that now takes it, once each 76 years, closer to the

It is more probable that [Halley's Comet] has been diverted into its present path, from one of much longer period, at some much less distant time, and has not yet worn out. Its orbit does not now come near any of the great planets, so that its capture can be no recent event; but the accumulation of ordinary small perturbations during a few thousand revolutions would probably suffice to shift the orbit enough to remove all traces of the original point of capture.
—Henry Norris Russell,
    The Solar System and Its Origin
    (New York, 1935)

*Or very nearly the same comet: the nucleus has lost many meters of ice since the perihelion passage of 1682.

When will a comet sweep the sky
Seizing and binding these evil leaders?
—From *Autumn Days in the
State of K'uei*,
by Tu Fu (712-770 A.D.)
Translated by Heather Smith and
Xie Yong

An apparition of Halley's Comet occurred in 760, and was recorded in the Chinese chronicles.

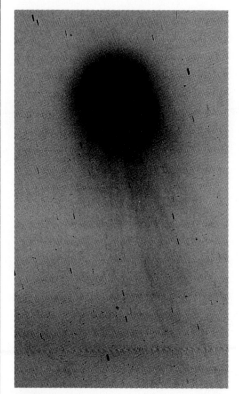

Halley's Comet in 1910 as photographed at the Helwan Observatory, Egypt. Courtesy National Aeronautics and Space Administration.

It would have been a great satisfaction to know that everyone who saw this wonderful object did so with the same feeling of elation and wonder—one would almost say veneration—with which the average astronomer regarded this beautiful and mysterious object stretching its wonderful stream of light across the sky.
  —E.E. Barnard, the preeminent
  observational astronomer of
  the time, on the 1910 apparition
  of Comet Halley

Sun than Venus is, and further from the Sun than Neptune. Every 76 years, more or less, for the last ten thousand, this comet has swept by, captivating the inhabitants of the Earth and then vanishing into the night.

Seventy-six years is a few generations or less, and so Halley's Comet is a kind of metronome beating out the rhythm of human progress or decline. Its 1910 apparition was the first since the invention of the airplane and the last before nuclear weapons. We have lately arranged the means of our self-destruction, and there is a real question of how many humans will be left the next time Halley's Comet comes by the Earth, in July 2061.

The dangers that we face are part of the process, now well underway, of the unification of the planet—in language, culture, science, and commerce. They are both driven by the identical technological advances—this critical and delicate time coincides with the widespread availability of nuclear weapons. At the present rate of change, it seems likely that in the period between now and 2061, the turning point for the human species will have been reached.

If we survive until then, our passage to the next apparition of Halley's Comet should be comparatively easy. That perihelion passage will be in March 2134, when the comet will make an unusually close encounter with the Earth. It will come within 0.09 Astronomical Units or 14 million kilometers, less than half the distance of the 1910 encounter. It will then be brighter than the brightest star. If there are those to do the commemorating, the years 2061 and 2134 should be celebrated for the courage, intelligence, and common purpose of a species forced by urgent necessity to come to its senses.

As everyone knows, contending inclinations war within each of us—toward understanding, creativity and growth, and toward chauvinism, violence and fear. This battle, on which the fate of the world truly rests, can be illustrated even by a mote of dust: Consider the history of a certain speck of matter dancing in the air before you—formed, it may be, from material ejected billions of years ago by a star on the other side of the Milky Way Galaxy; wandering for eons in the interstellar dark; gathered up into the forming solar nebula some five billion years ago; adhering to other similar grains in a lump of cosmic matter that eventually formed one of trillions of comets, probably in the vicinity of Uranus and Neptune; ejected out to the Oort Cloud; perturbed into the inner solar system, arriving a few thousand years ago; spewed off into the cometary tail, still encased in a particle of ice; wandering as a microplanet in interplanetary space; accidentally intercepted by the Earth a few years ago; gently settling through the Earth's atmosphere; and

now bobbing in the air, illuminated by a sunbeam, and giving no hint of its epic cosmic voyage.

Now consider another mote of dust, a little larger but with a similar history: It streaks into the Earth's atmosphere and burns up, ionizing the surrounding air, while it itself is reduced to molecules and atoms. That bit of matter makes a meteor trail, glowing brightly for a moment, perhaps to the accompaniment of expressions of delight by onlookers below. But if you think hard enough, you can find a use even for a shooting star. The trail of ions that a meteor briefly leaves will reflect very high frequency (VHF) radio waves. Since at any given moment there are enormous numbers of meteor trails in the atmosphere—most too faint to be seen with the naked eye—they collectively provide a kind of reflecting surface surrounding the Earth, off which radio waves of the proper frequencies can be bounced. Since the duration of an individual ion trail is less than a second, the message must be sent very fast. This has led to a new field of technology called Meteor Burst Communications.

And why would anyone go to such lengths, when there are perfectly adequate means of communication at hand? Because, after the imminent arms race in anti-satellite weaponry gets going, communications satellites would be among the early casualties in a nuclear war. Meteor Burst Communications has been developed so that a nuclear war can be fought. The comets have been enlisted. For the first time since they were thought to be warnings sent by an angry God, comets have practical value. But they have not been singled out. The entire worldwide enterprise of human knowledge is being drawn upon for similar services, and something approaching half the scientists on Earth now work for the various national military establishments.

Every encounter of Halley's Comet has been an occasion to express hopes and fears. Almost ritually, it has been a time to pray. So for this apparition of Halley's Comet, we offer a prayer of our own: We live on a fragile planet, whose thoughtful preservation is essential if our children are to have a future. We are only custodians for a moment of a world that is itself no more than a mote of dust in a universe incomprehensively vast and old. May we therefore learn to act, before all else, for the species and the planet.

Then, for innumerable nights to come, there will be humans to witness the grandeur of the comets that grace the skies of Earth.

* * *

2003

[The Comet of 1680] was observed with exquisite skill by Flamsteed and Cassini: And the mathematical science of Bernoulli, Newton, and Halley investigated the laws of its revolutions. At [its next return], in the year two thousand three hundred and fifty-five, their calculations may perhaps be verified by the astronomers of some future capital in the Siberian or American wilderness.
—Edward Gibbon,
*The Decline and Fall of the
Roman Empire*, Chapter 43 (1777)

That Gibbon, writing on the eve of the American Revolution, would foresee that the present territories of the United States and the Soviet Union would one day be the centers of astronomical discovery is almost as impressive as calculating the periodic return of comets, which he so clearly admires. But by the twenty-fourth century, the new centers of astronomical learning may be on Mars—or even on the comets themselves.

Halley's Comet, 1910. A photograph taken at the Observatorio Nacional Argentino. Courtesy Ruth S. Freitag, Library of Congress.

The Comet as a global cultural event. Courtesy Ruth S. Freitag, Library of Congress.

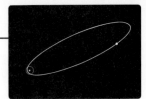

2004

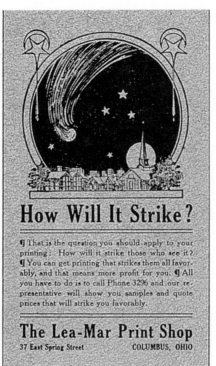

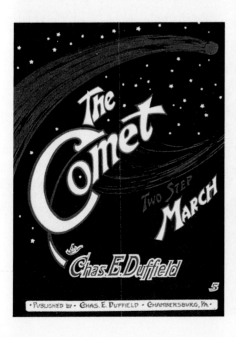

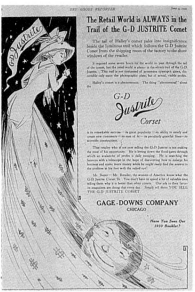

HALLEY'S COMET.

WILLIAM II.—"The end of the world! Impossible! I have given no such order."
—*Pasquino* (Turin).

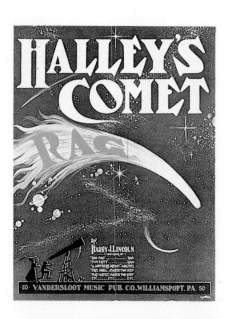

2005

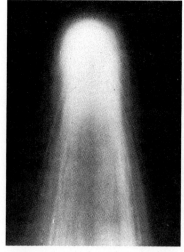

Halley's Comet in 1910. Lowell Observatory photograph.

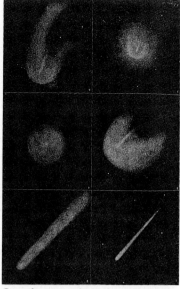

Six forms taken by Halley's Comet during the 1835 apparition as drawn by John Herschel. From Amédée Guillemin's *The Heavens* (Paris, 1868).

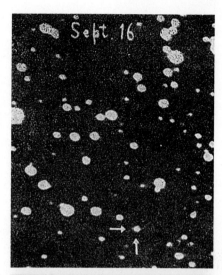

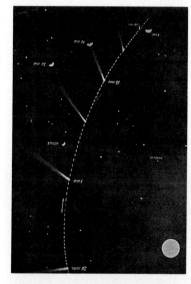

The positions of the tail of Halley's Comet during the 1910 apparition. A drawing by Lucien Rudaux for the French magazine *L'Illustration*. Courtesy Ruth S. Freitag, Library of Congress.

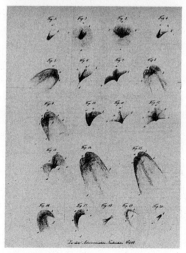

Twenty sketches of Halley's Comet drawn by Heinrich Schwabe in 1835. Courtesy Ruth S. Freitag, Library of Congress.

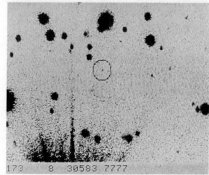

Two recovery photos of Halley's Comet as it was first seen in September 1909 and October of 1982. (Caltech photo).

The Comet of 1812

It was clear and frosty. Above the dirty, ill-lit streets, above the black roofs, stretched the dark starry sky. Only looking up at the sky did Pierre cease to feel how sordid and humiliating were all mundane things compared with the heights to which his soul had just been raised. At the entrance to the Arbat Square an immense expanse of dark starry sky presented itself to his eyes. Almost in the center of it, above the Prechistenka Boulevard, surrounded and sprinkled on all sides by stars but distinguished from them all by its nearness to the earth, its white light, and its long uplifted tail, shone the enormous and brilliant comet of 1812—the comet which was said to portend all kinds of woes and the end of the world. In Pierre, however, that comet with its long luminous tail aroused no feeling of fear. On the contrary he gazed joyfully, his eyes moist with tears, at this bright comet which, having traveled in its orbit with inconceivable velocity through immeasurable space, seemed suddenly—like an arrow piercing the earth—to remain fixed in a chosen spot, vigorously holding its tail erect, shining and displaying its white light amid countless other scintillating stars. It seemed to Pierre that this comet fully responded to what was passing in his own softened and uplifted soul, now blossoming into a new life.

—Leo Tolstoy,
War and Peace,
VIII, 22

[This is actually a reference to the Great Comet of 1811. It could still be seen with the naked eye in early 1812, but just barely. It was, however, a splendid object in the late fall of 1811.]

Comets approach the Sun, flicker a few hundred times, and die like moths around a flame. But a vast repository of them waits at the periphery of the Solar System. When the present configuration of continents is unrecognizably altered, when the Earth is engulfed by the expanding Sun, when, in its dotage, our star feebly illuminates the charred remains of this planet—then, even then, the skies will still be brightened as young comets, newly arrived from the interstellar dark, make their wild perihelion passages. When the rest of the solar system is dead, and the descendants of humans long ago emigrated or extinct, the comets will still be here.

2006

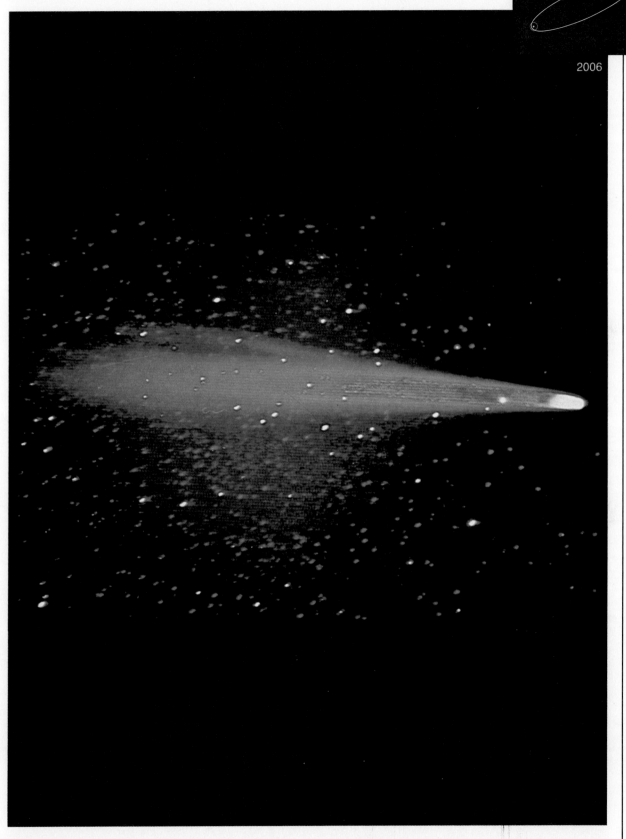

A computer color-enhanced photograph of Halley's Comet in 1910. Courtesy National Optical Astronomy Observatory/Lowell Observatory.

APPENDICES
COMETARY ORBITS AND METEOR SHOWERS

Appendix 1

Orbits of Selected Long-Period Comets

Comet	Name	Perihelion Distance (A.U.)
1811 I	Great Comet	1.035
1844 III	Great Comet	0.251
1858 VI	Donati	0.578
1861 II	Great Comet	0.822
1881 III	Great Comet	0.735
1882 II	Great Comet	0.008
1908 III	Morehouse	0.945
1910 I	Great Comet	0.129
1937 IV	Whipple	1.734
1943 I	Whipple-Fedtke-Tevzadze	1.354
1957 III	Arend-Roland	0.316
1957 V	Mrkos	0.355
1962 VIII	Humason	2.133
1965 VIII	Ikeya-Seki	0.008
1969 IX	Tago-Sato-Kosaka	0.473
1973 XII	Kohoutek	0.142
1976 VI	West	3.277
1980b	Bowell	3.364

Note that perihelion passage for these long-period comets varies from very close to the Sun (0.008 A.U.) to the middle of the asteroid belt, between the orbits of Jupiter and Mars (1.4 to 5 A.U.). There must be many long-period comets with more distant perihelia that are unknown to observers on Earth.

SOURCE: B. Marsden and E. Roemer, in *Comets*, L. Wilkening, ed., University of Arizona Press, 1982.

Appendix 2

Orbits of Selected Short-Period Comets

Periodic Comet	Perihelion Passage	Inclination in degrees	Perihelion Distance (A.U.)	Eccentricity	Period (years)
Oterma	1983 June	1.94	5.4709	0.2430	19.4
Crommelin	1984 Feb.	29.10	0.7345	0.9192	27.4
Giacobini-Zinner	1985 Sept.	31.88	1.0282	0.7076	6.59
Halley	1986 Feb.	162.23	0.5871	0.9673	76.0
Whipple	1986 June	9.94	3.0775	0.2606	8.49
Encke	1987 July	11.93	0.3317	0.8499	3.29
Borelly	1987 Dec.	30.32	1.3567	0.6242	6.86
Tempel 2	1988 Sept.	12.43	1.3834	0.5444	5.29
Tempel 1	1989 Jan.	10.54	1.4967	0.5197	5.50
d'Arrest	1989 Feb.	19.43	1.2921	0.6246	6.39
Perrine-Mrkos	1989 March	17.82	1.2977	0.6378	6.78
Tempel-Swift	1989 Apr.	13.17	1.5884	0.5391	6.40
Schwassman-Wachmann 1	1989 Oct.	9.37	5.7718	0.0447	14.9
Kopff	1990 Jan.	4.72	1.5851	0.5430	6.46
Tuttle-Giacobini-Kresak	1990 Feb.	9.23	1.0680	0.6557	5.46
Honda-Mrkos-Pajdusakova	1990 Sept.	4.23	0.5412	0.8219	5.30
Wild 2	1990 Dec.	3.25	1.5779	0.5410	6.37
Arend-Rigaux	1991 Oct.	17.89	1.4378	0.6001	6.82
Faye	1991 Nov.	9.09	1.5934	0.5782	7.34

The inclinations of the comets shown here vary from less than 2° (almost exactly in the ecliptic plane which includes the Earth and the planets), to 162° for Comet Halley; since a 90° inclination would mean a comet orbiting the Sun perpendicular to the ecliptic, a 162° inclination means a comet whose orbit is inclined some 180 − 162 = 18° from the ecliptic, but traveling in the opposite sense to the direction in which the planets revolve about the Sun. The eccentricity of the cometary orbits shown varies from about 0.04 (very close to a circle) to 0.97 for Comet Halley, which is extremely elongated.

SOURCE: Marsden and Roemer, 1982.

Appendix 3

Major Naked-Eye Meteor Showers, Late Twentieth Century

Shower	Shower Maximum Approximate Date	Single Observer Hourly Rate	Speed of Encounter with Earth (kilometers per second)	Normal Duration to 1/4 Strength of Maximum (days)	Associated Comet
Quadrantids	Jan. 3	40	41	1.1	?
Lyrids	April 22	15	48	2	1861 I
Eta Aquarids	May 4	20	65	3	Halley
S. Delta Aquarids	July 28	20	41	7	?
Perseids	Aug. 12	50	60	4.6	1862 III
Orionids	Oct. 21	25	66	2	Halley
S. Taurids	Nov. 3	15	28	—	Encke
Leonids	Nov. 17	15	71	—	1866 I
Geminids	Dec. 14	50	35	2.6	(1983 TB)?
Ursids	Dec. 22	15	34	2	Tuttle

Data based on table by P.M. Millman in *Observer's Handbook* of the Royal Astronomical Society of Canada (1985).

Appendix 4
Further Information

In addition to works listed in the Bibliography, further information about comets can be obtained from:

The Planetary Society
65 North Catalina Avenue
Pasadena, CA 91106
(818) 793-5100

The Astronomical Society of the Pacific
1290 24th Avenue
San Francisco, CA 94122
(415) 661-8660

The International Halley Watch
Jet Propulsion Laboratory
California Institute of Technology
4800 Oak Grove Drive
Pasadena, CA 91109
(818) 354-4321

Other resources for those who wish to dig more deeply include local planetariums and popular science publications such as *Sky and Telescope* or *Astronomy*. Periodic news on cometary findings is provided by *Comet News Service: A Quarterly Review and Irregular Bulletin*, P.O. Box TDR #92, Truckee, CA 95734.

Comets as depicted in many cultures: (A) Chinese painting on silk, 168 B.C. (B) From the Bayeux Tapestry, recording the appearance of Halley's Comet in spring 1066. (C) From the Eadwine Psalter, illuminated manuscript on vellum, c. 1145, by the monk Eadwine, based on accounts of Halley's Comet that appeared around this date. (D) From the fresco *The Adoration of the Magi* by Giotto, c. 1304. The gold pigments have been lost, revealing the red adhesive by which they were applied to the plaster. (E) From the Nuremberg Chronicles, a woodblock illustration published in 1493. It recounts the appearance of Halley's Comet in 684 A.D. (F) An Aztec depiction of a bright comet seen by the Emperor Montezuma II in the 1500's. From *Historia de las Indias de Nueva España* by Diego Duran. (G) A Turkish representation of the Great Comet of 1577. (H) A comet form according to Pliny, taken from Hevelius' *Cometographie*, 1640–80's. (I) One of several comet forms shown in Hevelius *Cometographie*, 1640–80's. (J) Detail of a probable comet from the hand guard in a Japanese wrought-iron sword with pierced design, Tsuba, c. 1700. (K) Illustration based on Halley's Comet photograph obtained in May 1910. From the Southern Observing Station of the Lick Observatory on Cerro San Cristóbal in Santiago, Chile. (L) Comet Bradfield, observed at approximately 1700 Greenwich Mean Time on January 10, 1980. Illustration computer enhanced by Goddard Space Center.

BIBLIOGRAPHY
Books on Comets Mainly for General Audiences

Asimov, Isaac. *Asimov's Guide to Halley's Comet.* Walker, New York, 1985.

Brandt, John C., ed. *Comets: Readings from Scientific American.* W. H. Freeman, San Francisco, 1981.

Brandt, John C. *Introduction to Comets.* Cambridge University Press, Cambridge, 1981.

Calder, Nigel. *The Comet Is Coming!* Viking Press, New York, 1980.

Chapman, Robert D., and John C. Brandt. *The Comet Book: A Guide for the Return of Halley's Comet.* Jones and Bartlett, Boston, 1984.

Comets: Career Oriented Modules to Explore through Pictures in Science. National Science Teachers Association, 1984.

Dahlquist, Raf, and Theresa Dahlquist. *Mr. Halley and His Comet.* Polestar, Canoga Park, CA, 1985. A charmingly illustrated children's book in rhyme.

Flaste, Richard, Holcomb Noble, Walter Sullivan, and John Noble Wilford. *The New York Times Guide to the Return of Halley's Comet.* Times Books, New York, 1985.

"Halley's Comet." *The Planetary Report,* 5, 3, May/June, 1985.

Halley Watch Amateur Observers' Manual for Scientific Comet Studies. Hillside, NJ: Enslow Publishers, 1983.

Krupp, E.C. *The Comet and You.* Macmillan, New York, 1985. For children.

Littman, Mark, and Donald K. Yeomans. *Comet Halley—Once in a Lifetime.* American Chemical Society, Washington, D.C., 1985.

Moore, Patrick. *Comets.* Scribner's, New York, 1976.

Moore, Patrick, and J. Mason. *The Return of Halley's Comet.* W. W. Norton, New York, 1984.

Mumford, George. *Everyone's Complete Guide to Halley's Comet.* Sky Publishing Co., Cambridge, 1985.

Pasachoff, Jay M., and Donald H. Menzel. *A Field Guide to the Stars and Planets.* Chapter 11, Houghton Mifflin, Boston, 1983.

Rahe, Jurgen, Bertram Donn, and Karl Würm. *Atlas of Cometary Forms: Structures Near the Nucleus.* NASA Special Publication 198. U.S. Government Printing Office, Washington, D.C. 1969.

Reddy, F. *Once in a Lifetime: Your Guide to Halley's Comet.* Astromedia, Milwaukee, 1985.

Richter, Nikolaus Benjamin. *The Nature of Comets.* Methuen, London, 1963.

The Royal Institution Library of Science: Astronomy. Vols. 1 and 2, Bernard Lovell, ed. Elsevier, New York, 1970. A compilation of Friday-evening discourses on astronomy at the Royal Institution, London, from the middle nineteenth century, including several interesting early talks on meteors and comets.

Seargent, David A. *Comets: Vagabonds of Space.* Doubleday, New York, 1982.

Stasiuk, Garry, and Dwight Gruber. *The Comet Handbook.* Stasiuk Enterprises, 1984.

Wilkening, L., ed. *Comets.* University of Arizona Press, Tucson, 1982. While technical, this collection of review papers is the single most comprehensive reference on all aspects of comets now in existence.

Representative Technical Papers Arranged by Chapters

CHAPTER 2

Dreyer, J. L. E. *A History of Astronomy from Thales to Kepler.* Cambridge University Press, Cambridge, 1906.

Hasegawa, Ichiro. "Catalogue of Ancient and Naked-Eye Comets." *Vistas in Astronomy 24,* 59, 1980.

Hellman, C. Doris. *The Comet of 1577: Its Place in the History of Astronomy.* Columbia University Press, New York, 1944.

Lagercrantz, Sture. "Traditional Beliefs in Africa Concerning Meteors, Comets, and Shooting Stars." In *Festschrift fur Ad. E. Jensen.* Klaus Renner Verlag, Munich, 1964.

Leon-Portilla, Miguel, ed. *The Broken Spears: The Aztec Account of the Conquest of Mexico.* Beacon Press, Boston, 1962.

Sarton, George. *A History of Science,* Vol. 1. Harvard University Press, Boston, 1952.

Stein, J., S.J. "Calixte III et la comète de Halley." *Specola Astronomica Vaticana II.* Tipografia Poliglotta Vaticana, Rome, 1909.

Thorndike, Lynn. A *History of Magic and Experimental Science*, Vol. 4. Columbia University Press, New York, 1934.

Wen wu. "Ma Wang Tui po shu 'T'ien wen ch'i hsiang tsa chan' nei jung chien shu" and "Ma Wang Tui Han ts'ao po shu chung to hui hsing t'u," Wen wu ch'u pan she, Beijing, Vol. 2, pp. 1–9, 1978.

Yoke, Ho Peng. "Ancient and Mediaeval Observations of Comets and Novae in Chinese Sources." *Vistas in Astronomy 5*, 127, 1962.

Ze-zong, Xi. "The Cometary Atlas in the Silk Book of the Han Tomb at Mawangdui." *Chinese Astronomy and Astrophysics 8*, 1, 1984.

CHAPTER 3

Barker, Thomas. *Of the Discoveries Concerning Comets.* Whiston and White, London, 1757.

Eddington, Arthur Stanley. "Halley's Observations on Halley's Comet, 1682." *Nature 83*, 373, 1910.

Freitag, Ruth S. *Halley's Comet: A Bibliography.* Library of Congress, Washington, D.C., 1984.

MacPike, Eugene, ed. *Correspondence and Papers of Edmond Halley.* Clarendon Press, Oxford, 1932.

Ronan, Colin. *Edmond Halley: Genius in Eclipse.* Doubleday, New York, 1969.

Stephenson, F. R., K. K. C. Yao, and H. Hunger. "Records of Halley's Comet on Babylonian Tablets." *Nature 314*, 587, 1985.

Walter, David L. "Medallic Memorials of the Great Comets and the Popular Superstitions Connected with Their Appearance." Scott Stamp and Coin Company, New York, 1893.

Westfall, Richard S. *Never at Rest: A Biography of Isaac Newton.* Cambridge University Press, Cambridge, 1980.

White, Andrew Dickson. "A History of the Doctrine of Comets." *Papers of the American Historical Association 2*, 2. G. P. Putnam, New York, 1887.

CHAPTER 4

Lalande, J. J. "Madame Nicole-Reine Etable de la Briere Lepaute." In *Astronomical Bibliography with a History of Astronomy between 1781 and 1802.* Paris, 1803.

Paulsen, Friedrich. *Immanuel Kant: His Life and Doctrine.* English translation by J. E. Creighton and Albert Lefevre. Frederick Ungar, New York, 1963. [Original printing 1899]

Wright, Thomas. *An Original Theory of the Universe.* [Original printing 1750] Macdonald, London, 1971. This edition has a very useful introduction, written by Michael A. Hoskin, to both Thomas Wright and his book.

CHAPTER 6

Feynman, Richard P., Robert B. Leighton, and Matthew Sands. *The Feynman Lectures on Physics.* Addison-Wesley, Reading, Mass. 1963. Our discussion of the physics of ice is partly based upon these remarkable lectures.

Hallett, John. "How Snow Crystals Grow." *American Scientist 72*, 582, 1984.

Patterson, W. S. B. *The Physics of Glaciers.* Pergamon Press, Oxford, 1969.

Whipple, F. L. "A Comet Model I. The Acceleration of Comet Encke." *Astrophysical Journal 111*, 375, 1950.

——. "A Comet Model II. Physical Relations for Comets and Meteors." *Astrophysical Journal 113*, 464, 1951.

CHAPTER 7

Fanale, F., and James Salvail. "An Idealized Short-Period Comet Model: Surface Insolation, H_2O Flux, Dust Flux, and Mantle Evolution," *Icarus 60*, 476, 1984.

Hughes, David W. "Cometary Outbursts: A Brief Survey." *Quarterly Journal of the Royal Astronomical Society 16*, 410, 1975.

CHAPTER 8

Khare, B. N., and Carl Sagan. "Experimental Interstellar Organic Chemistry: Preliminary Findings." In *Molecules in the Galactic Environment*, M. A. Gordon and L. E. Snyder, eds. John Wiley, New York, 1973.

Metz, Jerred. *Halley's Comet, 1910: Fire in the Sky.* Singing Bone Press, St. Louis, 1985.

Mitchell, G. F., S. S. Prasad, and W. T. Huntress. "Chemical Model Calculations of C_2, C_3, CH, CN, OH, and NH_2 Abundances in Cometary Comas." *Astrophysical Journal 244*, 1087, 1981.

Swings, P. "Le Spectre de la Comète d'Encke 1947 i." *Annales d'Astrophysique 11*, 1, 1948.

Wood, John, and Sherwood Chang, eds. *The Cosmic History of the Biogenic Elements and Compounds.* NASA Special Publication 476, 1985.

CHAPTER 9

Barnard, E. E. "On the Anomalous Tails of Comets." *Astrophysical Journal 22*, 249, 1905.

Biermann, L. "Solar Corpuscular Radiation and the Interplanetary Gas." *Observatory 77*, 109, 1957.

Biermann, L., and R. Lüst. "Comets: Structure and Dynamics of Tails." Chapter 18 of *The Solar System*, Vol. 4. G. P. Kuiper and B. M. Middlehurst, eds. University of Chicago Press, Chicago, 1963.

Henry, George E. "Radiation Pressure." *Scientific American 196*, 99–108, 1957.

Van Allen, James A. "Interplanetary Particles and Fields." In *The Solar System*, A Scientific American book. W. H. Freeman, San Francisco, 1975.

CHAPTER 10

Donnelly, Ignatius. *Ragnarok: The Age of Fire and Gravel*. University Books, New York, 1970 (original publication, 1883).

Goblet d'Alviella, Comte Eugène. *The Migration of Symbols*. University Books, New York, 1959. [Original printing, 1891]

Goldsmith, Donald, ed. *Scientists Confront Velikovsky: Evidence Against Velikovsky's Theory of 'Worlds in Collision."* W. W. Norton, New York, 1977.

Nuttall, Zelia. *"The Fundamental Principles of Old and New World Civilizations: A Comparative Research Based on a Study of the Ancient Mexican Religious, Sociological and Calendrical Systems." Archaeological and Ethnological Papers of the Peabody Museum*. Harvard University, Vol. 2, 1901.

Wilson, Thomas. *The Swastika: The Earliest Known Symbol, and Its Migrations; With Observations on the Migration of Certain Industries in Prehistoric Times*. Smithsonian Institution, Washington, D.C., 1896.

CHAPTER 11

Chebotarev, G. A. "On the Dynamical Limits of the Solar System." *Soviet Astronomy—AJ 8*, 787, 1965.

Oort, J. H. "The Structure of the Cloud of Comets Surrounding the Solar System, and a Hypothesis Concerning Its Origin." *Bulletin of the Astronomical Institutes of the Netherlands 11*, 91, 1950.

Oort, J. H. "Empirical Data on the Origin of Comets." Chapter 20 in G. P. Kuiper and B. M. Middlehurst, eds., *The Solar System*, Vol. 4. University of Chicago Press, Chicago, 1963.

Öpik, E. "Note on Stellar Perturbations of Nearly Parabolic Orbits." *Proceedings of the American Academy of Arts and Sciences 67*, 169, 1932.

Russell, Henry Norris. *The Solar System and Its Origin*. Macmillan, New York, 1935.

Van Woerkom, A. J. J. "On the Origin of Comets." *Bulletin of the Astronomical Institutes of the Netherlands 10*, 445, 1948.

CHAPTER 12

Biermann, L., and K. W. Michel. "On the Origin of Cometary Nuclei in the Presolar Nebula." *The Moon and the Planets 18*, 447, 1978.

Fernandez, Julio A. "Mass Removed by the Outer Planets in the Early Solar System." *Icarus 34*, 173, 1978.

Fernandez, J. A., and W.-H. Ip. "On the Time Evolution of the Cometary Influx in the Region of the Terrestrial Planets." *Icarus 54*, 377, 1983.

Goldreich, P., and W. R. Ward. "The Formation of Planetesimals." *Astrophysical Journal 183*, 1051, 1973.

Helmholtz, H. "On the Origin of the Planetary System." Lecture delivered by H. Helmholtz in Heidelberg and Cologne, 1871. Published in *Popular Articles on Scientific Subjects* by H. Helmholtz. D. Appleton and Company, New York, 1881.

Safronov, V. S. "Evolution of the Protoplanetary Cloud and the Formation of the Earth and the Planets." English-language version of the original Russian book published by Israel Program for Scientific Translations, Jerusalem, 1972.

Wetherill, George W. "Evolution of the Earth's Planetesimal Swarm Subsequent to the Formation of the Earth and Moon." *Proceedings of the Eighth Lunar Science Conference*, p. 1, 1977.

CHAPTER 13

Ball, Robert. *The Story of the Heavens*, rev. ed. Cassell and Company, London, 1900.

Keesing's Contemporary Archives. May 21–28, 17425–17429, 1960.

Von Humboldt, Alexander. "Events of the Night of Eleventh November, 1799." *Personal Narrative of Travels to the Equinoctial Regions of America During the Years 1799 to 1804*, Vol. 1. George Bell and Son, London, 1889.

CHAPTER 14

Gold, Thomas, and Steven Soter. "Cometary Impact and the Magnetization of the Moon." *Planetary and Space Sciences 24*, 45, 1976.

Kerr, Richard A. "Could an Asteroid Be a Comet in Disguise?" *Science 227*, February 22, 1985.

Marsden, B. G. "The Sungrazing Comet Group." *Astronomical Journal 72*, 1170, 1967.

Michels, D. J., N. R. Sheeley, R. A. Howard, and M. J. Koomen. "Observations of a Comet on Collision Course with the Sun." *Science 25*, 1097, 1982.

Öpik, Ernst. "The Stray Bodies in the Solar System. Part 1. The Survival of Cometary Nuclei in the Asteroids." *Advances in Astronomy and Astrophysics 2*, 219, 1963.

———."The Stray Bodies in the Solar System. Part 2. The Cometary Origin of Meteorites." *Advances in Astronomy and Astrophysics 4*, 301, 1966.

Shoemaker, Eugene M., and Ruth F. Wolfe. "Cratering Timescales for the Galilean Satellites." Chapter 10 in *The Satellites of Jupiter*, David Morrison, ed. University of Arizona Press, Tucson, 1982.

Shul'man, L. M. "The Evolution of Cometary Nuclei." In G. A. Chebotarev et al., eds., *The Motion, Evolution of Orbits, and Origin of Comets*. D. Reidel, Holland, 1972.

Turco, R.P., O.B. Toon, C. Park, R.C. Whitten, J.B. Pollack, and P. Noerdlinger. "An Analysis of the Physical, Chemical, Optical and Historical Impacts of the 1908 Tunguska Meteor Fall." *Icarus 50*, 1, 1982.

Wetherill, George W. "Occurrence of Giant Impacts During the Growth of the Terrestrial Planets." *Science 228*, 877, 1985.

CHAPTER 15

Alvarez, Luis W., Walter Alvarez, Frank Asaro, and Helen V. Michel. "Extraterrestrial Cause for the Cretaceous-Tertiary Extinction." *Science 208*, 1095, 1980.

Gould, Stephen Jay. "Sex, Drugs, Disaster, and the Extinction of the Dinosaurs." *Discover*, March 1984, p. 67.*

Hills, J. G. "Comet Showers and the Steady-State Infall of Comets from the Oort Cloud." *Astronomical Journal 86*, 1730, 1981.

National Museum of Natural Sciences and National Research Council of Canada, Syllogeus Series Number 12. "Cretaceous-Tertiary Extinctions and Possible Terrestrial and Extraterrestrial Causes: Proceedings of a Workshop Held in Ottawa, Canada, 16, 17 November, 1976." National Museums of Canada, Ottawa, Canada, March 1977.

Officer, Charles B., and Charles L. Drake. "Terminal Cretaceous Environmental Events." *Science 227*, 1161, 1985.

Pollack, James B., Owen B. Toon, Thomas P. Ackerman, Christopher P. McKay, and Richard P. Turco. "Environmental Effects of an Impact-Generated Dust Cloud: Implications for the Cretaceous-Tertiary Extinctions." *Science 219*, 287, 1983.

Sepkoski, J. John, Jr. "Mass Extinctions in the Phanerozoic Oceans: A Review." *Geological Society of America, Special Paper 190*, 283, 1982.

———. "A Kinetic Model of Phanerozoic Taxonomic Diversity. III. Post-Paleozoic Families and Mass Extinctions." *Paleobiology 10*, 246, 1984.

———. "Phanerozoic Overview of Mass Extinction." In *Pattern and Process in the History of Life*, D.M. Raup and D. Jablonski, eds. Springer-Verlag, Berlin, 1986.

Shoemaker, Eugene M. "Asteroid and Comet Bombardment of the Earth." *Annual Review of Earth and Planetary Sciences 11*, 461, 1983.

Steel, Rodney, and Anthony Harvey, eds. *The Encyclopedia of Prehistoric Life*. McGraw-Hill, New York, 1979.

"The Fossil Record and Evolution: Readings from *Scientific American*." W. H. Freeman, San Francisco, 1982.

Urey, Harold C. "Cometary Collisions and Geological Periods." *Nature 242*, 32, 1973.

CHAPTER 16

Alvarez, Walter, Frank Asaro, Helen V. Michel, and Luis W. Alvarez. "Iridium Anomaly Approximately Synchronous with Terminal Eocene Extinctions." *Science 216*, 886, 1982.

"A Talk with Eugene Shoemaker." Interview by Charlene Anderson, *The Planetary Report* 5 (1) 7, January/February, 1985.

Davis, Marc, Piet Hut, and Richard A. Muller. "Extinction of Species by Periodic Comet Showers." *Nature 308*, 715, 1984.

Gould, Stephen Jay. "Continuity." *Natural History*, April 1984, p. 4.*

*Collected in Stephen Jay Gould, *The Flamingo's Smile: Reflections in Natural History*. W.W. Norton, New York, 1985.

——. "The Cosmic Dance of Siva." *Natural History,* August 1984, p. 14.*

Hills, J. G. "Dynamical Constraints on the Mass and Perihelion Distance of Nemesis and the Stability of Its Orbit." *Nature 311,* 636, 1984.

Hoffman, Antoni. "Patterns of Family Extinction Depend on Definition and Geological Timescale." *Nature 315,* 659, 1985.

Rampino, Michael, and Richard Stothers. "Geological Rhythms and Cometary Impacts." *Science 226,* 1427, 1984.

——. "Terrestrial Mass Extinctions, Cometary Impacts and the Sun's Motion Perpendicular to the Galactic Plane." *Nature 308,* 709, 1984.

Raup, David M., and J. John Sepkoski, Jr. "Periodicity of Extinctions in the Geologic Past." *Proceedings of the National Academy of Sciences of the U.S.A. 81,* 801, 1984.

Schwartz, Richard D., and Philip B. James. "Periodic Mass Extinctions and the Sun's Oscillations about the Galactic Plane." *Nature 308,* 712, 1984.

Smoluchowski, R., J.N. Bahcall, and M.S. Matthews, eds. *The Galaxy and the Solar System.* University of Arizona Press, Tucson, 1985.

Thaddeus, Patrick, and Gary A. Chanan. "Cometary Impacts, Molecular Clouds, and the Motion of the Sun Perpendicular to the Galactic Plane." *Nature 314,* 73, 1985.

Whitmire, Daniel, and Albert A. Jackson. "Are Periodic Mass Extinctions Driven by a Distant Solar Companion?" *Nature 308,* 713, 1984.

CHAPTER 17

Bar-Nun, A., A. Lazcano-Araujo, and J. Oro. "Could Life Have Evolved in Cometary Nuclei?" *Origins of Life 11,* 387, 1981.

Forster, T. *Atmospheric Causes of Epidemic Diseases.* London, 1829.

Hobbs, R. W., and J. M. Hollis. "Probing the Presently Tenuous Link between Comets and the Origin of Life." *Origins of Life 12,* 125, 1982.

Hoyle, Fred. "Comets—A Matter of Life and Death." *Vistas in Astronomy 24,* 123, 1980.

Hoyle, Fred, and Chandra Wickramasinghe. *Diseases from Space.* Harper, New York, 1979.

Irvine, W. M., S. B. Leschine, and F. P. Schloerb. "Thermal History, Chemical Composition and Relationship of Comets to the Origin of Life." *Nature 283,* 748, 1980.

Oro, J. "Comets and the Formation of Biochemical Compounds on the Primitive Earth." *Nature 190,* 389, 1961.

Sales-Guyon de Montlivault, E.-J.-F. *Conjectures sur la réunion de la lune à la terre, et des satellites en général a leur planète principale; à l'aide desquelles on essaie d'expliquer la cause et les effets du déluge, la disparition totale d'anciennes espèces vivantes et organiques, et la formation soudaine ou apparition d'autres espèces nouvelles et de l'homme lui-même sur la globe terrestre.* Paris: Adrien Egron, 1821.

Shoemaker, Eugene M. "Asteroid and Comet Bombardment of the Earth." *Annual Review of Earth and Planetary Sciences 11,* 461, 1983.

Wallis, Max. "Radiogenic Melting of Primordial Comet Interiors." *Nature 284,* 431, 1980.

CHAPTER 18

Gay, Peter. *Education of the Senses.* Oxford University Press, New York, 1984.

"Halley's Comet from a Balloon." *Aeronautics 6,* 204, June 1910.

Morrison, David, et al. *Planetary Exploration Through the Year 2000: A Core Program.* National Aeronautics and Space Administration, U.S. Government Printing Office, Washington, D.C., 1983.

Neugebauer, M., et al., eds. *Space Missions to Comets.* NASA Conference Publication 2089, 1979.

Newburn, Ray L., and Jurgen Rahe. "The International Halley Watch." *Journal of the British Interplanetary Society 37,* 28, 1984.

Newsletters of the International Halley Watch. Periodicals produced by the NASA Jet Propulsion Laboratory, Publications 410, since August 1, 1982.

Planetary Society Fact Sheets on Halley Comet Spacecraft, available from The Planetary Society, 65 North Catalina Avenue, Pasadena, CA 91106.

"Report of the Comet Rendezvous Science Working Group." NASA Technical Memorandum 87564, 1985.

Sagdeev, R. Z., et al. "Cometary Probe of the Venera-Halley Mission." *Advances in Space Research 2,* 83, 1983.

Wilford, John Noble. "U.S. and Soviet Cooperating on Collection of Comet Dust." *New York Times,* December 21, 1984.

*Collected in Stephen Jay Gould, *The Flamingo's Smile: Reflections in Natural History.* W.W. Norton, New York, 1985.

Yeomans, Donald K., and John C. Brandt. "The Comet Giacobini-Zinner Handbook: An Observer's Guide." NASA/JPL Publication 400–254, 1985.

CHAPTER 19

Dyson, Freeman. "The World, the Flesh, and the Devil." Third J. D. Bernal Lecture, delivered at Birkbeck College, London. Reprinted in C. Sagan, ed., *Communication with Extraterrestrial Intelligence (CETI)*, MIT Press, Cambridge, MA, 1973.

Gaffey, Michael J., and Thomas B. McCord. "Mining Outer Space." *Technology Review* 79 (7), 51, June 1977.

Harwit, Martin, and E. E. Salpeter. "Radiation from Comets near Neutron Stars." *Astrophysical Journal 186*, L37, 1973.

MIT Student Project in Systems Engineering, "Project Icarus." MIT Press, Cambridge, MA, 1968.

CHAPTER 20

Arkin, William M., and Richard W. Fieldhouse. *Nuclear Battlefields: Global Links in the Arms Race.* Ballinger, Cambridge, MA, 1985.

Cruikshank, D. P., W. K. Hartmann, and D. J. Tholen. "Colour, Albedo, and Nucleus Size of Halley's Comet." *Nature 315*, 122, 1985.

Newburn, Ray L., and Donald K. Yeomans. "Halley's Comet." *Annual Review of Earth and Planetary Science 10*, 297, 1982.

Comet Index